HUMAN BRAIN EVOLUTION

HUMAN BRAIN EVOLUTION

The Influence of Freshwater and Marine Food Resources

Edited by

STEPHEN C. CUNNANE
Research Center on Aging
Departments of Medicine and Physiology and Biophysics
Université de Sherbrooke

KATHLYN M. STEWART
Canadian Museum of Nature

WILEY-BLACKWELL

A JOHN WILEY & SONS, INC., PUBLICATION

Published by John Wiley & Sons, Inc., Hoboken, New Jersey
Published simultaneously in Canada

Library of Congress Cataloging-in-Publication Data:

Cunnane, Stephen C.
Human brain evolution : the influence of freshwater and marine food resources / Stephen C. Cunnane and Kathlyn M. Stewart.
p. cm.
Includes index.
ISBN 978-0-470-45268-4 (cloth)
1. Brain–Evolution. 2. Human evolution. 3. Cognition. 4. Seafood. 5. Nutritional anthropology. I. Stewart, Kathlyn Moore. II. Title.
QP376.C857 2010
612.8′2–dc22

2009047979

Printed in the United States of America

CONTENTS

FOREWORD: EVOLUTION, ENCEPHALIZATION, ENVIRONMENT

Phillip V. Tobias

Stephen Cunnane of the University of Sherbrooke and Kathy Stewart of the Canadian Museum of Nature, Ottawa, have kindly invited me to write some thoughts by way of a foreword to their new book. This book takes a novel approach to the challenging question of the evolution of the human brain. Not content with the hard (and not-so-hard) facts culled from the fossil record, they and the team of contributors they have recruited look beyond these facts to seek out the impact of the environment, whose overarching force must have played a vital role in the several millions of years since the higher primate brain began to expand, absolutely and relatively, and to reorganize and reconstruct the telltale stigmata of cerebral localization.

About 2.6–2.5 million years ago (mya), there were marked climatic changes in Africa. These were attendant upon uplift of the southern and eastern parts of the continent; this great landmass, which comprises almost a quarter of the earth's habitable land surface, became cooler and drier. At that time and very shortly afterward, eight major changes are apparent in the palaeontological and archaeological record of Africa. They are as follows:

1. The retreat of the great wet forest of middle Africa, and the opening and spread of the savanna.
2. A number of animal extinctions.
3. The appearance of a number of new species and genera in the African fossil record.
4. The earliest stone tools appear in the archaeological archives.
5. The first appearance of a new genus of hominins, *Homo*, represented by the species *H. habilis* or by *H. rudolfensis.*
6. The development in early *Homo* of the first truly human kind of foot.
7. The first signs of appreciable enlargement of the hominin brain, as compared with the smaller brains of African great apes and australopithecines.
8. The earliest detectable appearance of the main speech areas of the cerebrum, namely Broca's and Wernicke's areas, on endocranial casts of *H. habilis*.

Were these all coincidences? It is more parsimonious and more likely that some or all of these phenomena were interlinked. Further, these eight phenomena were causally related to the tectonic and climatic changes in Africa, from a warmer and wetter climate in a more low-lying terrain before 2.6–2.5 mya, to a cooler and drier climate in more elevated terrain at the time of early *Homo*. It is obviously more satisfying and rigorous if logical causal steps linking these phenomena can be established, that is, the much, much earlier oxygenation of the atmosphere which made possible the survival and the flourishing of the aerobic eukaryotes; that was surely more than just a coincidence! As described in some detail, it is my contention that the causal pathway by which the tectonic and climatic changes of 2.6–2.5 mya influenced the development of the hominid brain involved not only selectively advantageous gene mutations, but also the intermediation of water.

Edible plants and animals abound in both marine and freshwater environments. Many specially-adapted types of plants and animals thrive along shorelines of rivers and lakes and along beaches and the shores of estuaries, both freshwater and saltwater, and on tidal and intertidal interfaces between them. These living creatures provide today, and would have provided in earlier times, an abundant source of foodstuffs (see Chapters 8 and 9). Much has been written in recent years about this subject, but we can trace the roots of the discourse back more than two hundred years. In 1798, Thomas Robert Malthus, an English clergyman and economist, published *An Essay on the Principle of Population: As It Affects the Future Improvement of Society.* (Malthus's sub-title hints at an early adumbration of the controversial eugenics movement, almost 100 years before Francis Galton introduced the term "eugenics" in 1883!) Malthus argued that whereas human populations increase by geometrical progression, their means of subsistence increase only by arithmetical progression. Hence, population size is necessarily limited by the means of subsistence. Two centuries later, the crucial role of food in human evolution has been the subject of uncounted important studies, many of which are reflected in the present volume.

While these thoughts on water, food, and mankind were being actively canvassed, the background for the development of the new hypotheses was being furnished by fossil evidence from South and East Africa. Going back to Raymond Dart and Robert Broom, there had been a long-standing hypothesis that early human evolution occurred in a savanna environment. I grew up with this paradigm and it had acquired an aura of sanctity! From the 1980s onward, studies of the mammalian and plant species accompanying early australopithecine remains from Makapansgat and Sterkfontein in South Africa, and from Ethiopia, had shown that by and large, these were not savanna species but forms associated with woodlands and even forests. At the Pliocene time depths represented in these deposits, the evidence torn from the ancient rocks told a story that seemed to be incompatible with the Savanna Hypothesis.

To this palaeontological and palaeobotanical evidence, data were added from the study of living forms, including humans. These studies showed that today's mammalian savanna dwellers are not characterized by such features as hairlessness, the great number and the nature of sweat glands and their distribution, the degree of development of subcutaneous fat, excessive urination, the absence of sun-reflecting fur, or poor water drinking capacity and retention. Yet these characteristics are hallmarks of modern mankind. When this anatomical, histological, biochemical, and physiological evidence was added to that of Africa's fossil-bearing deposits, it was clear that the savanna paradigm for early human evolution was passé. At a lecture at University College London in November 1995, I declared that all erstwhile savanna-supporters – including myself – must swallow their former assertions: "Open the window and throw out the Savanna Hypothesis, for it is no longer tenable!," I exclaimed. "The Savanna Hypothesis is dead; we are back to square one!"

With the Savanna Hypothesis put out to grass and perhaps even liquidated, the aquatic influences in human structure, function, and evolution were open to study with liberated and uncluttered minds. Michael Crawford and colleagues (see Chapters 2–4) had already been drawing attention to the special significance of the long-chain, polyunsaturated fatty acid – *docosahexaenoic acid* (DHA). They showed that DHA was necessary for the development of the large brain characteristic of hominids. There is a relative lack of DHA in savanna food. Crawford suggests that this would explain the "degenerative evolution" of the brains of truly savanna species and would be another reason why hominids are unlikely to have evolved their large brains on the savannas. On the other hand, the aquatic food chain has an abundant supply of DHA. Of necessity and convenience, early hominids would have made use of the aquatic food chain thereby making possible the spectacular evolution of the brain and brain size. The claim that the human brain depended on nutrients in the aquatic food chain furnished independent evidence to support the importance of water in human evolution.

DHA is not the only polyunsaturated fatty acid related to brain development and function. Two others are *arachidonic acid* (AA) and *eicosapentaenoic acid* (EPA). DHA and AA both comprise about 8% of the dry weight of the human brain, but EPA is present in much smaller amounts. These three fatty acids play important roles in the brain: both structural and functional roles are subserved by AA, while a structural role is concentrated in DHA and a functional role in EPA. Theoretically, humans could make AA, EPA, and DHA from their precursors (linoleic acid and alpha-linolenic acid), but the capacity for this conversion in humans seems to be inadequate to meet the needs of the developing brain. Also, the incorporation of DHA, AA, and EPA into brain phospholipids may be inadequate in the presence of much saturated fat in the diet. Our brain, therefore, needs a direct dietary supply of DHA, AA, and EPA. It is important to note that much of the AA, EPA, and DHA in the food supply originate from microalgae which grow in water. Hence, aquatic food chains are especially rich in these three important brain fatty acids. As David Horrobin (2001), who also studied polyunsaturated essential fatty acids and brain function, commented, "Perhaps that is why humans love, and pre-humans seem to have loved, water so much. They needed to eat water-based creatures to obtain the AA, EPA and DHA to grow their brains." Perhaps, also, I should like to add, that is why our mothers and grandmothers used to urge us to eat our fish, "otherwise our brains would not grow." They must have had some secret knowledge!

The endocranial capacities of *H. habilis* reveal a mean volume of 640 cm^3 (Tobias, 1997), which exceeds the mean value of 451 cm^3 for *Australopithecus africanus* by 189 cm^3. That is, when the values for fossil samples reasonably attributed to these two taxa are compared, the *H. habilis* value for mean absolute endocranial capacity exceeds that for *A. africanus* by 42%. When the absolute capacity values are related to estimates of body size, values may be obtained for *relative brain size*. These, too, show that *H. habilis* was significantly more encephalized than *A. africanus*. When these *encephalization quotients* (EQs) are expressed as percentages of the modern *H. sapiens'* relative brain size, the value for *A. africanus* falls at 46% and that of *H habilis* at 53%. That is, of all fossil hominin series for which EQs are determinable, *H. habilis* is the earliest for which the EQ is over half of that in modern *H. sapiens*; all of the determinable australopithecine species have EQs definitely below 50%.

It is a striking fact that this substantial increase in inferred brain size first becomes evident with the appearance of *H. habilis*. The earliest examples of the latter species occur in the fossil record after the tectonic and climatic changes of 2.6–2.5 mya. The drying and cooling of large parts of Africa would undoubtedly have affected the water resources available to the early hominins. The drying up of many streams might be expected to have set water supplies at a premium. Moreover, the tectonic elevation would have increased runoff and induced reversals of the direction of flow of some rivers. All of these direct and indirect sequelae of tectonic uplift, added to the general climatic desiccation, would have made water a more precious commodity to those living in the affected areas.

It goes without saying that fresh water had always been essential for survival. Under the more strained conditions just described, we may reasonably infer that an even closer proximity between man and water than before would have been of intense survival value. I envisage that such close relationship would have involved not only water for drinking and keeping cool, but also increased dependence upon aquatic food resources. At a time when selective pressures for larger brains must have been strong, the "brain foods" required to sustain brain development were acquired by increased, probably culturally influenced, foraging on aquatic plants and animals. By such a route, it is not difficult to envisage a causal link, or a set of links, between the more exacting challenges for survival and the selectively determined and palaeontologically testified increase in absolute and relative brain size. Water and edible aquatic organisms provided the crucial catalyst when early *Homo* confronted this evolutionary bottleneck. While many species did not survive the crisis of 2.6–2.5 mya, the hominins, some of whom were armed with genetic mutations for larger brains, a penchant for water foods, and stone culture, won through to become the diverse larger-brained species of Pleistocene mankind.

Modern humans need fresh water for drinking. In 1968, L.S.B. Leakey declared that, in order to seek early human remains, it was necessary to find a site that was near water. Prior to that, in his book, *Water, Weather and Prehistory*, Robert Raikes described in 1967 how, in his hydrological surveys in Baluchistan and the eastern Mediterranean, he learned to read the antiquity or recency of fresh water springs from the adjacent presence or absence of suitably ancient stone tools! Water is also essential for keeping cool. Third, as this book makes abundantly clear, water is a source of aquatic plant and animal foodstuffs. These studies have shown *inter alia* the important role that the aquatic food chain plays in providing an abundance of DHA and of AA, and their significance for the development and healthy functioning of the brain.

Waterways have been both deterrents to, and facilitators of, the dispersal of humans throughout Africa, across the Old World, and even into the New World. It has been suggested that the movement of peoples out of Africa tended to follow beaches, shorelines, and river banks, while "island hopping" was a likely means of peopling islands and landmasses beyond stretches of water. Strolling along the beach or swimming or floating or rafting would have been sufficient to carry mankind out of Africa – as Stringer suggested in 2000 and as I had proposed in 1998 in *Water and Human Evolution*. When much water was bound up on land in the Ice Ages, sea levels were lower than they are today. Previously submerged land bridges and insular stepping-stones appeared. At such times, it would have been possible to walk dry-footed from Tripoli and Tunisia to Malta, Sicily, and Sardinia; from South Korea to South Japan and from the Sakhalin Island/Peninsula to Hokkaido, North Japan; from Malaysia to Sumatra, Java, Madura, and Bali; and from Siberia to Alaska.

As noted by Morwood et al. (1998), the crossing by humans – and elephants (*Stegodon*) – into the Indonesian islands of Wallacia, especially Flores, nearly 1 mya raises interesting questions of the antiquity of simple watercraft and of hominids swimming, perhaps together with those splendid swimmers, the elephants. The presence of hominid remains in Iberia, both northern and southeastern Spain, certainly at 1 mya (Aguirre and Carbonell, 2001) and possibly at 1.5 mya (Gibert et al., 1989; Tobias, 1995) raises in an acute form the problem of the route by which these earliest Europeans reached Iberia (Tobias, 2002). The long journey through the Levantine Corridor around the eastern Mediterranean, and the land bridge from North Africa through Malta, Lampedusa, and Sicily to Peninsular Italy, or from North Africa to Sardinia and Corsica, and thence to Peninsular Italy, are several possible routes by which hominids might have migrated out of Africa, then moved westward and finally southward across the Pyrenees to Iberia. By analogy to the crossing from the southeast Asian Sunda Shelf to Flores, there is a reasonable likelihood of a water crossing with "island hopping" from Morocco and Ceuta to Spain. It has been estimated that at times of lower sea levels, with the emergence of a few islands that are at present submerged, and of a small peninsula hanging off the southern shelf of Iberia, the maximum required water crossing would not have exceeded 5 km!

Watercraft or rafts or floating tree trunks might have enabled some adventurous members of early *Homo* to cross. In this context, I propose that swimming has been a hominid activity for over a million years.

The presence of *Stegodon* along with stone tools in Flores across the deep strait of the Wallace Line, and the occurrence of the Algerian elephantid *Mammuthus* in Iberia of the early Pleistocene, raise this question: when one considers the close and intimate relationship that exists between elephants and their mahouts in the Indian subcontinent in recent centuries, it is interesting to speculate as to whether a relationship existed between the movement of early humans and early elephantids across such straits as that between the Sunda Shelf and Flores. If man and mammoths or man and *Stegodon* had a close relationship over a long period of time, it is not impossible that in some way, the *Stegodon* might have facilitated the crossing by humans of the Flores Straits 900,000 years ago. Moreover, could those North African mammoths have been associated in any way with the crossing by humans, from Ceuta and Morocco into Iberia, as long ago as 1.5 mya? At any rate, the possibility should certainly not be dismissed that human beings have been able to swim, or at least paddle with floats such as bladders, for a very long time.

A fifth way in which water is thought to have affected human evolution is an old proposal that mankind evolved some of its distinctive features in an aquatic environment (the Aquatic Ape Hypothesis). Although the idea is commonly ascribed to Sir Alister Hardy (1960), he was preceded by Wood Jones in England in 1929, G.L. Sera in Italy in 1938, Max Westenhöfer of Germany in 1942, and B. Henneberg. Supporters of the Aquatic Ape Hypothesis have listed many features, including bipedalism, voluntary breath holding, hairlessness, the distribution, excessive number, and nature of sweat glands, subcutaneous fat, excessive urination, absence of sun-reflecting fur, poor water drinking capacity, and water retention.

Until recently, the evolution of early hominids in the savanna has been a strongly held, prevailing hypothesis. Yet some of these human characteristics would have made us hopeless savanna dwellers. On the other hand, a number of our features align us with marine mammals, even including face-to-face copulation. Hence the suggestion by Hardy and others was, modestly, that human ancestors must have spent more time in the water in the early days of human evolution. Under these circumstances, it was proposed, humans lost most body hair, developed the layer of subcutaneous fat, and other features. Hardy's work was largely ignored by his contemporaries, but Elaine Morgan (1982, 1990, 1997), Marc Verhaegen (Verhaegen et al., 2002), Michael Crawford, Stephen Cunnane, Leigh Broadhurst, and others have revived interest in the fundamentally sound merits of aquatic diets and habitats, especially for the brain. My own paper, "Water and Human Evolution," (Tobias, 1998) gave a spurt to the resurgence of active interest.

In sum, it is widely accepted that the competing savanna hypothesis is no longer tenable, since I amassed much evidence against it at University College London in 1995. Therefore, I believe that scientists now have a duty to re-examine the evidence for a closer link between hominids and aquatic environments.

(i) The role of waterways in hominid development highlights a real problem that needs to be addressed. We need new investigations such as by fresh, open-minded research students and post-doctoral fellows.

(ii) I am not yet convinced that all of the traits included in the original Aquatic Ape Hypothesis can be reasonably attributed to that hypothesis. Research on those traits should be updated.

(iii) We should not telescope too many phases and characteristics of hominid evolution into a single, over-arching hypothesis.

(iv) Above all, let us keep our thought processes open to changes of paradigms, and especially to the change which would be necessitated with growing evidence of the role of waterways in hominid evolution.

(v) Finally, the role of water, while long appreciated and emphasized by ecologists, has been sadly neglected by human evolutionists.

This volume edited by Stephen Cunnane and Kathy Stewart is a puissant move away from the heavy, earthbound view of hominid evolution and a move toward a greater emphasis upon the role of water and waterways in hominid development, survival, and diversification. I hereby express my personal tribute and admiration to them for conceiving and editing this volume. As always my warm thanks are extended to Mrs. Felicity Krowitz.

Phillip V. Tobias, FRS
University of the Witwatersrand
Johannesburg, South Africa

REFERENCES CITED

AGUIRRE, E. and CARBONELL, E. 2001. Early human expansions into Eurasia: The Atapuerca evidence. *Quaternary International* **75**:11–18.

GIBERT, J., CAMPILLO, D., EISENMANN, V., et al. 1999. Spanish Late Pliocene and early Pleistocene hominid, Paleolithic and faunal finds from Orce (Granada) and Cueva Victoria (Murcia). *Human Evolution* **14**: 29–46.

HARDY, A. 1960. Was man more aquatic in the past? *The New Scientist* **7**:642–645.

HORROBIN, D.F. 2001. *The Madness of Adam and Eve: How Schizophrenia Shaped Humanity*. London: Bantam Press.

MORGAN, E. 1982. *The Aquatic Ape*. London: Souvenir Press.

MORGAN, E. 1990. *The Scars of Evolution*. London: Souvenir Press.

MORGAN, E. 1997. *The Aquatic Ape Hypothesis*. London: Souvenir Press.

MORWOOD, M.J., O'SULLIVAN, P., AZIZ, F., et al. 1998. Fission-track ages of stone tools and fossils on the east Indonesian island of Flores. *Nature* **392**:173–176.

STRINGER, C. 2000. Coasting out of Africa. *Nature* **405**: 24–27.

TOBIAS, P. V. 1995. The bearing of fossils and mitochondrial DNA on the evolution of modern humans, with a critique of the "mitochondrial Eve" hypothesis. *South African Archaeological Bulletin* **50**:155–167.

TOBIAS, P. V. 1997. Evolution of brain size, morphological restructuring and longevity in early hominids. In *Principles of Neural Ageing*, eds. Dani Su, A. Hori, and G.F. Walter, pp. 153–174. Amsterdam: Elsevier.

TOBIAS, P. V. 1998. Water and human evolution. *Out There* **35**:38–44.

TOBIAS, P.V. 2002. An Afro-European and Euro-African human pathway through Sardinia, with notes on humanity's world-wide water traversals and proboscidean comparisons. *Human Evolution* **17**:157–173.

VERHAEGEN, M., PUECH, P.-F. and MUNRO, S. 2002. Aquarboreal ancestors? *Trends in Ecology and Evolution* **17**:212–217.

INTRODUCTION

Kathlyn M. Stewart and Stephen C. Cunnane

Paleoanthropologists and others have traditionally viewed human brain capacity as a matter of fine-tuning over time by natural selection. However, there is no evidence that early hominins, including most australopithecines, "had acquired cognitive capacities significantly different from those of today's apes" (Tattersall, Chapter 1). It is not until about 2.5 Ma that the appearance of larger-brained hominins, with rough stone tools, signaled a cognitive advance. Long periods of "adaptive stasis" followed by cognitive advances and increased brain size also characterize the later evolution of *Homo*. With environmental changes in habitat and food resources being a primary driver in evolution, it follows that at several points in time, such changes contributed significantly to major changes in hominin morphology and behavior. This volume is founded on the premise that continued expansion of the hominin brain required significantly increased abundance and quality of food sources.

Much has been written about the high energy requirements of the human brain, and the need for high-quality food sources to fuel the adult brain. The adult human brain consumes about 23% of the body's energy requirement – much more than in other similarly sized mammals. There is clear evidence today that low intakes of specific dietary nutrients can be limiting for brain development and function. For instance, it is now well established that deficient intake of the omega-3 fatty acid docosahexaenoic acid (DHA) increases the risk of subnormal brain development. In fact, to insure normal growth of the fetal and infant brain, human babies need two dietary long-chained polyunsaturated fatty acids (PUFA): DHA and arachidonic acid (AA).

AA is found in many foods, with significant amounts in egg yolk, meat, seafood, and fish. Dietary sources of preformed DHA, however, are more or less confined to freshwater and marine fish and shellfish, and a few organs, notably the brain. Edible fats in the soft tissues of mammals and other land-based animals are less accessible, contain much lower amounts of long-chain PUFA, and are not consistently available. Given the human brain's specific requirements for these high-quality fatty acids, as well as iodine and iron, the earliest hominins must have evolved in an environment that could provide food resources with abundant availability of these nutrients (referred to as *brain-selective nutrients*). Otherwise, it seems empirically evident that what limits human brain development and function today would surely also have limited its evolution. This volume focuses on the biochemical and nutritional requirements of encephalization of the human brain, and provides several lines of evidence as to how hominins exaptively fulfilled these requirements.

Others have also argued that hominins evolved in an aquatic, wetlands, or shore-based environment (e.g., Hardy, 1960; Morgan, 1982, 1990, 1997; Verhaegen et al., 2002). However, they have focused more on morphological and physiological adaptations to aquatic environments and have said little about the implications of such an environment for the hominin brain. Most of the research studies dealing with the modern human brain and its high energy requirements have been published in journals and books focusing on metabolism or nutrition. Only occasionally has there been any peer-reviewed crossover into the paleoanthropological literature. Similarly, the morphology, ecology, and behavior of early hominins are not common topics in biochemical publications. Rare peer-reviewed exceptions have synthesized data and hypotheses from both fields and have focused on the importance of fish and aquatic invertebrates in underpinning the growth, development, and evolution of the human brain (Broadhurst et al., 1998).

In February 2006, a symposium was organized by the editors of this volume, and held at the Annual Meeting of the American Association for the Advancement of Science (AAAS), in St. Louis, Missouri. The aim was to have a discussion linking the fields of nutrition and metabolism of the brain, and morphology, behavior, and ecology of early hominins. The hope was to further stimulate this interdisciplinary crossover, thereby fostering a better understanding of the essential prerequisites for evolution of the human brain. The papers presented at that meeting (by Ian Tattersall, Michael Crawford, Stephen Cunnane, and Kathlyn Stewart) are published here, along with those of other experts in these disciplines.

The contributors to the present volume come from several fields – paleoanthropology, nutrition, neurochemistry, archaeology, and paleobiology – providing a multidisciplinary approach to the complex and constantly changing topic of the evolution of the hominin brain. In Chapter 1 of the volume, Ian Tattersall provides the conceptual underpinning for the volume with an overview of the evolution of the hominin brain and, more particularly, the development of human cognitive thought. He indicates that the emergence of human cognition was a process of long adaptive stasis interrupted by short periods of cognitive advance. These cognitive advances were not linear and directed, but were acquired indirectly as part of changing human behaviors; in other words, through the process of *exaptation*. One of these exaptive changing behaviors was the move toward consuming a higher quality diet. In particular, Tattersall links the emergence of *Homo sapiens* and the capacity for symbolic thought with evidence of opportunistic exploitation of coastal marine resources.

Michael Crawford in Chapter 2 discusses the role of lipids, in particular DHA and AA, in brain development, focusing on the evolution of the eye and neural systems in vertebrates. Crawford indicates how variety in lipid structure, with literally thousands of options when the various fatty acids are paired up in structural lipids, helped support evolutionary diversity. Nevertheless, DHA is the one constant in the visual and nervous systems in all organisms, attesting to its singular importance for signal transmission in these systems. One of his most compelling arguments about the impact of DHA for brain evolution involves brain size vis-à-vis body size. In mammals consuming a terrestrial diet low in DHA and AA, as body size increased from very small rodents to massive savanna herbivores, *with just one exception*, relative to the rest of the body, brain size did the reverse – it shrank. The exception is humans. The question is why.

Stephen Cunnane (Chapter 3) suggests that the cognitive abilities of the human brain were able to evolve because hominins not only accessed a reliable long-term supply of brain-selective nutrients, but because body fat evolved in the fetus and infant. Among primates, neonatal body fat is unique to humans. Infant body fat not only provides a store of fatty acids for energy metabolism in general but is also an important reserve of DHA. The adult human brain requires about 23% of the body's total energy needs, but in the infant, this figure is an astounding 74%. Infant body fat is crucial as a source of fatty acids converted to ketones used by the brain whenever glucose is less available. Cunnane suggests humans are still evolutionarily dependent on shore-based food resources, as exemplified by the widespread occurrence of iodine and iron deficiencies. These problems particularly affect cognitive development, and occur in inland and mountainous regions, but are uncommon in coastal areas or where fish and seafood are common in the diet.

Tom Brenna in Chapter 4 notes that "Humans are the species of the runaway brain." He details how DHA is different from two other long-chain PUFA (the two docosapentaenoic acids) that are structurally very similar to but cannot functionally replace DHA. He notes that for normal development, the disproportionately large human brain requires a disproportionate intake of DHA, sufficient amounts of which must be acquired preformed exogenously, that is, through the diet. Thus, he concludes that humans must have inhabited environments that provided a "ready-made" dietary source of DHA.

Frits Muskiet and Remko Kuipers (Chapter 5) examine the role of DHA and AA from the perspective of dietary differences in present day "traditional" populations in East Africa. As a proxy for Paleolithic diets, they compare the fatty acid composition of breast milk from women from marine and/or freshwater fish-eating societies in eastern Tanzania, with that from women who do not eat fish. Their findings indicate that the higher DHA in breast milk contrasts with the populations that do not consume fish.

Crawford, Cunnane, Brenna, and Muskiet and Kuipers have all provided arguments for the need for brain-selective nutrients, particularly DHA and iodine, for normal development of the neonatal and infant brain, for maintenance of the adult brain and, by association, for human brain evolution. But what mechanism(s) actually triggered the encephalization process? Sebastiano Venturi and Michel Bégin in Chapter 6 discuss the importance of iodine and thyroid hormone in brain development and function throughout human evolution. They also point to the important role iodine plays in defending fragile lipid molecules such as DHA and AA from oxidation. Similar to other studies (Crockford, 2003, 2008), they suggest that thyroid hormone and iodine were important components of a biological control mechanism that could potentially coordinate a suite of physiological, morphological, and behavioral changes, including the encephalization process.

These chapters provide persuasive evidence of the importance of brain-selective nutrients, particularly DHA and AA, in the successful development of the human brain and, by inference, the encephalization of the hominin brain. The importance of freshwater and/or marine fish and invertebrates are repeatedly emphasized as excellent sources of these essential nutrients. However, through morphological and behavioral analogy with extant apes, the diet of the earliest hominins is usually reconstructed as being based primarily on fruits and leaves. The chapters by Erlandson, Stewart, Shabel, and Parkington provide data and inferences on the shift from a plant-based diet to aquatic and marine foods, which are much more abundant in brain-selective nutrients.

Jon Erlandson, in his overview of fish procurement by hominins in Chapter 7, suggests that coastal marine foraging or fishing has been given a minor role in hominin evolution, due in large part to high present-day sea levels which have drowned coastal evidence of such activities. Non-archaeological and archaeological evidence alike increasingly points to the use of coastal (and freshwater; see Chapter 8) dispersal routes used by hominins. Dispersal along these routes implies exploitation of the available aquatic foods. Increasing evidence of fish and shellfish procurement by Neanderthals and anatomically modern *H. sapiens* indicates that fish and shellfish were of high importance to hominins long before the appearance of intensive fishing and fishing implements about 10,000 years ago.

Fossil evidence for consumption of fish and aquatic invertebrates by early hominins is presented in Kathlyn Stewart's Chapter 8. She suggests that early hominins exploited freshwater environments and consumed wetlands vegetation, supplemented by occasional consumption of invertebrates, small fish, and larger slow-moving fish such as catfish. Enhanced climatic change in the Plio-Pleistocene, particularly in the dry seasons, intensified hominin use of wetlands and wetlands resources, with increasing exploitation of fish and aquatic invertebrates. Enhanced dry periods resulted in a scarcity of terrestrial foods and forced hominins to rely on aquatic food sources. Stewart suggests that periods of aridity, increased consumption of freshwater resources, and isolation in areas of potable water produced circumstances favorable to the encephalization process.

Also providing evidence for hominin exploitation of foods at the water/land interface is Alan Shabel's "durophage-ecotone" model in Chapter 9. This model reconstructs the robust australopithecines as opportunistic consumers of hard-shelled foods obtained at the land/water interface, including eggs, crabs, and mollusks, as well as nuts. These foods all contain the brain-selective nutrients linked to encephalization. The model is based on the craniodental traits, particularly cranial capacity, of the carnivoran species that forage for hard-shelled prey at the land/water ecotone. The suite of features that characterizes the skulls of the wetland durophages (hard-object eaters) is also exhibited by the robust australopithecines.

John Parkington in Chapter 10 examines the emergence of *H. sapiens* in the Middle Paleolithic in Africa. He lays to rest the old paradigm that presented southern Africa as a "backwater," where archaic hominins and stone tools existed alongside *H. sapiens* in Europe, with its flourishing art and artifacts. He presents evidence arguing for the emergence of anatomically modern humans in southern Africa between about 260,000 and 110,000 years ago. Somewhat toward the end of this period, sites emerged containing technologically "innovative" stone assemblages, and other modified bone and bead artifacts, and ochre. Almost all of these innovative assemblages are associated with shell middens near the southern African coast, indicating exploitation of coastal food resources. Parkington suggests that these anatomically modern humans in coastal southern Africa, with their innovative technologies and behaviors, were systematically consuming shellfish and other marine food resources; this diet may have been a key factor facilitating their encephalization.

In short, following Tattersall's overview of the development of cognitive thought in humans, the first half of this volume focuses specifically on the biochemical and nutritional requirements of encephalization of the human brain, best acquired exaptively through consumption of fish and/or shellfish. The papers in the second half provide multidisciplinary evidence on the exploitation of initially, freshwater, and later, marine fish and shellfish, by successive hominin taxa. The increased consumption of fish and shellfish, the increasing climatic aridity, and the isolation of hominin groups in areas of potable water produced circumstances favorable to the encephalization process.

We invite you, the reader, to explore the evidence gathered together here, for the first time, and form your own opinion as to the role freshwater and marine foods could have played in human brain evolution.

ACKNOWLEDGMENTS

The editors would like to express their gratitude to the contributors whose expertise and insights into the links between brain evolution and diet made this volume possible. This volume builds on the work of others before us, and we gratefully acknowledge in particular the foundation work by Alister Hardy and Elaine Morgan. We are also very grateful to Karen Chambers at Wiley-Blackwell for her ongoing support and enthusiasm in bringing this volume to fruition. Notwithstanding the patience and professionalism of other Wiley-Blackwell staff, we take full responsibility for errors and shortcomings.

REFERENCES

Broadhurst, C.L., Cunnane, S.C., and Crawford, M.A. 1998. Rift Valley lake fish and shellfish provided brain-specific nutrition for early *Homo*. *British Journal of Nutrition* **79**:3–21.

CROCKFORD, S.J. 2003. Thyroid rhythm phenotypes and hominid evolution: A new paradigm implicates pulsatile hormone secretion in speciation and adaptation changes. *Comparative Biochemistry and Physiology (Part A)* **135**:105–129.

CROCKFORD, S.J. 2008. *Rhythms of Life: Thyroid Hormone and the Origin of Species*. Victoria: Trafford Publishing.

HARDY, A. 1960. Was man more aquatic in the past? *The New Scientist* **7**:642–645.

MORGAN, E. 1982. *The Aquatic Ape*. London: Souvenir Press.

MORGAN, E. 1990. *The Scars of Evolution*. London: Souvenir Press.

MORGAN, E. 1997. *The Aquatic Ape Hypothesis*. London: Souvenir Press.

VERHAEGEN, M., PUECH, P.-F., and MUNRO, S. 2002. Aquarboreal ancestors? *Trends in Ecology and Evolution* **17** (5):212–217.

CONTRIBUTORS

Michel E. Bégin, Research Center on Aging, Health and Social Services Centre, Sherbrooke Geriatrics University Institute, Sherbrooke, Québec, Canada J1H 4C4; Email: beginme@videotron.ca

J. Thomas Brenna, Division of Nutritional Sciences, Cornell University, Ithaca, NY 14853; Email: jtb4@cornell.edu

Michael A. Crawford, Institute of Brain Chemistry and Human Nutrition, London Metropolitan University, London N7 8DB; Email: michael@macrawf.demon.co.uk

Stephen C. Cunnane, Research Center on Aging and Departments of Medicine and Physiology and Biophysics, Université de Sherbrooke, Sherbrooke, Québec, Canada; Email: stephen.cunnane@usherbrooke.ca

Jon M. Erlandson, Department of Anthropology and Museum of Natural and Cultural History, University of Oregon, Eugene, OR 97403-1224; Email: jerland@uoregon.edu

Remko S. Kuipers, Laboratory Medicine, University Medical Center, Groningen, The Netherlands; Email: remkokuipers@hotmail.com

Frits A.J. Muskiet, Laboratory Medicine, University Medical Center Groningen (UMCG), RB Groningen, The Netherlands; Email: f.a.j.muskiet@lc.umcg.nl

John Parkington, Department of Archaeology, University of Cape Town, Rondebosch 7701, Capetown, South Africa; Email: john.parkington@uct.ac.za

Alan B. Shabel, Department of Integrative Biology, University of California, Berkeley, CA 94720; Email: shabel@berkeley.edu

Kathlyn M. Stewart, Paleobiology, Canadian Museum of Nature, Ottawa, K1P 6P4, Canada; Email: kstewart@mus-nature.ca

Ian Tattersall, Division of Anthropology, American Museum of Natural History, Central Park West at 79th Street, New York, NY 10024; Email: iant@amnh.org

Sebastiano Venturi, Tre Genghe n. 2; Department of Health, Marche Region, 61016-Pennabilli (PU), Italy; E-mail: venturis@freeweb.org

CHAPTER 1

MACROEVOLUTIONARY PATTERNS, EXAPTATION, AND EMERGENCE IN THE EVOLUTION OF THE HUMAN BRAIN AND COGNITION

Ian Tattersall

INTRODUCTION

In 1950 the late Ernst Mayr, one of the principal architects of the Evolutionary Synthesis, sketched human evolution as a simple, unilinear affair involving the gradual transformation over time of a minimal number of species within a simple lineage (Mayr, 1950). Imbued with authority by the seductive reductionism of the Synthesis, a view of the evolutionary process which preached that virtually all evolutionary phenomena ultimately boiled down to gradual within-lineage gene frequency shifts under the guiding hand of natural selection, Mayr's formulation rapidly took over the paleoanthropological mind-set (Tattersall, 1995, 2009).

This takeover was bolstered by the unassailable observation that *Homo sapiens* is the only hominid species in the world today, something that had already conditioned hominid paleontologists and others to think of human evolution as a process of steady refinement of a single entity over the eons, a matter of gradual progress from primitiveness to perfection. For, while not eliminating the possibility of extinct collateral relatives, projecting a single species back into the past inevitably results in the perception of a central "main line" of evolution. Additionally, as the product of a species that prides itself above all on its cognitive differences from the rest of the living world, this perception was amplified by the knowledge that the average size of hominid brains has fairly steadily increased over time – or at least, as we now realize, over the last 2 million years of human evolution – and the strong temptation has been simply to join up the dots.

NATURAL SELECTION

What makes the idea of natural selection so compelling is that this process appears on the face of it to be an inevitable consequence of the production of more offspring than survive to reproduce. And in concert with the various influences just adumbrated, this fact has resulted in a strong tendency among paleoanthropologists to look upon the process by which a nonsymbolic, nonlinguistic ancestor became transformed into a symbolic, linguistic descendant, as little more than a matter of generation-by-generation fine-tuning. After

Human Brain Evolution, Edited by Stephen C. Cunnane and Kathlyn M. Stewart

all, to smart members of a smart species, any incremental increase in intelligence inevitably looks like an obvious advantage that should continuously feed back into itself via reproductive gain. But is this a valid perspective? There are good reasons, both theoretical and empirical, for thinking not.

Let us start with the concept of natural selection itself. The basic notion here is that "fitter," or better adapted, individuals will outreproduce the less fit, and thus that their heritable advantages will become more common in the evolving population. However, this reasoning really only works when we think of beneficial characteristics as evolving in isolation. For, in reality, each organism is an integrated whole made up of huge numbers of usually polygenic traits that are typically specified by variably regulated pleiotropic genes. And natural selection can, by its very nature, only vote up or down on the reproductive success of the whole individual, and not on that of its individual features (Tattersall, 1999). Still less can it focus on particular genes. Yet we still tend to speak of such things as "the evolution of the brain" or "the evolution of the foot" or the "evolution of the gut," as if these structures somehow existed separately from the entire organisms in which they are and were embedded, and as if each had an evolutionary history independent not only of them, but also of the other traits that together make up the functioning whole.

Of course, some attributes may well make an absolutely critical contribution to reproductive success and may thus be individually subject to strong natural selection. But it is entirely probable that such features are largely limited to those that directly relate to the reproductive process itself. It is, for example, very likely to be no accident that testis size is consistently greatest in those primate species with polygamous mating systems (Harcourt et al., 1981). Chimpanzee males compete vigorously for females, and it is very highly plausible that their remarkably large testes result directly from a history of sperm competition. Gorilla males, on the other hand, equally plausibly have smaller testes because an excess of metabolically expensive reproductive tissue is unnecessary where single males monopolize groups of females for extended periods of time. Similarly, most well-documented cases of sexual selection seem to be convincingly attributable to pressures for the origin and maintenance of the structures concerned.

Still, most animal tissues and activities are not devoted to reproduction per se but rather to economic ends (Eldredge, 2004), and neither affect nor reflect reproduction except in indirect and complex ways. If an individual is not economically successful, it is highly unlikely to be reproductively successful and in this context, it may plausibly be argued that reproductive structures and activities are little more than a veneer imposed on the basic economic machinery.

The upshot is that natural selection of the kind that promotes population change, in contrast to population stability, probably typically acts to propagate a very large and diverse subgroup of reproductively superior individuals whose economic performance is not necessarily any more than adequate. After all, evolution is constrained in ways that preclude it from being a process of structural optimization (Gould, 2002); and what succeeds most of the time is merely what works. Reproductive success often will not involve being the best, but simply being good enough.

MACROEVOLUTION

Local diversification within any species is a critically important evolutionary phenomenon; and, in conjunction with speciation, it provides a crucial link between microevolutionary and macroevolutionary processes. However, most of the time, rather than being directional, it seems to work within (to misappropriate a term from Eldredge, 2003) what might be termed a "sloshing bucket" principle, whereby genetic and phenotypic frequencies move back and forth within the containing wall of the species, the occasional overflow being lost. This is what might be expected in a world in which, as the paleoclimatic and faunal/floral records show, environments have typically changed on timescales far too short to be tracked by gradual generation-to-generation natural selection. If your habitat changes, you are much more likely to become at least locally extinct, or to migrate to a more congenial environment, than to adapt *in situ* (Eldredge and Gould, 1972).

Of more routine evolutionary importance, then – and certainly in promoting macroevolutionary trends and patterns – is economic competition among species as wholes. As just suggested, environmental changes will generally result in the movement of floras and faunas – and it is, after all, of very little use to be the best adapted individual of your species – whatever that may in practice mean – if your entire species is being outcompeted into extinction by a new arrival, closely related or otherwise.

There are, then, many reasons to suppose a priori that sustained directional change within lineages should be at least a rare evolutionary phenomenon. This expectation is borne out in the hominid case by examination of both the fossil and the archaeological records. In neither case do we find evidence of continuous modification over time; instead, in both, substantial change is highly intermittent – with the significant exception, in very recent times, of technological development. Interestingly, despite the similarity in overall pattern, there is no synchronicity between major innovations in the biological and technological realms. This only seems counterintuitive, though, until one realizes that it is uniquely *within* a species that anatomical novelties and technological inventions can occur (Tattersall, 1998).

PATTERNS IN HUMAN EVOLUTION

The African origins of the hominid family itself are still somewhat mysterious, but recent finds strongly suggest that the hominid family tree was bushy from the very start, some 7 million years ago (Fig. 1.1). In other words, the early history of Hominidae was pretty conventional, in the sense that the dominant signal emerging from our family's fossil record is consistently one of evolutionary experimentation, of an ongoing exploration of the many ways that there evidently are to be hominid.

The one common feature claimed for all of the very early fossil hominids is terrestrial bipedality (Gibbons, 2006). But while it was clearly the precursor of modern human upright bipedalism, the locomotion of the "bipedal apes" of the period from about 6 to 2 million years ago can hardly be viewed as "transitional" between ancestral arboreality and modern striding terrestriality. The "bipedal ape" body plan involved a unique combination of characteristics that would have been useful in an arboreal context even with the key characters of terrestrial bipedality. And this "eat-your-cake-and-have-it" structure was evidently a stable and successful one, remaining essentially unchanged for several million years even as a host of ancient australopith species came and went.

Despite their innovative body structure, from nowhere during this period of adaptive stasis is there any evidence that our ancient precursors had acquired cognitive capacities significantly different from those of today's apes. This is not to belittle the capacities of the archaic hominids: cognitively, today's apes are very complex beings (de Waal, 2005), and doubtless so were our distant ancestors. But at about 2.6 million years ago, we do begin to find striking indications, in the form of the appearance of crude but effective stone tools, of a recognizable advance in cognitive abilities. For it takes an insight into the fracturing properties of brittle materials well beyond what any ape has yet displayed, even

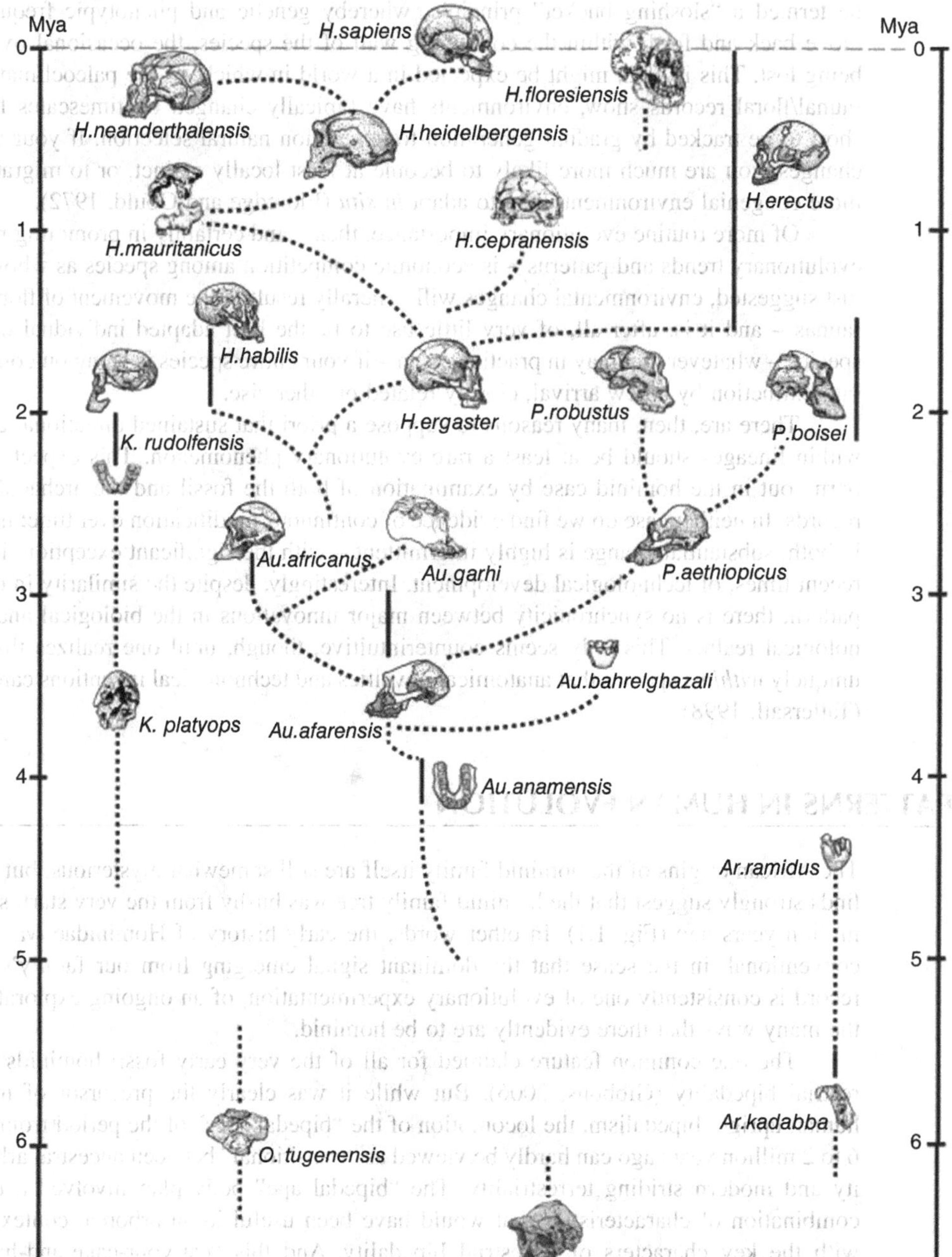

Figure 1.1 One possible scheme of phylogenetic relationships within the family Hominidae, with time on the vertical axis. Solid lines show stratigraphic ranges. Many of the details of relationship are entirely conjectural, and this diagram shows most importantly that typically several different hominid species have coexisted at any one point in time. It is very much the exception that *Homo sapiens* is the lone hominid in the world today. © Ian Tattersall.

after long and intensive coaching, to hit one cobble with another at exactly the angle and force necessary to detach a sharp cutting flake (Schick and Toth, 1993). The exact identity of the first stone toolmaker is not known, but almost certainly, he or she was of archaic body build, and possessed a brain not much larger than one would expect in an ape of similar size. The earliest stone toolmakers were, in other words, australopiths.

The earliest stone toolmakers simply sought an *attribute*: a sharp cutting edge. What the sharp flake actually looked like was of no importance. But following an extended period of stasis, at about 1.6 million years ago toolmakers began to make "hand axes," implements of a standardized and symmetrical shape that evidently corresponded to a "mental template" that existed in their minds before toolmaking began (Schick and Toth, 1993). Significantly, this indirect evidence of another cognitive advance was bequeathed us by a new kind of hominid that had, significantly, already been around for several hundred thousand years *before* this innovation was made.

These creatures, who appeared on the scene at a little under 2 million years ago, were the first hominids of essentially modern body size and build: the earliest members of a morphologically coherent genus *Homo* (Wood and Collard, 1999). With brains perhaps a little bigger than those of the bipedal apes, but still at best not much more than half the size of ours today, these early upright striders were the first hominids to be truly emancipated from the forest edge and woodland habitats to which their precursors were largely confined. And, still wielding only crude stone tools, they rapidly spread far beyond Africa – a development made possible, it seems, by the new body form alone (Tattersall, 1997).

Following the invention of the hand ax, there is once again a long wait for the next major technological innovation, wherein a stone "core" was carefully shaped until a single blow would detach a more or less finished tool. This too surely signifies another notch-up in cognitive complexity (Klein, 1999). And again, this invention came a long time after a new kind of hominid had shown up in the fossil record, at about 600,000 years ago in Africa and shortly thereafter in Eurasia. It was hominids of this new species, *Homo heidelbergensis*, that some 200,000 years later apparently introduced such important novelties as the building of shelters (de Lumley and Boone, 1976), the regular domestication of fire in hearths (de Lumley, 1986), and the careful shaping of wooden throwing spears (Thieme, 1997). There is, however, nothing in the archaeological record of these hominids that convincingly suggests they indulged in symbolic activities of any kind.

Perhaps the most accomplished practitioners of prepared-core toolmaking were the Neanderthals, *Homo neanderthalensis*. It is this species, which flourished in Europe and western Asia following about 200,000 years ago, which provides us with the best yardstick by which to measure the uniqueness of our own species, *H. sapiens*. But while the Neanderthals had brains as large as ours, invented the burial of the dead, and probably took care of disadvantaged members of society, they too left little behind them to suggest unequivocally that they possessed symbolic consciousness (Klein, 1999). That is to say, their material products embodied nothing to suggest that they processed information about the world in the uniquely modern human way that involves deconstructing the environment into a mass of discrete mental symbols that can be recombined to pose "what if?" questions.

SYMBOLIC COGNITION

Beginning some 40,000 years ago (and paralleled by similar processes that apparently took place in eastern Asia at about the same time), the Neanderthals were entirely evicted from their vast territory, in not much more than 10,000 years, by arriving *H. sapiens* whose existences were very clearly drenched in symbol (White, 1982, 1986). These early European *H. sapiens*, known as the Cro-Magnons, created astonishing art on the walls of caves. They carved exquisite figurines. They decorated everyday objects and made notations on plaques of bone. They played music on bone flutes, and without question sang and danced as well. In short, they were *us*. And the material record they left behind is distinguished most notably from that of their Neanderthal contemporaries by its clear indications of a symbol-based mode of cognition.

Significantly, though, the Cro-Magnons were not the first creatures who *looked* just like us. The highly characteristic bony anatomy that distinguishes modern *H. sapiens* had its roots in Africa as long as 160–200,000 years ago (White et al., 2003; McDougall et al., 2005), long before we find the possible earliest intimations of symbolic behaviors on that continent at about 100–80,000 years ago (Deacon and Deacon, 1999; Henshilwood et al., 2003, 2004). Similarly, while anatomically modern *H. sapiens* appeared in the Levant at a little under 100,000 years ago (Valladas et al., 1988), these people made stone tools that were virtually indistinguishable from those made by the Neanderthals who continued to persist alongside or in alternation with them.

The final eviction of the Neanderthals from the Levant came only following the appearance there of stone tools equivalent to those the Cro-Magnons brought with them into Europe. This suggests that neither hominid species had an overall competitive advantage as long as the behaviors of both could be described as the most sophisticated extrapolations yet of the trends toward increasing brain size and cognitive complexity – in both lineages – that had preceded them. But once *H. sapiens* began to behave in a "modern" way, the Neanderthals were faced with an entirely unanticipated phenomenon. With the advent of that phenomenon, the rules of the game changed entirely, and our species became an irresistible force in Nature, intolerant of competition from close relatives, and with the ability to indulge that intolerance.

There is nothing in the record I have just briefly summarized to suggest that the acquisition of modern symbolic consciousness marked the culmination of a gradual trend through time, under the beneficent supervision of natural selection. Certainly, the acquisition of modern human cognition was *based* on what had gone before and could not have happened without it. But the event itself marked a qualitative leap, rather than a small final step in an ongoing process of refinement. Despite clear evidence of sporadically increasing cognitive complexity among hominids over the last 2 million years, the nature of modern hominid cognition was not predicted by what preceded it; for, rather than being an incremental improvement, it represented a radical departure from tradition. Various recent studies have pointed to hints in the earlier record of aspects of behavior that we commonly associate with modern humans; but very likely, these straws in the wind merely point to a complex cognition that was nonetheless not symbolic.

In this context, it is important not to be misled by the undoubted increase in the average size of hominid brains over the past 2 million years or so (Tattersall, 2008). Superficially, it certainly looks as if crude plots of average hominid brain sizes over time, such as that shown in Fig. 1.2, indicate a pretty consistent enlargement. This is, of course, true as far as it goes; and clearly, the extraordinary predisposition of members of the genus *Homo* to an increasing brain size certainly tells us something of fundamental importance about this group – something that we will have to understand before we can fully comprehend the evolutionary dynamic that was at work. However, it has to be remembered that not only is there no demonstrable link between raw brain size and "intelligence," but that such enlargement was clearly *not* the product of evolution within a single steadily modifying lineage. To the contrary, hominid history has been a story of vigorous evolu-

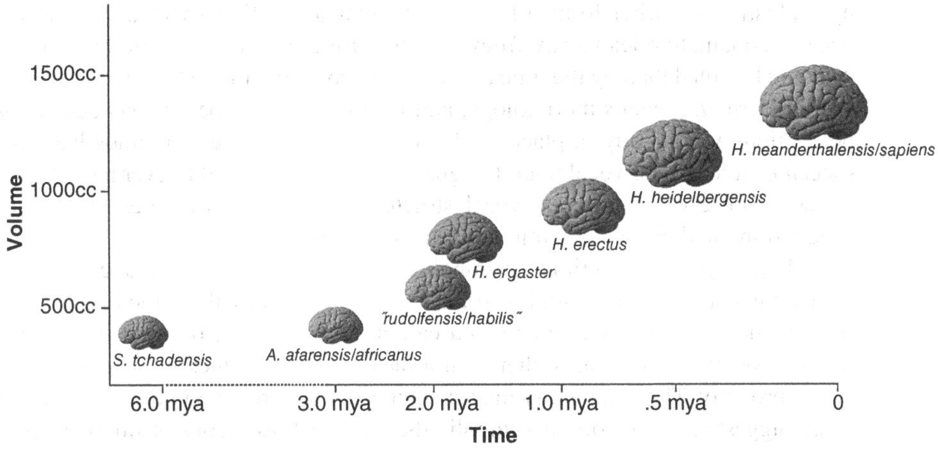

Figure 1.2 A crude plot of average hominid brain sizes against time. Following an initial flatlining this plot appears to indicate a consistent enlargement of the hominid brain over the last 2 million years. However, it is important to bear in mind that these brain volumes are averaged across an unknown number of different lineages within the genus *Homo*, and that it is likely that what the plot reflects is the preferential success over this period of larger-brained hominid species, rather than steady increase within a single lineage. Illustration by Gisselle Garcia. ©Ian Tattersall.

tionary experimentation, with several hominid species typically in existence at any one time: a diversity that is clearly evident from the hominid phylogeny depicted in Fig. 1.1. What we are thus seeing in the apparently inexorable increase in the amount of metabolically expensive brain tissue over the Pleistocene is almost certainly a product of the preferential survival of larger-brained species in multiple lineages within the genus *Homo* (Tattersall, 2004, 2008).

In any event, large brains by themselves are clearly not enough to assure symbolic consciousness. Neanderthals had brains of modern human size; but, cognitively sophisticated as they doubtless were, they failed to leave convincing evidence of symbolic behaviors, certainly in pre-contact times. Perhaps even more remarkably, this was also true for the first anatomically modern humans. The earliest potential *H. sapiens* fossils in Africa are associated with relatively unsophisticated stone artifacts, and the Early Moderns of the Levant some 90,000 years ago (Valladas et al., 1988) made stone tools just like those of the Neanderthals, with little if any sign of symbolic behaviors (Klein, 1999). There was thus a very considerable time lag between the acquisition of modern anatomy and the expression of modern behavior patterns, putatively as long ago as 75,000 years in Africa, and most dramatically expressed in Europe following about 40,000 years ago.

It has been eloquently argued that an enabling genetic change, whose effects were limited to brain activity, may have occurred and spread within *H. sapiens* in the period following about 50,000 years ago (Klein and Edgar, 2002). More likely, though, the neural capacity that underwrites the faculty for symbolic thought emerged with the substantial biological reorganization that accompanied the emergence of *H. sapiens* as a (highly) distinctive anatomical entity, at some point over 150,000 years ago (Tattersall, 2004, 2008). That potential must then have lain undiscovered and dormant for many millennia, until it was released by a cultural, rather than a biological, innovation.

The most plausible cultural releaser of this kind is the invention of language. This unique way of transferring knowledge, so intimately connected with the unique human way of processing information about the world, was clearly founded on a basis supplied by sophisticated earlier forms of vocal communication. Equally clearly, however, it represents a qualitative leap away from any other form of communication we know of. And it should be noted that, by the time that we have any good inferential evidence for language use, modern *H. sapiens* morphology, and thus the vocal apparatus necessary for speech production, was already in place – whether or not any earlier hominids had possessed a speech-enabled upper vocal tract. Language as we know it could never have been invented in the absence of appropriate vocal structures, but those structures had initially been acquired in another context entirely (Tattersall, 2008).

Language is the ultimate symbolic activity, involving as it does the creation of intangible symbols in the mind, and their recombination to allow the asking of questions such as the "what if?" one mentioned earlier. What is more (even if it is in essence an interior conduit to thought, rather than a means of communication), language is a communal property. This externality makes language more credible in the role of releaser than other suggested facilitators of symbolic thought, such as theory of mind (Dunbar, 1998) which are internalized.

EXAPTATION AND EMERGENCE

Whatever it may have been that ushered in the beginnings of symbolic thought in a species *H. sapiens* that, like all other organisms, had lived until that point in a concrete external world rather than in one constantly mentally remade, it is clear that symbolic thought itself cannot have been propelled into existence by natural selection. Indeed, natural selection is *not* a creative force; it can only exert itself on variations that come into existence spontaneously. In this sense, any useful novelty has to arise not as an *adaptation* but as an *exaptation* (Gould and Vrba, 1982): as a feature that is not acquired in the context of any function to which it might eventually be put. The vocal structures that make language possible are a prime example of this.

The bottom line here is that nothing arises *for* anything, and selection can only work with what is already there. Sometimes, novelties persist in populations for no better reason than that they do not get in the way. Looking at the matter from this perspective takes the origin of our vaunted human cognitive capacities out of the arena of gradual honing by natural selection, placing it instead in that of emergence (Tattersall, 2004, 2008). That is to say, an entirely new and unanticipated level of cognitive complexity was acquired as result of an entirely fortuitous coincidence of acquisitions. The whole, in other words, is greater than the sum of its parts.

Human cognition is without doubt the product of the human brain, which has had a long and accretionary evolutionary history reaching back to the very origin of the vertebrates. But although the record of vertebrate brain evolution yields a story of increasing cognitive complexity over hundreds of millions of years, nothing we know of in that history predicts symbolic thought. Evidently, at the origin of the anatomically distinctive species *H. sapiens*, a development that is exceedingly poorly documented in the fossil record (Tattersall and Schwartz, 2008), something was added to the ancestral brain that gave it an altogether new and unexpected potential. What that something was structurally is beyond my expertise to speculate; but recognizing the new capacity as a simple by-product of the larger biological reorganization that produced morphologically distinctive *H. sapiens* makes it altogether unsurprising that our species did not exploit this new potential immediately.

Indeed, a lag of this kind is entirely routine in terms of evolutionary process. Early bird relatives, for instance, possessed feathers for a very long time before co-opting them as essential components of the flight mechanism; and ancestral tetrapods initially developed their four limbs in an entirely watery context (Shubin, 2008).

Since the discovery some 100,000 years ago that the human brain could be used for symbolic manipulations, the history of our species has largely been one of finding new ways in which to deploy this cognitive capacity: a process that is, indeed, still under way. And because this emergent capacity of the human brain seems to be a general-purpose ability rather than a fine-tuned one, we can begin to see why the only species that as far as we know agonizes about its condition is the one whose condition is least well-defined. You can find an individual human being to illustrate both parts of any pair of behavioral antitheses you might care to mention; and neither state can be attributed to the ancient (and mythical) Environment of Evolutionary Adaptedness so beloved of evolutionary psychologists.

Much of what I have said about the appearance of modern human cognition also applies to earlier innovations in the human lineage. And this is important, because while we are naturally fascinated by the question of what the critical addition to human brain function was that gave rise to *H. sapiens* as we know it worldwide today, it remains true in this as in all other cases that the resulting capacity depends utterly on what was there before. What is more, the ancestral condition had its own accretionary history: a history that did not simply involve the acquisition of new structures as modifications of old ones, but that entailed the histories of the populations in which those innovations were embedded: populations that competed with others for ecological space and ultimately for survival. To build a complete picture, we therefore need to know not only *what* happened, but to understand the conditions that made it possible.

LARGE BRAINS AND AQUATIC RESOURCES

This volume deals with some of the prerequisites that underpinned the evolutionary enlargement of the hominid brain, in the sense that they made it possible for larger brains to be acquired and maintained among those hominid species that contrived to displace their smaller-brained brethren over the long span of the Pleistocene. The modern human brain and its normal cognitive functioning are known to be constrained by a host of dietary requirements, among them the availability of high-quality fatty acids (see Chapter 4). Resources must thus have been available, and exploited, to sustain the hominid brain enlargement that substantially preceded the acquisition of symbolic cognitive processes in our own lineage.

Brain enlargement to modern size actually occurred independently in at least two hominid lineages (those eventuating in *H. sapiens* in Africa and *H. neanderthalensis* in Europe); and a similar trend seems also to have characterized a third long-lived species (*Homo erectus* in eastern Asia), in which later members had significantly larger brains than early ones. Interestingly, the exploitation of coastal resources (including shellfish rich in omega-3 [ω3] fatty acids) has been imputed to Middle Stone Age presumed *H. sapiens* in southern Africa (Deacon and Deacon, 1999; Marean et al., 2007), and to late Neanderthals in Iberia (Finlayson, 2008; Stringer et al., 2008). According to Marean et al. (2007: 906–907), "The expansion to shellfish is one of the last additions of a new class of food to the human diet before the introduction of domesticates at the end of the Pleistocene." From the 165,000-year-old site PP13B at Pinnacle Point on the southern African coast, these authors recovered substantial evidence of shellfishing, and they believe that the high cognitive skills involved in shellfish-collecting and the presumably associated modern hunter-gathering would have provided "excellent contexts for stimulating symbolic expression through material culture" (Marean et al., 2007).

The supporting evidence for symbolic, as opposed to impressively complex, behaviors among the Pinnacle Point people is nonetheless rather inferential. Besides shellfishing itself, it consists of the use of ochre and of bladelet technology, both of them arguable as proxies for symbolic cognitive processes (Tattersall, 2008). On the other hand, it is very plausible to suggest that "shellfish may have been crucial to the survival of … [the Pinnacle Point] humans as they expanded their range to include coastlines." (Marean et al., 2007: 905). The implication here is that under certain circumstances, the availability of resources such as those the Pinnacle Point people exploited may be vital to the survival of large-brained hominids.

The case of the Iberian Neanderthals is different in one critical respect. There is good evidence for both shellfishing and the exploitation of marine mammals at sites in Gibraltar (Finlayson, 2008; Stringer et al., 2008). However, at the same time, there is a complete absence of any direct cultural evidence suggesting that the hominids living at the sites were symbolic. Thus, while it is entirely reasonable for Finlayson (2008) to propose that the availability of coastal resources was key to late Neanderthal survival in Iberia in the deteriorating climates of Isotopic Stage 3, it cannot be argued that the evidence from this region shows any link between shellfishing and an identifiable increase in cultural or cognitive complexity. Clearly, the existence of the energy-hungry large modern human brain, like that of the Neanderthals, must from the very beginning have been underwritten both by the availability and the exploitation of dietary sources yielding the nutritional elements necessary to sustain it (see also Chapter 3). But equally clearly, the ability to acquire such resources does not necessarily imply the possession of symbolic cognition.

REFERENCES

de Lumley, H. 1986. Les civilisations du Paleolithique inferieur en Languedoc mediterraneen et en Roussillon. In *La Prehistoire Francaise*, Vols. 1–2, pp. 852–874. Paris: CNRS.

de Lumley, H. and Boone, Y. 1976. *Les structures d'habitat au Paléolithique inférieur*. In *La Préhistoire Francaise*, Vol. 1, ed. H. de Lumley, pp. 625–643. Paris: CNRS.

de Waal, F.B.M. 2005. A century of getting to know the chimpanzee. *Nature* **437**:56–59.

Deacon, H. and Deacon, J. 1999. *Human Beginnings in South Africa: Uncovering the Secrets of the Stone Age.* Cape Town: David Philip.

Dunbar, R. 1998. Theory of mind and the evolution of language. In *Approaches to the Evolution of Language: Social and Cognitive Bases*, ed. J.R. Hurford, M. Studdert-Kennedy, and C. Knight. Cambridge: Cambridge University Press.

Eldredge, N. 2003. The sloshing bucket – How the physical realm controls evolution. In *Evolutionary Dynamics*, ed. J.P. Critchfield and P. Schuster, pp. 3–32. New York: Oxford University Press.

Eldredge, N. 2004. *Why We Do It: Rethinking Sex and the Selfish Gene.* New York: W.W. Norton.

Eldredge, N. and Gould, S.J. 1972. Punctuated equilibria: An alternative to phyletic gradualism. In *Models in Paleobiology*, ed. T.J.M. Schopf, pp. 82–115. San Francisco: Freeman Cooper.

Finlayson, C. 2008. On the importance of coastal areas in the survival of Neanderthal populations during the Late Pleistocene. *Quaternary Science Reviews*, **27**: 2246–2252, doi:10.1016/jquascirev.2008.08.033.

Gould, S.J. 2002. *The Structure of Evolutionary Theory.* Cambridge, MA: Belknap/Harvard University Press.

Gould, S.J. and Vrba, E. 1982. Exaptation – A missing term in the science of form. *Paleobiology* **8**:4–15.

Gibbons, A. 2006. *The First Human: The Race to Discover Our Earliest Ancestors.* New York: Doubleday.

Harcourt, A.H., Harvey, P.H., Larson, S.G., and Short, R.V. 1981. Testis weight, body weight and breeding system in primates. *Nature* **293**:55–57.

Henshilwood, C., d'Errico, F., Vanhaeren, M., van Niekerk, K., and Jacobs, Z. 2004. Middle Stone Age shell beads from South Africa. *Science* **304**:404.

Henshilwood, C., d'Errico, F., Yates, R., Jacobs, Z., Tribolo, C., Duller, G.A., Mercier, N., Sealy, J.C., Valladas, H., Watts, I., and Wintle, A.G. 2003. Emergence of modern human behavior: Middle Stone Age engravings from South Africa. *Science* **295**: 1278–1280.

Klein, R. 1999. *The Human Career.* 2nd ed. Chicago: University of Chicago Press.

Klein, R. and Edgar, B. 2002. *The Dawn of Human Culture.* New York: Wiley.

Marean, C.W., Bar-Matthews, M., Bernatchez, J., Fisher, E., Goldberg, P., Herries, A.I.R., Jacobs, Z., Jerardino, A., Karkanas, P., Minichillo, T., Nilssen, P.J., Thompson, E., Watts, I., and Williams H.M. 2007. Early human use of marine resources and pigment in South Africa in the Middle Pleistocene. *Nature* **449**: 905–909.

Mayr, E. 1950. Taxonomic categories in fossil hominids. *Cold Spring Harbor Symposiums on Quantitative Biology* **15**:109–118.

McDougall I., Brown, F.H., and Fleagle, J.G. 2005. Stratigraphic placement and age of modern humans from Kibish, Ethiopia. *Nature* **433**:733–736.

Schick, K.D. and Toth, N. 1993. *Making Silent Stones Speak: Human Evolution and the Dawn of Technology.* New York: Simon and Schuster.

Shubin, N. 2008. *Your Inner Fish: A Journey into the 3.5-Billion-Year History of the Human Body.* New York: Pantheon.

Stringer, C.B., Finlayson, J.C., Barton, R.N.E., Fernandez-Jalvo, Y., Caceres, I., Sabin, R.C., Rhodes, E.J., Currant, A.P., Rodriguez-Vidal, J., Giles-Pacheco, F., and Riquelme-Cantal, J.A. 2008. Neanderthal exploitation of marine mammals in Gibraltar. *Proceedings of the National Academy of Sciences U S A* **105**:14319–14324.

Tattersall, I. 1995. *The Fossil Trail: How We Know What We Think We Know About Human Evolution.* New York: Oxford University Press.

Tattersall, I. 1997. Out of Africa again ... and again? *Scientific American* **276** (4):46–53.

Tattersall, I. 1998. The origin of the human capacity. *James Arthur Lecture Series (American Museum of Natural History)* **68**:1–27.

Tattersall, I. 1999. The abuse of adaptation. *Evolutionary Anthropology* **7**:115–116.

Tattersall, I. 2004. What happened in the origin of human consciousness? *Anatomical Record (New Anatomist)* **267B**:19–26.

Tattersall, I. 2008. An evolutionary framework for the acquisition of symbolic cognition by *Homo sapiens. Comparative Cognition and Behavior Reviews* **3**:99–114.

Tattersall, I. 2009. *The Fossil Trail: How We Know What We Think We Know About Human Evolution.* 2nd ed. New York: Oxford University Press.

Tattersall, I. and Schwartz, J.H. 2008. The morphological distinctiveness of *Homo sapiens* and its recognition in the fossil record: Clarifying the problem. *Evolutionary Anthropology* **17**:49–54.

Thieme, H. 1997. Lower Palaeolithic hunting spears from Germany. *Nature* **385**:807–810.

Valladas, H., Reyss, J.L., Joron, J.L., Valladas, G., Bar-Yosef, O., and Vandermeersch, B. 1988. Thermoluminescence dating of Mousterian "Proto-Cro-Magnon" remains from Israel and the origin of modern man. *Nature* **331**:614–616.

White, R. 1982. Rethinking the Middle/Upper Paleolithic transition. *Current Anthropology* **23**:169–192.

White, R. 1986. *Dark Caves, Bright Visions: Life in Ice Age Europe.* New York: Norton.

White, T.D., Asfaw, B., DeGusta, D., Gilbert, H., Richards, G.D., Suwa, G., and Howell, F.C. 2003. Pleistocene *Homo sapiens* from Middle Awash, Ethiopia. *Nature* **423**:742–747.

Wood, B. and Collard, M. 1999. The human genus. *Science* **284**:65–71.

CHAPTER 2

LONG-CHAIN POLYUNSATURATED FATTY ACIDS IN HUMAN BRAIN EVOLUTION

Michael A. Crawford

INTRODUCTION – LIPIDS AND EVOLUTION

The earth's life history is usually divided sharply into two phases – an anaerobic and an aerobic phase. I propose adding a third phase, with each reflecting environmental chemistry as a driving force:

1. The *Precambrian era*: The stromatolite reef at Pilbara in Western Australia has been dated to 3.43–3.35 billion years old when it was submerged in a shallow sea (Allwood et al., 2006). Since then and for the next 2.5 billion years, the only fossils found in abundance are the *stromatolites*. This vast *Precambrian era* was dominated by *prokaryotes*, the *Cyanophyta* (blue green algae). Although photosynthetic, they show no intracellular detail.
2. The *Vendian and Cambrian eras:* Starting about 600 million years ago, oxygen tension in the atmosphere rose above the Pasteur point, at which point aerobic metabolism became thermodynamically possible (Holland, 2006). The 32 phyla we know today exploded into the fossil record in a short period of time thereafter and, in contrast to the prokaryotes, these fossils of *eukaryotes* provide considerable intracellular detail.
3. The *Cretaceous era*, during which flowering plants and the new reproductive system of the mammals evolved. By providing a new source of omega-6 *(ω-6) polyunsaturated fatty acids* (PUFA) from the parent fatty acid, *linoleic acid*, this new lipid chemistry ultimately made possible the evolution of the human brain.

Each of these phases is an example of a change in environmental chemistry. First, chemical evolution as a by-product of the supernova became biochemistry some 3 billion years ago. In the second phase, during the Vendian and Cambrian eras, biochemistry shifted up eightfold in efficiency with aerobic metabolism giving rise to the 32 phyla. The third phase during the Cretaceous was a lateral shift in the reproductive system that filled a gap in the requirements for the brain by providing ω6 PUFA required for mammalian reproduction to compliment the *ω3 PUFA – docosahexaenoic acid* (DHA) – with both PUFA classes being needed for optimal brain development.

Human Brain Evolution, Edited by Stephen C. Cunnane and Kathlyn M. Stewart

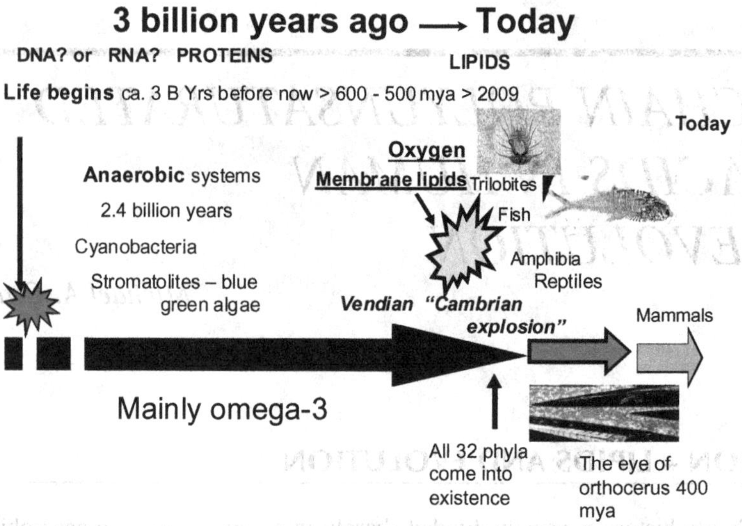

Figure 2.1 There was little change in the design of Precambrian prokaryote life forms during the Precambrian era despite ample opportunity for mutational change to provide for substantial modification during this 2.5-billion-year period. This prolonged stasis is powerful evidence for two forces in evolution – *natural selection and conditions of existence*. Insufficient atmospheric oxygen tension was a "condition of existence" that prevented advancement to more complex life forms. The *Cambrian explosion* is clear evidence of the importance of Darwin's conditions of existence.

THE EVOLUTION OF COMPLEX LIFE FORMS

The Precambrian era was occupied by prokaryote life. There was little change in the design of the life forms despite the 2.5-billion-year time period offering ample opportunity for the rate of mutational change to provide for substantial modification. This 2.5-billion-year stasis is powerful evidence for Darwin's view that there were two forces in evolution: "*natural selection and conditions of existence*." Of the two, he wrote, the latter was the most powerful. The absence of sufficient atmospheric oxygen tension was a "condition of existence," which did not permit advancement to more complex life forms. When the Pasteur point was breached at which point oxygen metabolizing systems became thermodynamically possible, all the current 32 phyla appeared in the fossil record with a remarkable suddenness (see Fig. 2.1). This *Cambrian explosion* is, therefore, clear evidence of the importance of Darwin's conditions of existence, and is consistent with the recent realization that the gene-centric obsession in evolution is misplaced in evolution (Crawford and Marsh, 1995); both the Precambrian stasis and the Cambrian explosion are clear and incontrovertible evidence of the power of the environment over gene mutations in affecting evolution.

In eukaryotic evolution, intracellular structures appear in the fossil record. This intracellular detail is made largely of membrane lipid bilayers and their embedded proteins. The organization of cellular structures was made possible by membrane lipids. It seems likely that extensive speciation was not only a product of the rise in the oxygen tension but was also due to the cell structural complexity in which the lipids would have played an important role in generating intracellular specialization and then speciation.

Was Increased Complexity of Lipids Responsible for Cell Structures and Specialization among the Eukaryotes?

The state of cellular lipids leading up to the Cambrian explosion can only be a topic of conjecture. Freshwater algae operate more anaerobically than seawater algae and synthesize little of the long-chain PUFA (20 and 22 carbons with 3–6 double bonds). Moreover, anaerobic systems, as in the gut flora or rumen, use unsaturated fatty acids as hydrogen acceptors, resulting in the production of trans isomers and saturated fatty acids. It is likely then that in the anaerobic period (Precambrian), long-chain PUFA would have been rare.

Oxygen is required in the desaturation reactions converting shorter- to long-chain PUFA. The synthesis of the six double bonds in DHA requires six oxygen atoms without including the energy requirement for the chain elongations. These lipids are today used for the organization of complex cellular structures, that is, those in the *reticular endothelium*, *mitochondrial electron transport systems*, *nuclear envelope*, and plasma membranes that accommodate receptors, transporters, signaling systems, and antioxidant enzymes. Oxidative metabolism brought with it the emergence of the PUFA, which would have provided a great wealth of novel architectural possibilities to evolve sophisticated organization and functional accommodation of proteins, a process not previously possible.

THE LANGUAGE OF LIPIDS

The language of DNA operates with four letter words, with several words forming a sentence that eventually is translated into a protein. The proteins are built with 20 words that are assembled into functional three-dimensional structures. Because of the molecular movement of the final assembly of a protein, it is an example of supramolecular chemistry which includes reversible *noncovalent associations*, *hydrogen bonding*, *metal coordination*, *π–π interactions*, and electrochemical effects involving *lipophilic* and *hydrophilic* assemblies. In that sense, a protein in a living cell exists in five dimensions: the fourth dimension being the *electrochemical profile* and the fifth dimension being the *van der Waals-type forces.*

The van der Waals equation in its first form (from Wikipedia) expresses relationships with lipids:

$$\left(p+\frac{a'}{v^2}\right)(v-b')=kT$$

Where:

p is the pressure of the fluid

v is the volume of the container holding the particles divided by the total number of particles k is Boltzmann's constant which is the physical constant relating energy at the particle level with temperature. The Boltzmann's constant is the gas constant R (pressure-volume measure) divided by the Avogadro constant (the number of atoms or molecules in one mole of a gas) N_A:

$$k=\frac{R}{N_A}.$$

T is the absolute temperature

a' is a measure for the attraction between the particles

b' is the average volume excluded from v by a particle

The interest in the van der Waals equation is that it coincidentally describes the properties of lipids in which a' will vary with the chain length and degrees of unsaturation, which

TABLE 2.1 Principal glycerophospholipid molecular species in the language of lipids associated with fatty acids acylated in sn-1 and sn-2 position of the glycerol backbone and sn-3 attached to choline, ethanolamine, serine, or inositol polar head groups = 352 words, sphingomyelin = 11 words, cholesterol = 1 word, total 364 basic words

sn-1			sn-2		
16:0	16:0	18:0	20:0	22:0	24:0
16:1	16:1	18:1	20:1	22:1	24:1
18:0		18:2	20:2	22:2	
18:1		18:3	20:3	22:3	
			20:4	22:4	
			20:5	22:5	
				22:6	

alters the pKa of the acid and will influence the polarity of the whole lipid. *b′* will also vary with physical chain length of the fatty acid (16–24 carbons) and degree of unsaturation (1–6 double bonds) as well as with lipid concentration as lipids form micelles and other macromolecular structures in an aqueous milieu. *T* and *p* are functional determinants of the degree of unsaturation. Hence, with the electrical properties of the polar head groups and *ceramides*, based on the same principles as described for proteins, lipids themselves can be considered to exist in five dimensions but varying in biological systems depending on temperature and pressure. Moreover, a number of isomeric structures exist for several of the monounsaturated fatty acids and PUFA that give rise to an even greater variety of fatty acid combinations. In addition to the most common *diacyl phosphoglycerides*, there are *choline* and *ethanolamine plasmalogens, cardiolipin, sphingolipid*, and *glycosphingolipid* species which include the many *galactocerebrosides* found in neural tissue, and *glucocerebrosides* present in muscle and other tissues, contributing another considerable number of words to the lipid dictionary. The language of DNA has 4 words (nucleotides) and that of proteins 20 words (amino acids). However, there are 364 common lipid words (molecular species) and several hundred organelle- and tissue-specific, less ubiquitously used, yet highly abundant lipid words. The greater the number of words in a language, the greater the potential for expression and subtlety of meaning (see Table 2.1).

The cell membrane is the first point of contact between the cell and its external environment. Cell behavior is known to respond to diet, temperature, and pressure. Moreover, throughout biology, the lipid composition of cells varies with temperature and pressure. With humans being homeothermic, the main variables become the diet and environment, and there is an abundance of evidence that diet, stress, exercise, drugs, and toxins affect the composition of the acyl groups, which are a major part of the alphabet of the lipid molecular species.

Membrane proteins are the receptors, signalers, and transporters. Thus, the environment talks to the proteins through the language of lipids and influences their behavior which in turn influences the cell nucleus. The specific interactions between the lipid and the protein may be chemical or electrical. It is a two-way phenomenon in that the protein will demand a multidimensional and thermodynamic fit with the lipid on the one hand, while on the other, the lipid can be manipulated by diet and is modified by environmental stress. The variation of the language alters the behavior of the cell by influencing protein function, as well as the signaling that affects genomics. In addition, there are several lipid-derived molecules actively involved in cellular signaling pathways as well as individual fatty acids which in the free form act as ligands for nuclear receptors (Chawla et al., 2001). Again, there is likely to be a stoichiometric relationship between the small amount of ligand released in the free form and the membrane concentration.

From an evolutionary perspective, the first 2.5 billion years passed with no significant change apparent in the prokaryotes. That stasis must say something about the stability or conservation of their DNA. So, in face of this rigidity, what caused the sudden about-face with the creation of the 32 phyla during the Cambrian explosion? How was it that the DNA and its proteins which had done little, if anything, for 2.5 billion years, suddenly got into top gear and life forms started to change so radically? One likely explanation is the enhancement of cellular energy production using oxygen. With metabolic events happening at greater speeds, there was probably more opportunity for the DNA to mutate. However, once the 32 known phyla were formed, no new phyla appear subsequently in the fossil record over the next 500–600 million years despite continued use of oxygen.

Another possibility not mutually exclusive to the first is that the changes occurring during the Cambrian explosion were a function of the dimensional leap of increased complexity of the lipids brought about by aerobic metabolism; that is, the simple lipid language of prokaryotes with only a few words was transformed into a dictionary with over 1,000 words. The introduction of order and organization by the membrane lipids was then subject to different environmental conditions in chemistry, temperature, and pressure, thereby altering the language of the lipids, including the way they talked in domains around the proteins and other membrane structures including *rafts* and *caveolae*.

The influence of environment as a rapid and powerful force in evolution is seldom given its full due but is exemplified by the fact that Western human populations have changed in shape, size, and disease patterns in just one century. With the genome being well conserved in time and the proteins dictated by the genome, there has to be a variable to explain evolution at a chemical and molecular level. The ability of the environment to effect change can be witnessed both in the Cambrian explosion and in the very recent changes in human paradigm of shape, size and chronic disease, in one century. There can be little doubt that the lipids are just that kind of variable.

DHA

In all vertebrates, docosahexaenoic acid (all-cis-docosa-4,7,10,13,16,19-hexaenoic acid – C22:6ω3, DHA) is the major PUFA constituent of the brain (Crawford and Sinclair, 1972; see also Chapter 4). There is a paucity of DHA in the land food chain that also contains competing saturated fatty acids and ω6 PUFA (Hibbeln, 2007). The brain first evolved using the marine food web some 500–600 million years ago. Since then, DHA has been conserved as the principle lipid-molecular component of visual and neural signaling membranes in the cephalopods, fish, amphibian, reptiles, birds, mammals, and humans. This is an example of extreme conservation. The preservation of DHA in neural signaling systems which occurred despite the genomic changes over 600 million years implies that DHA could actually have been dictating to the DNA rather than the more conventional view of evolution occurring the other way round.

The ω-6 docosapentaenoic acid (all-cis-docosa-4,7,10,13,16-pentaenoic acid – C22:5ω6, DPA) and the ω3 DPA (all-cis-docosa-7,10,13,16,19-pentaenoic acid – C22:5ω3) differ from DHA by the absence of just one double bond but neither replaced DHA during 600 million years of evolution (see Fig. 2.2). Fluidity cannot be the answer to the selection of DHA over either of the two DPAs because this structural difference between them is very small and certainly not enough to explain 600 million years of conservation

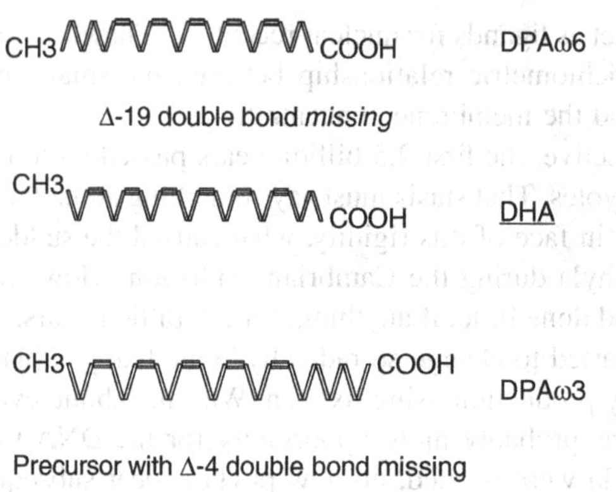

Figure 2.2 Docosahexaenoic acid (DHA, C22:6ω3) compared with two docosapentaenoic acids (DPAω6 and DPAω3; both with 22 carbons and 5 double bonds). The latter has only one double bond different from DHA yet neither replaced DHA in 600 million years of evolution.

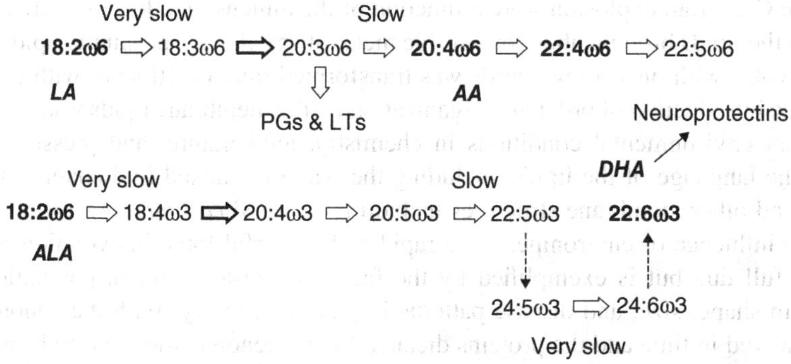

Figure 2.3 Synthesis of arachidonic acid (20:4ω6) and docosahexaenoic acid (22:6ω3) from their precursors, linoleic acid (18:2ω6) and α-linolenic acid (18:3ω3), respectively. Plant seeds are rich in linoleic acid and green leaves in α-linolenic acid. Meat, eggs, and human milk are rich in arachidonic acid, while fish, shellfish, eggs, and small mammals are rich in docosahexaenoic acid. Docosahexaenoic acid is the precursor to neuroprotectins while arachidonic acid is the precursor to eicosanoids and leukotrienes which, on balance, are commonly associated with inflammatory processes. Conversion of linoleic acid and α-linolenic acid downstream is very slow. Synthesis of docosahexaenoic acid involves peroxisomes.

(Bloom et al., 1999). The possibility that DHA has unique electrical properties not fully shared by either of the DPAs could provide the explanation (Crawford et al., 2008).

Nature's preference for DHA in the brain is strikingly demonstrated in large, vegetarian land mammals. The omega-3 (ω3) DPA is the dominant ω3 PUFA found in non-neural tissues and is thus abundantly available in the body, much more so than DHA (Crawford et al., 1969). However, neural membranes even in these large mammals still specifically conserved DHA. During evolution of the land mammals, this retention of DHA composition was associated with economy in brain size relative to the body, with a logarithmic *reduction* as they evolved larger bodies (Crawford et al., 1993).

The structure of the brain is 60% lipid. Neural cells have a particularly high membrane content of DHA. In different mammalian species, the content of both arachidonic acid (20:4ω6, AA) and DHA does not vary: it is brain size that varies (Crawford et al., 1976, 1993; Figs. 2.1 and 2.3). DHA is rapidly and selectively incorporated in neural membranes and is concentrated at synaptic signaling sites (Suzuki et al., 1997). It is the most unsaturated of cell membrane fatty acids (Jump, 2002). Some DHA can be synthesized from alpha (α)-linolenic acid, but the process is highly rate-limited (Sprecher, 1993, Sprecher et al., 1999; Plourde and Cunnane, 2007) and, moreover, α-linolenic acid is oxidized at a very rapid rate of 60% in 24h compared with only <5% for DHA (Leyton et al., 1987; Freemantle et al., 2006).

In 1972, Crawford and Sinclair first published evidence that DHA itself was an independent determinant of brain growth and evolution (Broadhurst et al., 2002). DHA deficiency studies in rodents (Sinclair and Crawford, 1972; Benolken et al., 1973; Galli and Socini, 1983; Weisinger et al., 1999; Catalan et al., 2002), chickens (Budowski et al., 1987), and primates (Fiennes et al., 1973; Neuringer et al., 1986), and visual and cognitive trials in human infants (Carlson and Werkman, 1996; Martinez and Vasquez, 1998; Birch et al., 2000) support the concept that DHA is essential to brain development and function. Moreover, in collaboration with Professor Pierre Budowski at the Hebrew University of Jerusalem, we described competition existing between ω6 and ω3 PUFA and showed that their balance is critical for brain development and structural integrity (Budowski and Crawford, 1985). We will return to the question of balance later when we consider the evolution of mammals and *Homo sapiens.*

EVOLUTION OF *HOMO SAPIENS*

Starting about 50 million years ago, certain mammals migrated to the coastal regions and then left the land and radiated into the marine habitat. With unlimited access to the DHA food web, the marine mammals retained a far higher brain-to-body weight ratio than is seen today in the large land mammals. For example, the dolphin (*Tursiops truncatus*) has a 1.8-kg brain, which compares to little more than 350g in the zebra *(Equus quagga)*, which has a similar body weight to the dolphin and is also a nonruminant. A coastal ecological niche would have provided the rich source of DHA, iodine, and other trace elements essential to the brain and in poor supply on land (see also Chapter 3). Such an environmental and dietary niche would have offered a substantial evolutionary advantage compared to that of the land food chain and so would have avoided the loss of relative brain capacity as occurred in all large land-based mammals.

In Fig. 2.4, we plot the approximate arithmetic decline of brain size in some land mammals. Because of the discrepancy between body size and brain size, some have used logarithmic plots to obtain straight lines to explore the relationship. This strategy of course means that one of the parameters is varying logarithmically to the other. In fact, in land mammals, brain size diminishes logarithmically with the increase in body size. Even with logarithmic correction, *H. sapiens* and the marine mammals do not fall on the straight line. Of the large mammals, the dolphin with about 1% of its body weight as brain comes the closest to *H. sapiens*. At just under 1.86%, *H. sapiens* has a brain/body weight ratio that would be totally exceptional if considered as a large, land-based mammal. Interestingly, all the very small mammals have brain/body weight ratios similar to or greater than *H. sapiens.*

The synthesis of DHA from the plant-based ω3 fatty acid – α-linolenic acid – is powerfully rate limited (Plourde and Cunnane, 2007; see also Chapter 4). As land-based mammals evolved larger and larger bodies, their increased growth velocity outstripped the ability to make DHA. These data suggest that the decrease in relative brain size in land mammals as their body size increased can be simply explained by the lack of DHA in the land food web. Single amino acids in the diet are incorporated directly into protein whereas the same is true for some dietary fatty acids. However, α-linolenic acid is of negligible structural importance (especially in the brain) and has to be chain elongated and

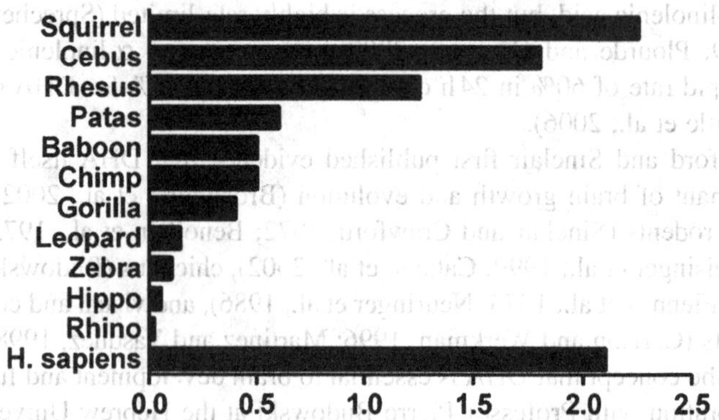

Figure 2.4 Approximate brain weight as a proportion of body weight in mammals declines markedly as body weight increases.

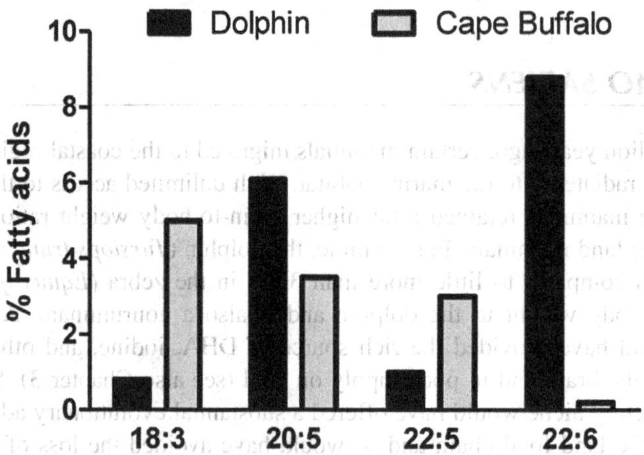

Figure 2.5 Polyunsaturated fatty acids in liver ethanolamine phosphoglycerides of the dolphin (*Tursiops truncatus*) compared to the Cape Buffalo (*Syncerus caffer*).

desaturated to DHA before it becomes available for membrane synthesis. Note that in Fig. 2.5, buffalo liver lipids are quite rich in α-linolenic acid, eicosapentaenoic acid (EPA), and even the ω3 DPA, but despite this wealth of precursor, the buffalo fails to synthesize significant DHA, leaving little to be transferred to the developing fetus. In this respect, the contrast with dolphin lipids is striking; the dolphin has access to ample DHA from the marine food chain and develops a much bigger brain than the large land mammals.

Both AA and DHA are needed for normal development and function of the brain (Crawford and Sinclair, 1972). In contrast to the land-based mammals, the difficulty the dolphin and other marine mammals have is in obtaining sufficient AA from the marine food chain to serve the needs of the brain and reproduction. Hence, AA supply could well be a constraint on brain evolution in the marine habitat and is also required for mammalian reproduction (Williams and Crawford, 1987). A littoral ecosystem would have provided an evolving hominin with abundant access to both AA and DHA and hence would have had the best of both worlds (Crawford et al., 1999; Broadhurst et al., 2002).

One needs to turn the clock back to 5–7 million years ago. For the previous 45 million years or so land-based mammals had consistently exploited the richness of coastal habitats and food resources. First, they would have eaten the seaweeds, mussels, cockles, oysters, and fish stranded in rocky pools when the tide receded. Slowly, they became more adventurous and began collecting food from the shoreline, wading, and then diving and learning to swim. The land to sea migration that led to the creation of the marine mammals finished about 7 million years ago. It may be just a coincidence, but that is about the time that the geneticists tell us that the primate that came down to *H. sapiens* separated from the great apes. Nature uses every possible ecological niche from the rigors of the Arctic to the deserts and high mountains, resulting in the polar bear, the desert rat, and the snow leopard, so it is unlikely that the richest food resource on the planet at the coastline would have remained empty. If hominids separated genetically, that meant they separated geographically; otherwise, there would still have been interbreeding. There is the classical view that hominins were on the savannas of Africa, but that would have committed them to a land-based food chain with little DHA, in which case hominids would have undergone the universal collapse of brain size as the big ruminants and primates evolved larger and larger bodies, which is not what happened.

One of the justifications for the savanna hypothesis was that hominins developed upright stance because of the need to see over the tall grasses to throw spears at animals. But this does not bear scrutiny. First, if hominins were making spears, they already had highly evolved brains. Second, one does not hunt animals by standing upright and declaring one's presence, but instead one crawls like the big cats. Darwin comments on the Aborigines' "usual manner of crawling after wild animals" and their ability to conceal themselves on almost bare ground. Another idea proposed as favoring hominin evolution on the savannas is that food could be acquired by "endurance running" to find carcasses (Bramble and Lieberman, 2004). This idea, although published in Nature, seems quite naïve. Dogs can outrun humans. They have an acute sense of smell that can detect a dead animal at long distances. Humans have a poorly developed sense of smell. Vultures can see dead or dying animals from great distances. Both dogs and vultures would beat any endurance running hominin to a carcass even if the latter knew where it was. Moreover, the fact that we lose water to keep the body temperature stable means any endurance runner would have had to carry water.

There is now good documentation of great apes knuckle-walking down to rivers and then standing perfectly upright when they enter the water. Indeed, at the end of a BBC-TV film celebrating Darwin's birth, Sir Richard Attenborough showed a film of apes walking upright in the river. He made the comment that this was how the chimpanzees took their first steps to become humans. The line that came to down to *H. sapiens* probably wandered down a river and came to an estuary. There they would have seen the seabirds feasting on shellfish. It would not take a bright little primate long to recognize the wealth of food ready for the picking. Even a pregnant woman would be able to feast off the seafood with little effort and in so doing would have fed her fetus with a rich source of DHA, iodine, and other trace essentials (Cunnane, 2005). Some people criticize the coastal origin by saying that humans could have obtained their DHA from brains and marrow of dead animals. Well, a small brain is not going to go far among the ladies even if it was still in an edible condition when they got it back after their endurance run. The evolution of humans at a coastal rather than a land-based hunting system is now well explained by the evidence on ω3 PUFA and, in particular, on DHA in neural gene expression (Barcelo Coblijn et al., 2003a,b; Kitajka et al., 2002, 2004; Yavin et al., 2009). DHA is by far the dominant ω3 fatty acid in the brain (Crawford et al., 1976). Preformed DHA is far superior for brain growth and development compared to its synthesis from plant-based α-linolenic acid, even in rodents (Sinclair,

1975). Lipid biochemistry therefore puts the evolution of *H. sapiens* firmly at the marine, riverine, and lacustrine coastlines with access to preformed DHA from the aquatic resources. Indeed, the concept that *H. sapiens* actually went through an aqueous phase was put forward by Sir Alistair Hardy (1960) and was followed up in several books written by Elaine Morgan (1995, 1997) and others (Crawford and Marsh, 1995; Cunnane, 2005).

Chris Stringer (2000) has suggested that *H. sapiens* populated the planet by migrating "out of Africa" around the coastlines. A coastal migration route would have clearly meant using shore-based components of the marine food chain. There is incontrovertible fossil evidence of use of the marine food web dated to a time close to the biological emergence of modern humans (Broadhurst et al., 2002; Marean et al., 2007; see also Chapter 8). There is also contemporary evidence of fishing people in the Rift Valley of Africa, with healthier cardiovascular health profiles than their inland cousins (Pauletto et al., 1996; Crawford et al., 1999), and contemporary evidence of the Moken and other sea dwellers living around the coast of Asia with a healthy lifestyle, possibly as a remnant of this migration (Gislén et al., 2003).

In a simple yet devastating comment for the savanna hypothesis, Phillip Tobias noted at a McCarrison Society conference at the Zoological Society of London ("A New Light on Human Origins," September 22, 2000) that wherever humans were evolving, they had to have water to drink. Previously in his review of the 1998 Dual Congress of Paleoanthropologists and Biologists in Sun City, South Africa, he listened to the evidence on the coastal origins of *H. sapiens* and the need for the proper nutrition of the brain, and wrote about the demise of the savanna hypothesis and the inherent plausibility of coastal origins of humans saying that "... we were all profoundly and unutterably wrong" (Tobias, 1998). It takes a great man to concede in such a manner that much of what he believed in during his academic career was misplaced.

DHA AND NEURAL PATHWAYS?

A key characteristic of dietary ω3 PUFA and brain DHA deficiency is reduced learning capacity and behavioral pathology. We were the first to describe the behavioral pathology in an ω3 PUFA-deficient primate (Fiennes et al., 1973). A similar condition is seen in Dr. Joe Hibbeln's work at the NIH USA (Hibbeln et al., 2004, 2007). The selective uptake of DHA by the synapse has been shown by Suzuki et al. (1997). Brain DHA depletion as a consequence of dietary ω3 PUFA deficiency is associated with loss of learning capacity, a feature reported consistently by many authors since it was first described over 30 years ago (Lamptey and Walker, 1976; Bourre et al., 1989). The loss of brain DHA from studies on cognition in older rats would be expected to be associated with increased memory loss (Yamamoto et al., 1991). So how could DHA be involved in the learning process?

The signaling function of the photoreceptor is more easily studied than the synapse and its function has been well described. The electrical activity of the retina is directly related to the presence of DHA (Benolken et al., 1973). The memory or visual impairment in ω3 PUFA deficiency involves a specific deficit of DHA as it is the only significant ω3 PUFA in the brain (see Fig. 2.6; Anderson and Maude, 1972; Neuringer et al., 1986; Birch et al., 2000). The brain turns over and conserves its lipid constituents rather than importing them. The lipid chemistry of the photoreceptor is dominated by DHA which fits tightly with the receptor molecule – *rhodopsin* (Bazan, 2007). The photoreceptor sheds its DHA receptor discs when exposed to light, a process that involves loss of DHA which is then recovered (Wheeler et al., 1975; Rodriguez de Turco et al., 1992; Wiegand et al., 1995). However, no recycling process is 100% efficient, so there is some continual loss and with that loss, functional decline of vision with ageing.

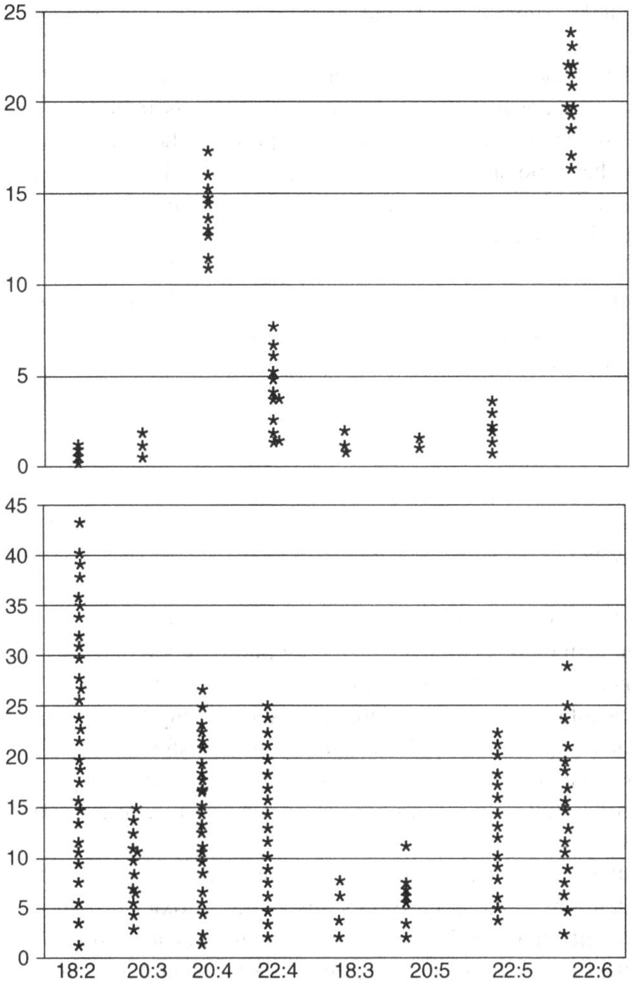

Figure 2.6 The composition of polyunsaturated fatty acids in inner membrane phospholipids varies widely in mammalian liver but not in the brain (n = 42 species; from Crawford et al., 1976).

So what is the role of brain DHA in learning and memory of the subject learnt? Let us assume the letter A is seen on a teach-yourself typing screen. The response of putting the left hand's small finger on the second end key on the left of the 4th row of the PC keyboard requires the correct visualization of A, its recognition requiring a motor response, transmission to a motor section of the brain, identification of the hand and then the small finger, the left hand, and then the transmission of the message in 3D to the small finger. That neural pathway has to be learnt so that when the photoreceptors and then the brain call for the letter A, the correct response is elicited. Learning requires repetition. In the repetition process, the synapses fire and reconstitute. With remodeling of the pre- and postsynaptic membranes and/or selective uptake of DHA, the synapses in the pathway will be enriched. Such a process could be expected to facilitate the release of neurotransmitters. The more enriched the synapse, the better its function which is the converse of the ω3 PUFA deficiency experiment that depresses learning ability. Repetition enriches a neural pathway just as water takes the path of least resistance when flowing down a hill or a pathway is formed in a country when people repeatedly walk in the same direction. This idea is supported by the evidence on the interaction between exercise and learning on synaptic systems underlying learning and memory (Chung et al., 2008; Chytrova et al., 2009).

This concept of memory is not independent of other similar concepts of protein activation and synthesis except that the evidence on memory is mostly published with respect to ω3 PUFA deficiency. One would expect the proteins that are encoded by DNA to be robustly built to the same specification regardless of diet. However, the lipids and lipid composition is subject to environmental inputs and variation. It might be argued that protein-calorie malnutrition can also lead to poor learning capacity. However, protein-calorie malnutrition also involves PUFA deficiency (Smit et al., 2004) as the foods linked to it are deficient of both protein and PUFA (Crawford, 1971). This proposed function of a DHA-enriched synaptic pathway would facilitate conduction of a signal and the establishment and function of a neural pathway.

A COMMENT ON AA

It is often reported that AA can be synthesized from linoleic acid without difficulty. In the early isotope studies by James Mead (Steinberg et al., 1956), the recovery of isotope from linoleic acid in AA was very small, a point reiterated in a recent review on human PUFA metabolism (Plourde et al., 2007). Similarly, Andrew Sinclair demonstrated a strong preference for uptake of preformed AA into the developing rat brain compared to its synthesis from linoleic acid (Sinclair, 1975). The human placenta is extremely rich in AA (Bitsanis et al., 2005), so the placenta might be considered as a super pump transferring AA to the fetus. Indeed, the human placenta is much more active in transferring AA than DHA to the fetus. There is also evidence from myographic studies of vascular function that AA acts as an endothelium-derived relaxant, a property not observed with DHA or with EPA (Crawford et al., 2003).

However, the inflammatory and thrombotic mediators derived from AA are almost always the focus of attention, to the point of labeling AA as a toxic substance. However, this view distorts the reality that much of the time, vasodilatory and anti-adhesive *eicosanoids* are being synthesized from the rich source of AA in the endothelium. Indeed, the Food and Drug Administration commented on the problem of aggressive pharmaceutical suppression of COX2 as adversely affecting *prostacyclin* synthesis and hence contributing to the excess of deaths in patients being so treated (e.g., Ray et al., 2002; James and Cleland, 2004).

There are some locations in the brain with much more AA and its elongation product than DHA. Serhan et al. (2008) make the point in a recent paper:

> The popular view that all lipid mediators are pro-inflammatory arises largely from the finding that *non-steroidal anti-inflammatory drugs* block the biosynthesis of *eicosanoids*. The resolution of inflammation was widely held as a passive event until recently, with the characterization of novel biochemical pathways and lipid-derived mediators that are actively turned on in resolution and that possess potent anti-inflammatory and pro-resolving actions. A lipid-mediator informatics approach was employed to systematically identify new families of endogenous local-acting mediators from ω3 PUFA (EPA and DHA) in resolving exudates, which also contain aspirin-triggered *lipoxins* generated from AA.

Figure 2.7 illustrates the reason why there is a question regarding the capability of the preterm infant to efficiently synthesize AA from linoleic acid. While there are many issues

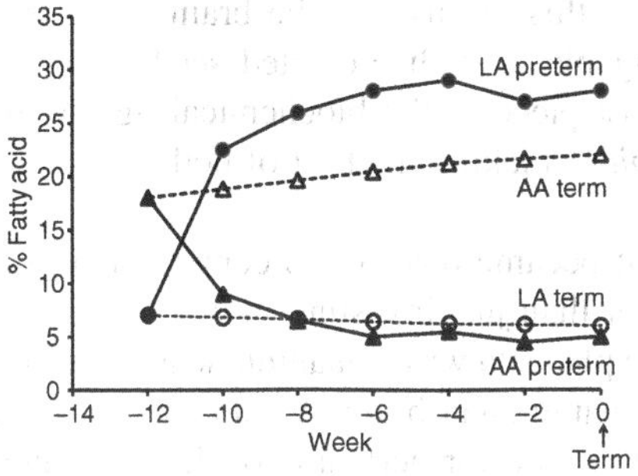

Figure 2.7 Inefficient conversion of linoleic acid to arachidonic acid in the preterm infant (from Leaf et al. 1992). These results show that, postnatally in plasma choline phosphoglycerides, arachidonic acid drops precipitously to a third of the values at birth despite linoleic acid rising three- to fourfold during the same period. Data are for the 12 weeks preterm up to normal expected term birth (Week 0).

affecting the metabolism of the preterm infant and stable isotope data indicate that there is some conversion, the reality of the situation demonstrates that conversion is sufficiently limited that it cannot meet the demand for growth. In comparison to the infant which received most ω6 PUFA as linoleic acid, the fetus benefits from direct and selective transfer of AA by the placenta with the return of linoleic acid to the maternal circulation (Crawford et al., 2003).

The known functions of AA in the brain involve production of *eicosanoids* and related compounds, *endocannabinoids*, protein kinase C activation, *peroxisomal proliferator-activated receptors*, *inositol phosphoglycerides* as well as several other signaling pathways (Hindenes et al., 2000). As it is hard to induce AA deficiency, it may have other functions yet to be discovered. There is a need for further research on the functions of dihomo-γ-linolenic acid (20:3ω6), AA, and docosatetraenoic acid (22:4ω6) in the brain.

It is not even properly understood how AA or DHA reach the brain. Analysis of plasma free fatty acids and triglycerides reveal little AA or DHA although their transport to the brain is thought to be from this pool via a fatty acid binding protein. It is difficult to see how concentrations of 0.5% are amplified to over 20% for DHA in some lipids of the brain or from the same level to over 18% for AA across the placenta. Studies by Lagarde et al. (2001) in rats suggest that *lysophosphatidylcholine* may be a preferred carrier of DHA for the brain. Bill Connor's group in Portland, Oregon, have presented evidence that red cells contribute to the DHA-rich phosphoglycerides crossing the placenta (Ruyle et al., 1990).

THE THIRD PHASE OF EARTH'S LIFE HISTORY – AA AND REPRODUCTION IN MAMMALS

DHA is not alone – AA also plays a part in the human brain evolution story. It presents yet another key example of the conditions of existence with evolution directed by chemistry in the emergence of the mammals and the flowering plants and their protected seeds. Fish require ω3 PUFA for reproduction, but the mammals also came to require ω6 PUFA for their reproduction. During the Cretaceous period, there was an explosion of the flowering plants with protected seeds which followed the collapse of the giant reptiles and the giant ginkos, ferns, and the like. Hence, one can envisage the use of AA in the adherence of the fertilized egg to the endometrium rather than ejecting it with an eggshell. Hence, prior to the Cretaceous, the ω3 PUFA dominance of the land and sea food webs would have resulted in a constraint on brain development because of a paucity of ω6 PUFA because, critically for this discussion, the brain requires both AA and DHA. The evolution of the flowering plants with protected seeds, rich in the ω6 PUFA – linoleic acid – put in place the last piece of the biochemical jigsaw that led to an advancement of brain size, with *H. sapiens* making the best of both worlds from land and sea.

With the evolution of the placenta, it became possible to continuously perfuse the products of conception with nutrients rather than just investing in one squirt of nutrients as in a shelled egg. The development of the placenta was a quantum leap forward in creating opportunities for the evolution of the mammalian brain. AA dominates vascular and placental lipids (Bitsanis et al., 2005) and is vital for perfusion of the fetus and feeding the high energy demands of fetal brain growth.

DARWIN AND THE CONDITIONS OF EXISTENCE

A description of the environmental determinism of DHA in neural signaling systems and the requirement by mammals for ω6 PUFA for reproduction would not be complete without referring to Darwin's treatise in this 150th celebratory period. In the *Origin of Species* (Darwin, 1868), he stated that there were two forces in evolution, *natural selection* and the *conditions of existence*. Of the two, he said, *the latter was the more powerful*. However, based on experiments in which he cut off the tails of breeding rodents and observed that subsequent generations still produced tails, Weismann (1893) rejected this view claiming that natural selection was sufficient and that there was no need for the conditions of existence. That set in train the present paradigm of the *modern synthesis*, and genomic determinism within which DNA is seen as the sole dictator of difference and evolution, including the notion of the "selfish gene" (Dawkins, 1999). Why Weismann needed to do this experiment on mutilation is difficult to understand today because, after all, English farmers had been docking the tails of sheep for centuries, and the Jews had been cutting off foreskins for 3,000 years or more.

Nonetheless, Darwin spent much of the later part of his life searching for what he called *pangenes* that he thought were responsible for translating environmental influences on gene function. His failure served the "all-sufficiency of natural selection" camp, which successfully excluded the conditions of existence down to the present time though without justification. This "all-sufficiency of natural selection" is still the dominant paradigm for many evolutionary biologists. Darwin's "pangenes" are now in evidence in epigenetic effects and the response of plasma membrane receptors to nutrients influencing gene expression (Chawla et al., 2001; Anderle et al., 2004; Kitajka et al., 2004; Corella and Ordovas, 2005). Epigenetic effects affecting gene expression during early development were in evidence following the Dutch food shortage during World War II, in which low birth weight was transmitted to a second generation (Stein et al., 2006). Another example is prenatal programming (Barker, 2004) resulting in adult risk to heart disease, diabetes, and stroke from poor maternal/fetal nutrition. Over evolutionary timescales, these epigenetic effects are plausibly cemented together just as the inability to make AA makes the cat an obligatory carnivore (Rivers et al., 1975).

Darwin's original view on conditions of existence is totally consistent with the remarkable conservation of DHA in neural and visual signaling systems over the past 500–600 million years. Despite wide-ranging changes in the genetic code and great evolutionary changes, DHA has been rigorously conserved. It is as though DHA has been instructing the genes to do its bidding rather than the conventional view which is the other way round. Apart from vindicating Darwin's concept that conditions of existence play a crucial role in evolution, DHA raises basic questions in biology that enhance our understanding of the relationship between environment, the genes, and function.

The functionality of natural selection is beyond question and is readily identified in animals, plants, and insects. It has one drawback in that it does not fit with degenerative loss of organ size, that is, the loss of relative brain size in all land-based mammals as they evolved larger bodies (Crawford et al., 1993). Most importantly, natural selection does not predict such a degenerative process since it is based on randomness. However, Darwin's conditions of existence offer predictive value, which is the true hallmark of science. The strength of the emerging evidence on ω3 PUFA-rich marine food consumption in pregnancy affecting childhood intelligence and behavior measured at 8 years of age is a reminder that *H. sapiens* is also subject to Darwin's conditions of existence (Hibbeln et al., 2007). The recent change in opinion by some leading geneticists about the significance of the gene-centric paradigm that dominated the research scene and investments throws us back onto the importance of the environmental conditions as amply witnessed by the key advents of evolution referred to by Stephen J. Gould as *punctuated evolution* (Eldredge and Gould, 1997; Otsuka, 2008).

Peter Ward (2009) has presented an awakening from the natural selection dogma, which calls upon the ingenuity of people to explain away one of the most influential paradigms of the twentieth century – the extinctions supposedly caused by dramatic extraterrestrial impacts or massive volcanic eruptions. Ward argues that it was changing environmental conditions, brought on by life itself, that led to these extinctions. These are essentially chemical arguments and, like mine, support Darwin's *conditions of existence.* The only difference is that Ward talks about reasons for extinction whereas I am discussing the converse, the conditions for synthesis (Crawford and Marsh, 1995). Both concepts are mutually interactive and, unlike the all-sufficiency of natural selection, offer predictions as to the future. The rise in brain disorders to overtake all other burdens of ill-health in Europe (Andlin-Sobocki et al., 2005) is a serious warning about the vulnerability of brain function of the greatest significance which we ignore at our peril.

IMPLICATIONS

Today in the world, there are 1.6 billion people at risk of iodine deficiency, a sure progenitor for mental retardation but seldom seen in the fishing communities. In Europe, the cost of treatment of brain disorders have now overtaken all other burdens of ill-health (Andlin-Sobocki et al., 2005), and mental ill-health is predicted by the Global Forum of Health (www.globalforumhealth.org) to be in the top three burdens of ill-health worldwide by 2020. There is compelling evidence that the reasons are related to the declining intake of seafoods and their replacement by land foods (Hibbeln, 2002; Hibbeln et al., 2004, 2007). Solving this problem and preventing a further rise in disorders of the brain may well require a new paradigm with a focus on the nutritional requirements for the brain. The savanna hypothesis provided no predictions for the future other than the complacency of having been selected through evolution. It said nothing about the biochemical dependence of the brain on nutrition, the richest source for which is in the aquatic food webs (see also Chapter 3). By contrast, with its dependence on specific lipids and trace elements for the brain, human evolution on the coasts and shorelines explains how the brain evolved, arrived, and, equally importantly, warns us of the future without this unique nutritional resource. The alarming rise in brain disorders, which, collectively, have now overtaken all other burdens of ill-health in Europe, is a sign of Darwin's conditions of existence, this time operating negatively. We now have the enormous challenge to deliver a solution that will arrest this decline which has already started. This means agriculturalizing the oceans and enhancing the responsible development, use, and consumption of seafoods worldwide.

CONCLUSION

The unique and specific use of DHA in neural signaling systems over a 600-million-year stretch of evolution is compelling evidence for its essentiality. It is now known to be involved in neural receptor domains, gene expression with derivatives providing protection from oxidative stress in the brain, and resolution of injury.

DHA is the most biosynthetically limiting of the brain-selective fatty acids. It therefore needs to be obtained preformed for optimum human nutrition, especially during pregnancy and lactation when the fetal and infant brain is rapidly developing. DHA is poorly represented in the land food chain whereas the richest source is the marine food web where the brain first evolved. In the face of the rising brain disorders, the implications of hominin cerebral expansion relying on brain-selective nutrients as opposed to the savanna hypothesis and its variants, which depend on competition with savanna predators, are important to the future of humanity.

ACKNOWLEDGMENTS

I wish to express my appreciation to the many colleagues who have helped formulate these ideas, including Myer Bloom Physics – Vancouver, Javier Caraveo-Patiño – Mexico, Laurence Harbige – Greenwich, Ivan Golfetto – Caracas, Venezuela, Lucilla Poston – St. Thomas' Hospital, London, John Parkington – Capetown, Hiramitsu Suzuki – National Food Research Institute (NFRI) Japan, Ephraim Yavin and Annette Brand-Yavin who assisted with ideas on the language of lipids, Yiqun Wang – Institute of Brain Chemistry and Human Nutrition (IBCHN) London, Ole Mouristsen – MEMPHYS Denmark, the Letten Foundation for encouragement and support, and Dr. Joseph Hibbeln for his many discussions on the importance of seafood and mental health. We also thank David Marsh for drawing to attention Darwin's notion of pangenes and conditions of existence persisting into his sixth edition. I am also especially grateful to Catherine Lehane for reading the manuscript, correcting the text, and organizing the references. The author has no conflicts of interest to declare.

NOTES

1. See *Science*, 2002, vol. 296, April 12, pp. 233 and 340.
2. Seventy kilograms is considered a standard for men, so, at a brain weight of 1.3 kg, the ratio to the body is 1.86%; at 1.4 kg it is 2%.

REFERENCES

Allwood, A.C., Walter, M.R., Kamber, B.S., Marshall, C.P., and Burch, I.W. 2006. Stromatolite reef from the Early Archaean era of Australia. *Nature* **441**:714–718.

Anderle, P., Farmer, P., Berger, A., and Roberts, M.A. 2004. Nutrigenomic approach to understanding the mechanisms by which dietary long chain fatty acids induce gene signals and control mechanisms involved in carcinogenesis. *Nutrition* **20**:103–108.

Anderson, R.E. and Maude, M.B. 1972. Lipids of ocular tissues: The effects of essential fatty acid deficiency on the phospholipids of the photoreceptor membranes of rat retina. *Arch Biochem Biophys* **151**:270–276.

ANDLIN-SOBOCKI, P., JONSSON, J., WITTCHEN, H.-U., AND OLESEN, J. 2005. Cost of disorders of the brain in Europe. *Eur J Neurol* **12** (Suppl. 1):1–27.

BARCELO COBLIJN, G., HOGYES, E., KITAJKA, K., PUSKAS, L.G., ZVARA, A., HACKLER, L. JR., NYAKAS, C., PENKE, Z., AND FARKAS, T. 2003a. Modification by docosahexaenoic acid of age-induced alterations in gene expression and molecular composition of rat brain phospholipids. *Proc Natl Acad Sci U S A* **100**:11321–11326.

BARCELO COBLIJN, G., KITAJKA, K., PUSKAS, L.G., HOGYES, E., ZVARA, A., HACKLER, L. JR., AND FARKAS, T. 2003b. Gene expression and molecular composition of phospholipids in rat brain in relation to dietary n-6 to n-3 fatty acid ratio. *Biochim Biophys Acta* **1632**:72–79.

BARKER, D.J. 2004. The developmental origins of chronic adult disease. *Acta Paediatr Suppl* **93** (446):26–33.

BAZAN, N.G. 2007. Homeostatic regulation of photoreceptor cell integrity: Significance of the potent mediator neuroprotectin D1 biosynthesized from docosahexaenoic acid. The Proctor Lecture. *Invest Ophthalmol Vis Sci* **48** (11):4866–4881.

BENOLKEN, R.M., ANDERSON, R.E., AND WHEELER, T.G. 1973. Membrane fatty acids associated with the electrical response in visual excitation. *Science* **182** (118): 1253–1254.

BIRCH, E.E., GARFIELD, S., HOFFMAN, D.E., HOFFMAN, D.R., UAUY, R., AND BIRCH, D.G. 2000. A randomised trial of early dietary supply of long chain polyunsaturated fatty acids and mental development in term infants. *Dev Med Child Neurol* **42**:174–181.

BITSANIS, D., CRAWFORD, M.A., MOODLEY, T., HOLMSEN, H., GHEBREMESKEL, K., AND DJAHANBAKHCH, O. 2005. AA predominates in the membrane phosphoglycerides of the early and term human placenta. *J Nutr* **135** (11): 2566–2571.

BLOOM, M., LINSEISEN, F., LLOYD-SMITH, J., AND CRAWFORD, M.A. 1999. Insights from NMR on the functional role of polyunsaturated lipids in the brain. *Magnetic Resonance and Brain Function – Approaches from Physics*, ed. B. Maraviglia. Proceedings of the 1998 Enrico Fermi International School of Physics, Enrico Fermi Lecture, Course #139, Varenna, Italy, pp. 1–27.

BOURRE, J.M., FRANCOIS, M., YOUYOU, A., DUMONT, O., PICIOTTI, M., PASCAL, G., AND DURAND, G. 1989. The effects of dietary alpha-linolenic acid on the composition of nerve membranes, enzymatic activity, amplitude of electrophysiological parameters, resistance to poisons and performance of learning tasks in rats. *J Nutr* **119** (12): 1880–1892.

BRAMBLE, D. AND LIEBERMAN, D. 2004. Endurance running and the evolution of *Homo*. *Nature* **432**:345–352.

BROADHURST, C.L., WANG, Y., CRAWFORD, M.A., CUNNANE, S., PARKINGTON, J., AND SCHMID, W.F. 2002. Brain-specific lipids from marine, lacustrine, or terrestrial food resources: Potential impact on early African *Homo sapiens*. *Comp Biochem Physiol B Biochem Mol Biol* **131**:653–673.

BUDOWSKI, P. AND CRAWFORD, M.A. 1985. Alpha-linolenic acid as a regulator of the metabolism of AA: Dietary implications of the ratio n-6:n-3 fatty acids. *Proc Nutr Soc* **44**:221–229.

BUDOWSKI, P., LEIGHFIELD, M.J., AND CRAWFORD, M.A. 1987. Nutritional encephalomalacia in the chick: An exposure of the vulnerable period for cerebellar development and the possible need for both ω6 and ω3 fatty acids. *Br J Nutr* **58**:511–520.

CARLSON, S.E. AND WERKMAN, S.H. 1996. A randomized trial of visual attention of preterm infants fed docosahexaenoic acid until two months. *Lipids* **31**:85–90.

CATALAN, J., MORIGUCHI, T., SLOTNICK, B., MURTHY, M., GREINER, R.S., AND SALEM, N. JR. 2002. Cognitive deficits in docosahexaenoic acid-deficient rats. *Behav Neurosci* **116**:1022–1031, 2078S–2083S.

CHAWLA, A., REPA, J.J., EVANS, R.M., AND MANGELSDORF, D.J. 2001. Nuclear receptors and lipid physiology: Opening the X-files. *Science* **294** (5548):1866–1870.

CHUNG, W.L., CHEN, J.J., AND SU, H.M. 2008. Fish oil supplementation of control and (n-3) fatty acid-deficient male rats enhances reference and working memory performance and increases brain regional docosahexaenoic acid levels. *J Nutr* **138** (6):1165–1167.

CHYTROVA, G., YING, Z., AND GOMEZ-PINILLA, F. 2009. Exercise contributes to the effects of DHA dietary supplementation by acting on membrane-related synaptic systems. *Brain Res* May 13, ahead of publication at time of submission.

CORELLA, D. AND ORDOVAS, J.M. 2005. Single nucleotide polymorphisms that influence lipid metabolism: Interaction with dietary factors. *Annu Rev Nutr* **25**:341–390.

CRAWFORD, M.A. 1971. Epidemiological interactions. In *Symposium on Mycotoxins in Human Health*, ed. I.F.H. Purchase, pp. 231–244. Proceedings of a Symposium in Pretoria, 1970. London: Macmillan.

CRAWFORD, M.A., BLOOM, M., BROADHURST, C.L., SCHMIDT, W.F., CUNNANE, S.C., GALLI, C., GHEBREMESKEL, K., LINSEISEN, F., LLOYD-SMITH, J., AND PARKINGTON, J. 1999. Evidence for the unique function of DHA during the evolution of the modern hominid brain. *Lipids* **34**:S39–S47.

CRAWFORD, M.A., BROADHURST, C.L., GALLI, C., GHEBREMESKEL, K., HOLMSEN, H., SAUGSTAD, L.F., SCHMIDT, F., SINCLAIR, A.J., AND CUNNANE, S.C. 2008. The role of docosahexaenoic and AAs as determinants of evolution and hominid brain development. In *Fisheries for Global Welfare and Environment*, ed. K. Tsukamoto, T. Kawamura, T. Takeuchi, T.D. Beard Jr., and M.J. Kaiser, pp. 57–76. 5th World Fisheries Congress 2008, Terrapub, Tokyo.

CRAWFORD, M.A., CASPERD, N.M., AND SINCLAIR, A.J. 1976. The long chain metabolites of linoleic and linolenic acids in liver and brain in herbivores and carnivores. *Comp Biochem Physiol* **54B**:395–401.

CRAWFORD, M.A., CUNNANE, S.C., AND HARBIGE, L.S. 1993. A new theory of evolution: Quantum theory. IIIrd International Congress on Essential Fatty Acids and Eicosanoids. American Oil Chemists Society, ed A.J. Sinclair and R. Gibson. Adelaide, pp. 87–95.

CRAWFORD, M.A., GALE, M.M., AND WOODFORD, M.H. 1969. Linoleic acid and linolenic acid elongation products in muscle tissue of *Syncerus caffer* and other ruminant species. *Biochem J* **115**:25–27.

CRAWFORD, M.A., GOLFETTO, I., BISTANIS, D., GHEBREMESKEL, K., MIN, Y., MOODLEY, T., POSTON, L., PHYLACTOS, A., CUNNANE, S., AND SCHMIDT, W. 2003. Arachidonic and docosahexaenoic acids in protection against central nervous system damage in preterm infants. *Lipids* **38** (4):303–315.

CRAWFORD, M.A. AND MARSH, D.E. 1995. *Nutrition and Evolution: Food in Evolution and the Future*. New Canaan, CT: Keats.

CRAWFORD, M.A. AND SINCLAIR, A.J. 1972. Nutritional influences in the evolution of the mammalian brain. In *Lipids, Malnutrition and Developing Brain*, ed. K. Elliot and J. Knight, pp. 267–292. A Ciba Foundation Symposium. Amsterdam: Elsevier.

CUNNANE, S.C. 2005. *Survival of the Fattest: The Key to Human Brain Evolution*, River Edge, NJ: World Scientific Publishing.

DARWIN, C. 1868. *The Origin of Species by Means of Natural Selection. Or the Preservation of Favoured Races in the Struggle for Life*. London: John Murray.

DAWKINS, C.R. 1999. *The Selfish Gene*. Oxford University Press.

ELDRIDGE, N. AND GOULD, S.J. 1972. Punctuated equilibria: an alternative to phyletic gradualism. In T.J.M. Schopf, ed., *Models in Paleobiology*. San Francisco, CA: Freeman, Cooper, pp. 82–115.

FIENNES, R.N.T.-W., SINCLAIR, A.J., AND CRAWFORD, M.A. 1973. Essential fatty acid studies in primates: Linolenic acid requirements of Capuchins. *J Med Primatol* **2**:155–169.

FREEMANTLE, E., TREMBLAY, S., VANDAL, M., TREMBLAY-MERCIER, J., BLACHÈRE, J.C., BÉGIN, M.E., WINDUST, T., BRENNA, J.T., AND CUNNANE, S.C. 2006. Omega-3 fatty acids, energy substrates and brain function during aging. *Prostagl Leukotri Essential Fatty Acids* **75**: 213–220.

GALLI, C. AND SOCINI, A. 1983. Dietary lipids in pre- and post-natal development. In *Dietary Fats and Health*, ed. E.G. Perkins and W.J. Visek, pp. 278–301. Champaign, IL: American Oil Chemists Society.

GISLÉN, A., DACKE, M., KRÖGER, R.H., ABRAHAMSSON, M., NILSSON, D.E., AND WARRANT, E.J. 2003. Superior underwater vision in a human population of sea gypsies. *Curr Biol* **13** (10):833–836.

HARDY, A. 1960. Was man more aquatic in the past? *The New Scientist*, March 17, pp. 642–645.

HIBBELN, J.R. 2002. Seafood consumption, the DHA content of mothers' milk and prevalence rates of postpartum depression: A cross-national, ecological analysis. *J Affect Disord* **69** (1–3):15–29.

HIBBELN, J.R. 2007. From homicide to happiness – A commentary on omega-3 fatty acids in human society. Cleave Award Lecture. *Nutr Health* **19** (1–2):9–19.

HIBBELN, J.R., DAVIS, J., STEER, C., EMMETT, P., ROGERS, I., WILLIAMS, C., AND GOLDING, J. 2007. Maternal seafood consumption in pregnancy and neurodevelopmental outcomes in childhood (ALSPAC study): An observational cohort study. *Lancet* **369** (9561):578–585.

HIBBELN, J.R., NIEMINEN, L.R., AND LANDS, W.E. 2004. Increasing homicide rates and linoleic acid consumption among five Western countries, 1961–2000. *Lipids* **39** (12):1207–1213.

HINDENES, J.O., NERDAL, W., GUO, W., DI, L., SMALL, D.M., AND HOLMSEN, H. 2000. Physical properties of the transmembrane signal molecule, sn-1-stearoyl 2-arachidonoylglycerol. Acyl chain segregation and its biochemical implications. *J Biol Chem* **275** (10): 6857–6867.

HOLLAND, H.D. 2006. The oxygenation of the atmosphere and oceans. *Philos Trans R Soc Lond B Biol Sci* **361** (1470):903–915.

JAMES, M.J. AND CLELAND, L.G. 2004. Applying a research ethics committee approach to a medical practice controversy: The case of the selective COX-2 inhibitor rofecoxib. *J Med Ethics* **30**:182–184.

JUMP, D.B. 2002. The biochemistry of n-3 polyunsaturated fatty acids. *J Biol Chem* **277** (11):8755–8758.

KITAJKA, K., PUSKAS, L.G., ZVARA, A., HACKLER, L. JR., BARCELO COBLIJN, G., YEO, Y.K., AND FARKAS, T. 2002. The role of n-3 polyunsaturated fatty acids in brain: Modulation of rat brain gene expression by dietary n-3 fatty acids. *Proc Natl Acad Sci U S A* **99**:2619–2624.

KITAJKA, K., SINCLAIR, A.J., WEISINGER, R.S., WEISINGER, H.S., MATHAI, M., JAYASOORIYA, A.P., HALVER, J.E., AND PUSKAS, L.G. 2004. Effects of dietary omega-3 polyunsaturated fatty acids on brain gene expression. *Proc Natl Acad Sci U S A* **101**:10931–10936.

LAGARDE, M., BERNOUD, N., BROSSARD, N., LEMAITRE-DELAUNAY, D., THIES, F., CROSET, M., AND LECERF, J. 2001. Lysophosphatidylcholine as a preferred carrier form of docosahexaenoic acid to the brain. *J Mol Neurosci* **16** (2–3):201–204; discussion 215–221.

LAMPTEY, M.S. AND WALKER, B.L. 1976. A possible essential role for dietary linolenic acid in the development of the young rat. *J Nutr* **106** (1):86–93.

LEAF, A.A., LEIGHFIELD, M.J., COSTELOE, K.L., and CRAWFORD, M.A. 1992. Factors affecting long-chain polyunsaturated fatty acid composition of plasma choline phosphoglycerides in preterm infants. *J Pediatr Gastroenterol Nutr* **14**:300–308.

LEYTON, J., DRURY, P.J., AND CRAWFORD, M.A. 1987. Differential oxidation of saturated and unsaturated fatty acids in vivo in the rat. *Br J Nutr* **57**:383–393.

MAREAN, C.W., BAR-MATTHEWS, M., BERNATCHEZ, J., FISHER, E., GOLDBERG, P., HERRIES, A.I., JACOBS, Z., JERARDINO, A., KARKANAS, P., MINICHILLO, T., NILSSEN, P.J., THOMPSON, E., WATTS, I., AND WILLIAMS, H.M. 2007. Early human use of marine resources and pigment in South Africa during the Middle Pleistocene. *Nature* **449** (7164):905–908.

MARTINEZ, M. AND VAZQUEZ, E. 1998. MRI evidence that docosahexaenoic acid ethyl ester improves myelination in generalized peroxisomal disorders. *Neurology* **51** (1): 26–32.

MORGAN, E. 1995. *The Descent of the Child*. Oxford: Oxford University Press.

MORGAN, E. 1997. *The Aquatic Ape Hypothesis*. London: Souvenir Press.

NEURINGER, M., CONNOR, W.E., LIN, D.S., BARSTAD, L., AND LUCK, S. 1986. Biochemical and functional effects of prenatal and postnatal omega 3 fatty acid deficiency on retina and brain in rhesus monkeys. *Proc Natl Acad Sci U S A* **83** (11):4021–4025.

OTSUKA, J. 2008. Theoretical approach to the large-scale evolution of multicellularity and cell differentiation. *J Theor Biol* **255** (1):129–136.

PAULETTO, P., PUATO, M., CAROLI, M.G., CASIGLIA, E., MUNHAMBO, A.E., CAZZOLATO, G., BON, G.B., ANGELI, M.T., GALLI, C., AND PESSINA, A.C. 1996. Blood pressure and atherogenic lipoprotein profiles of fish-diet and vegetarian villagers in Tanzania: The Lugalawa study. *Lancet* **348**:784–788.

PLOURDE, M. and CUNNANE, S.C. 2007. Extremely limited conversion of long chain omega 3 polyunsaturates in adults: Implications for their dietary essentiality and use as supplements. *Appl Physiol Nutr Metab* **32**:619–634.

RAY, W.A., STEIN, C.M., DAUGHERTY, J.R. HALL, K., ARBOGAST, P.G., AND GRIFFIN, M.R. 2002. COX-2 selective non-steroidal anti-inflammatory drugs and risk of serious coronary heart disease. *Lancet* **360**:1071–1073.

RIVERS, J.P.W., SINCLAIR, A.J., AND CRAWFORD, M.A. 1975. Inability of the cat to desaturate essential fatty acids. *Nature* **285**:171–173.

RODRIGUEZ DE TURCO, E.B., GORDON, W.C., AND BAZAN, N.G. 1992. Light stimulates in vivo inositol lipid turnover in frog retinal pigment epithelial cells at the onset of shedding and phagocytosis of photoreceptor membranes. *Exp Eye Res* **55**:719–725.

RUYLE, M., CONNOR, W.E., ANDERSON, G.J., AND LOWENSOHN, R.I. 1990. Placental transfer of essential fatty acids in humans: Venous-arterial difference for docosahexaenoic acid in fetal umbilical erythrocytes. *Proc Natl Acad Sci U S A* **87** (20):7902–7906.

SERHAN, C.N., CHIANG, N., AND VAN DYKE, T.E. 2008. Resolving inflammation: Dual anti-inflammatory and pro-resolution lipid mediators. *Nat Rev Immunol* **8** (5): 349–361.

SINCLAIR, A.J. 1975. Incorporation of radioactive polyunsaturated fatty acids into liver and brain of developing rat. *Lipids* **10** (3):175–184.

SINCLAIR, A.J. AND CRAWFORD, M.A. 1972. The incorporation of linolenic and docosahexaenoic acid into liver and brain lipids of developing rats. *FEBS Lett* **26**:127–129.

SMIT, E.N., MUSKIET, F.A., AND BOERSMA, E.R. 2004. The possible role of essential fatty acids in the pathophysiology of malnutrition. *Prostaglandins Leukot Essent Fatty Acids* **71** (4):241–250.

SPRECHER, H. 1993. Interconversions between 20- and 22-carbon n-3 and n-6 fatty acids *via* 4-desaturase independant pathways. IIIrd International Congress on Essential Fatty Acids and Eicosanoids. American Oil Chemists Society, ed. A.J. Sinclair and R. Gibson. Adelaide, pp. 18–22.

SPRECHER, H., CHEN, Q., AND YIN, F.Q. 1999. Regulation of the biosynthesis of 22:5n-6 and 22:6n-3: A complex intracellular process. *Lipids* **34**:S153–S156.

STEIN, A.D., ZYBERT, P.A., VAN DER PAL-DE BRUIN, K., AND LUMEY, L.H. 2006. Exposure to famine during gestation, size at birth, and blood pressure at age 59 y: Evidence from the Dutch famine. *Eur J Epidemiol* **21** (10): 759–765.

STEINBERG, G., SLATON, W.H. JR., HOWTON, D.R., AND MEAD, J.F. 1956. Metabolism of essential fatty acids. IV. Incorporation of linoleate into AA. *J Biol Chem* **220** (1): 257–264.

STRINGER, C. 2000. Palaeoanthropology. Coasting out of Africa. *Nature* **405** (6782):24–25, 27.

SUZUKI, H., MANABE, S., WADA, O., AND CRAWFORD, M.A. 1997. Rapid incorporation of docosahexaenoic acid from dietary sources into brain microsomal, synaptosomal and mitochondrial membranes in adult mice. *Internat J Vit Res* **67**:272–278.

TOBIAS, P.V. 1998. Water and human evolution. *Out There* **35**:38–44.

WARD, P. 2009. What will become of Homo sapiens? *Sci Am* **300**:68–73.

WEISINGER, H.S., VINGRYS, A.J., BUI, B.V., AND SINCLAIR, A.J. 1999. Effects of dietary n-3 fatty acid deficiency and repletion in the guinea pig retina. *Invest Ophthalmol Vis Sci* **40**:327–338.

WEISMANN, A. 1893. The all-sufficiency of natural selection. *Contemp Rev* **64**:309–338, 596–610.

WHEELER, T.G., BENOLKEN, R.M., AND ANDERSON, R.E. 1975. Visual membranes: Specificity of fatty acid precursors for the electrical response to illumination. *Science* **188**:1312–1314.

WIEGAND, R.D., KOUTZ, C.A., CHEN, H., AND ANDERSON, R.E. 1995. Effect of dietary fat and environmental lighting on the phospholipid molecular species of rat photoreceptor membranes. *Exp Eye Res* **60** (3):291–306.

WILLIAMS, G. AND CRAWFORD, M.A. 1987. Comparison of the fatty acid component in structural lipids from dolphins, zebra and giraffe: Possible evolutionary implications. *J Zool Lond* **213**:673–684.

YAMAMOTO, N., OKANIWA, Y., MORI, S., NOMURA, M., AND OKUYAMA, H. 1991. Effects of a high-linoleate and a high-alpha-linolenate diet on the learning ability of aged rats. Evidence against an autoxidation-related lipid peroxide theory of aging. *J Gerontol* **46** (1):B17–B22.

YAVIN, E., HIMOVICHI, E., and EILAM, R. 2009. Delayed cell migration in the developing rat brain following maternal omega 3 alpha linolenic acid dietary deficiency. *Neuroscience* **162**:1011–1022.

CHAPTER 3

HUMAN BRAIN EVOLUTION: A QUESTION OF SOLVING KEY NUTRITIONAL AND METABOLIC CONSTRAINTS ON MAMMALIAN BRAIN DEVELOPMENT

Stephen C. Cunnane

INTRODUCTION

The origins of the large brain and advanced cognitive capacities of humans are most commonly linked to toolmaking, starting in *Homo habilis* (Aeillo and Dean, 1990; Changeux and Chavaillon, 1995; Conroy, 1997). With sharp-edged stone axes and spear points for simple weapons, early hominins could acquire fresh meat and hence a higher quality diet which, in turn, helped to meet the growing nutrient and energy needs of the expanding brain. A "higher quality diet" is a nebulous term but generally refers to less intake of fibrous plant material and higher intake of meat obtained either by scavenging or predation. A higher quality diet is generally easier to digest and has a higher energy, protein, and fat content than a plant-based diet. Such an increase in the energy and protein density of the diet is widely viewed as having been specific to hominins and to have permitted the gradual threefold expansion of the relatively small australopithecine brain leading to the large present-day human brain. Of note, the term "higher quality diet" rarely refers to the *micronutrient* content of the diet (minerals, vitamins, and certain vitamin-like fatty acids) but the case will be made here that the micronutrient content of the diet was actually far more important for human brain evolution than its energy or protein content (Crawford and Marsh, 1989; Cunnane, 2005a).

Other antecedents of increasing brain sophistication in hominins that coincide roughly with the emergence of skilled hunting include the evolution of social networking, language, and culture. Indeed, language, culture, and advanced social structure are clearly associated with evolution of the highly advanced human brain and, for the most part, they are uniquely human attributes. Where opportunities lack for social interaction, that is, mother–infant bonding or play during childhood, human brain development clearly suffers. Still, without downplaying the indelible interdependence between higher cognitive performance, culture, language, and social organization in humans, several questions remain: Are they a sufficient explanation for the origins of the large human brain? Other woodland primates are highly social, so why only in hominins did the brain undergo such advanced changes? What is the mechanism explaining the link between culture, language, diet, and

Human Brain Evolution, Edited by Stephen C. Cunnane and Kathlyn M. Stewart

expanding cognitive performance as *Homo sapiens* emerged? In terms of the emergence of human behavior or cognitive capacity, what can be predicted about brain function in those consuming diets containing meat versus diets that contain little or no meat?

The fossil record is incomplete so it is difficult to clearly establish how closely advancements in brain size preceded, followed, or were coincident with the quality of stone tools or the evolution of language, symbolic thought, and culture. Chemical signatures in fossils also provide little hard proof of biochemical or neurological processes that could clearly establish the circumstances of human brain evolution. My view is that neurological sophistication, brain enlargement, hunting, language, and sociocultural skills are mutually interdependent. As such, one could not evolve very far without advances in the others. Small increases in brain size would facilitate small improvements in language, toolmaking, and culture, and *vice versa*, so how could one really be the trigger for the other? Crucially, what got the ball rolling? What catalyzed the process and propelled it so much further than in other primates? In keeping with the most fundamental tenet of natural selection, there was no particular purpose behind human brain evolution; there was no survival imperative, no social imperative, no need for more advanced cognitive skills; that is, there could be no specific purpose or outcome in mind.

Rather, the human brain evolved as an *exaptation* (Tattersall, 1998; see Chapter 1) – fortuitously, for better or for worse. There were advantages to human brain evolution but there were also disadvantages, including its vulnerability especially during development of the infant and child. I believe that to understand how human brain evolution occurred, we need to integrate two concepts into our thinking: (1) the vulnerability of the developing brain, specifically, the ongoing dependency of successful human brain development on an adequate dietary nutrient supply, and (2) the crucial role of fatness for healthy brain development in human babies and, hence, evolution of the human brain.

With respect to the first point, events in early human brain development lay the groundwork for the brain's functionality later in life. Optimal language development and social integration are closely and bidirectionally linked with healthy brain development. In turn, optimum brain development depends absolutely on a secure, nutrient-rich food supply permitting the mother–infant pair to successfully get through the first few vulnerable years after birth. The blueprint for mammalian brain development is programmed to a certain extent, but it is also malleable. In humans, this malleability is more apparent than in nonhuman species; it provides the potential for more sophisticated function than in other species, but this potential carries with it a greater risk of suboptimal development as well. Aside from the importance of the maternal–infant bond, successful brain development is vulnerable to insufficient dietary intake of a cluster of nutrients called *brain-selective nutrients* (Cunnane et al., 1993; see Brain-Selective Nutrients section). Brain-selective nutrients are unequivocally essential to the normal development of the human brain. Over a billion people today have suboptimal brain function simply because they have inadequate dietary intake of iodine and iron. Inadequate intake of omega-3 (ω3) fatty acids also increases the risk of slower neurological development and adult-onset psychiatric illness and cognitive decline. Therefore, hominin brain evolution and expansion could not have advanced very far without finding a solution to the vulnerability imposed by this important nutritional constraint. Equally importantly, the worldwide prevalence of suboptimal brain development in humans subsisting on diets providing inadequate amounts of brain-selective nutrients attests to the fact that the developmental vulnerability of the human brain was not eliminated and may actually have *increased* during evolution. Genetic mutations may well be involved but if human brain evolution was a purely genetic event, this nutritional vulnerability of human brain development would be highly unlikely.

The second point is that, among primates, healthy term human babies are unique in being born fat. Fatness is much more than an endearing feature in babies – it is critical to optimal development of the infant brain. At all ages, body fat is a fuel reserve when food intake is insufficient to meet energy requirements. Due to the brain's voracious energy needs during infancy, this role of baby fat is especially important in optimizing brain development. *Ketone bodies* (or simply, *ketones*) are generated from fatty acids stored in the fat reserves and are by far the most important fuel backing up glucose for the brain. The developing brain not only needs a guaranteed fuel reserve but must also synthesize a large part of its cellular structures *in situ*. Aside from their critical role backing up glucose as a brain fuel, ketones are also the main substrate for synthesis of brain lipids that are vital for brain function, principally cholesterol and some fatty acids. Ketones are not stored in the body but are produced from fatty acids released from body fat (see Baby Fat – The Reserve for Brain Lipids section). Hence, larger body fat reserves accumulating in the third trimester human fetus must have evolved more or less simultaneously with the evolving brain because they remain crucial for postnatal nourishment of the brain today. Hence, human brain evolution was tied directly to the unique evolution of fat babies (Cunnane, 2005a).

I will make the case here that human brain evolution required a nutritionally enriched habitat providing an abundant and accessible dietary supply of brain-selective nutrients. In my view, it happened at the land–water interface; that is, it was specifically a shore-based habitat. Such a habitat masked (and continues to mask) the human brain's developmental vulnerability because this habitat alone provided a selection of foods capable of meeting the brain's expanding energy and nutrient needs (Crawford and Marsh, 1989; Cunnane, 2005a). Shore-based foods include a large variety of plants, shallow water fish such as catfish, crustaceans, shellfish, amphibians, and eggs of birds nesting on or near shorelines. Shore-based habitats provide not only a secure, accessible, and abundant food supply, but they also provide the richest source of brain-selective nutrients available in any known ecosystem. Equally importantly, shore-based foods can be obtained without stone tools or other sophisticated technology.

BRAIN EVOLUTION IN HOMININS

Brain Size in Hominins

The average adult human weighs about 60 kg and the brain weighs about 1360 g (Allen et al., 2002; Table 3.1). Hence, for our size and compared to other animals of human weight, brain weight in humans is significantly above average. The australopithecines – *Australopithecus afarensis, A. garhi*, and *A. africanus* – all had brains about one-third the size of human brains. *A. afarensis* was much smaller than humans, but *A. africanus* and the *Paranthropus* species were similar in size to humans. Adjusting the brain weight for body weight corrects for differences in body weight but still shows that the *A. afarensis* brain was proportionally about one-third smaller than in humans (brain-to-body weight ratio of ~1.7% compared to ~2.3% in humans (Table 3.1). Hence, the threefold greater brain weight in humans compared to *A. afarensis* is more than the difference in their brain-to-body weight ratios that would be predicted from normal allometric scaling according to body size alone (Martin, 1981; Mink et al., 1981; Armstrong, 1983; Martin, 1996). The newborn human has a somewhat larger but less mature brain than a chimpanzee.

The concept of *encephalization quotient* (EQ) mathematically corrects for the fact that within hominins, brain size does not increase as fast as body size (Jerison, 1973). This formula has been adjusted from time to time, resulting in published EQ values that differ

TABLE 3.1 Brain and body dimensions of extinct and extant adult hominins, hominins, some primates, and living humans[a]

Species	Brain weight (g)	Brain-to-body weight ratio (%)	Relative EQ[b] (humans = 100)
	Hominins and hominins		
A. afarensis (3.6–2.8 mya)	455	1.7	41
	180[c]		
A. africanus (3.0–2.2 mya)	450	1.0	44
P. aethiopicus (2.7–2.3 mya)	405	1.1	44
P. boisei (2.3–1.3 mya)	510	0.9	46
P. robustus (1.8–1.5 mya)	520	1.1	50
H. rudolfensis (2.4–1.7 mya)	750	1.7	59
H. habilis (1.9–1.5 mya)	600	1.7	57
H. ergaster (1.8–1.4 mya)	855	–	–
Homo erectus (1.8–1.5 mya)	863	1.6	63
H. erectus (0.5–0.3 mya)	1000	1.6	63
H. heidelbergensis (600–200 kya)	1200	1.8	74
H. neanderthalensis (200–40 kya)	1450	1.9	75
H. sapiens (Modern) (100–10 kya)	1490	2.4	102
	Extant nonhuman primates		
Pan troglodytes– male	400	0.9	42
P. troglodytes– newborn	160	10.0	–
Pongo pygmaeus– male	400	0.5	32
Gorilla Gorilla– male	500	0.3	25
	Extant humans		
H. sapiens– male	1360	2.3	100
	–	2.7[d]	117
H. sapiens– newborn	380	10.9	–
		13.1[d]	–

[a] Data averaged from MacKinnon (1978); Cronin et al. (1981); Ho et al. (1981); Blumenberg (1983); Passingham (1985); Aeillo and Dean (1990); Kappelman (1996); Langdon (1997); Ruff et al. (1997); Falk (1998); Wood and Collard (1999); Allen et al. (2002); Pinker (2002); Carroll (2003); Leonard et al. (2003); and DeSilva and Lesnik (2008).

[b] Encephalization Quotient: Data are averaged from Ruff et al. (1997); Martin (1981); and Kappelman (1996), and standardized relative to living *H. sapiens* (100), as modified from Cunnane (2005a).

[c] Newborn (DeSilva and Lesnik, 2008).

[d] Corrected to lean body weight.

Mya, million years ago; kya, thousand years ago.

somewhat according to who reported them. Nevertheless, all studies show EQs increased an average two- to threefold when comparing australopithecines to more advanced hominins or living humans (Table 3.1). Thus, no matter how "brain capacity" in humans is expressed, it is two to three times more than in the earliest hominins.

To make comparisons easier, the EQ of *Homo sapiens* can be normalized to an arbitrary value of 100, against which the EQ of other hominins can then be referenced as a percentage of this "relative EQ." Thus, *A. afarensis* had a relative EQ of about 41, or a brain capacity about 41% of humans (Table 3.5). Brain size and relative EQ both increased by 40–45% in going from *A. afarensis* to *H. habilis*. Similarly, from *H. habilis* to

Homo erectus, brain size increased a further 40%, but relative EQ actually increased much less because body size in *H. erectus* was considerably more than *H. habilis*. Absolute brain size in *Homo heidelbergensis* was 33% more than in *H. erectus*, but body weight also increased so, again, relative brain weight changed less than absolute brain weight because relative EQ was 74 in *H. heidelbergensis* compared to 63 in *H. erectus*. *Homo neanderthalensis* had a still larger brain and body weight and thus had almost identical brain capacity to the somewhat earlier *H. heidelbergensis*. At 1490 g, early modern humans (Cro-Magnon) had the largest hominin brain size known but were physically somewhat smaller than Neanderthals so they experienced a 36% increase in EQ compared to Neanderthals. Compared to the Cro-Magnon, living humans not only have lost an average of about 130 g of brain weight but have also lost about 3 kg of body weight, resulting in both the brain-to-body weight ratio and the relative EQ of humans decreasing 2–3%.

Brain Size and Higher Cognitive Function

Brain size is clearly only a crude measure of cognitive potential. Although a small decrease actually occurred in brain size from early *H. sapiens* (Cro-Magnon) to present-day humans, there is no evidence that functionality of the brain has decreased. Neanderthals also had 10% larger brains than present-day humans but, again, there is no indication that they were any more intelligent. In humans, the cortex undergoes many more cycles of cell division than in other primates and even more so than in rodents (Hill and Walsh, 2005). This extended process of neurogenesis adds more layers of later-derived neurons to the human cortex than in primates. Specialization is a feature within and between cortical areas, with some such areas being greatly expanded in humans, that is, the prefrontal cortex. Nevertheless, the differences in proportion of brain regions between humans and other primates may not be as significant as commonly thought (Carroll, 2003).

Functional asymmetries are also more a feature of human than nonhuman cerebral cortex. Speech is a higher brain function peculiar to humans and is associated with asymmetrical enlargement of Broca's area in the left hemisphere. Asymmetry in Wernicke's area of the left temporal lobe may also be associated with speech and communication by gestures.

Body Fatness and Relative Brain Weight

Body fat usually contributes to a minimum of 15–20% of overall weight in healthy adult humans. Terrestrial animals in the wild rarely have more than 5% body fat so even adult humans of "normal fatness" are remarkably fat in comparison to other terrestrial mammals. The higher amount of body fat in humans leads to underestimating brain weight in humans relative to nonhuman primates. To compare like with like, about 15% of average adult human body weight should be removed so that normally lean nonhuman species can be compared to the corrected lean weight of humans. This correction leaves the average lean, present-day adult human weighing about 51 kg instead of 60 kg. Correcting for fatness does not change actual brain size, but it does raise the relative brain weight of modern humans to about 2.7% from 2.3% of body weight, which raises the corrected relative EQ in humans to about 117 (see Table 3.1). In turn, this means that increase in relative EQ during hominin evolution was proportionally more than without the correction for increasing body fat.

Full-term, healthy human babies weigh about 3500 g and have an average brain weight of 380 g, so relative brain weight in the human newborn is about 11%. However, 500–600 g or about 14% of newborn body weight is fat. Since other terrestrial species do not have fat babies, for more accurate comparison of relative brain sizes, the fat component of the body weight of human infants should be excluded. Thus, an average healthy newborn human has a corrected *lean* body weight of about 2900–3000 g and has a relative brain weight of 13% (Table 3.1). Also, in comparison, newborn chimpanzee infants are lean and have relative brain weights of about 10%, so lean human babies have relative brain weights about 30% larger than in chimpanzees.

Several points arise from these comparisons. First, in the newborn human, the brain is about one-third its size in the adult, but *relative* brain weight at birth is much higher than in the adult. This is true across species, not just in humans, and has important implications for human brain evolution. Second, at birth, the difference in relative brain weight in chimpanzees and humans is smaller in infants than it is in adults; that is, relative to body weight, the human infant brain is about 30% bigger than in chimpanzee infants while in human adults, the brain is about three times bigger than in adult chimpanzees. Although humans and chimpanzees are born with a relatively small difference in brain size, after birth, body growth increases more in chimpanzees while humans retain a significantly higher postnatal commitment to brain growth. Third, fatness is essential for healthy brain growth and neurological development in humans but apparently not in chimpanzees or other primates. Human infants who are born lean like chimpanzees, that is, premature or low birth weight infants, have a significantly higher risk of impaired brain development. In contrast, to my knowledge, spontaneous neurodevelopmental delay in chimpanzees or other primates is unknown (see Prematurity, Low Birth Weight, and Body Fat Stores section).

NEED FOR A NEW PARADIGM

For about 30 years now, an increasingly abundant and secure dietary energy supply has been recognized as probably having contributed to human brain evolution (Mace et al., 1981; Leonard and Robertson, 1994, 1996; Aeillo and Wheeler, 1995; Martin, 1996; Broadhurst et al., 1998; Teaford and Ungar, 2000; Leonard, 2002). However, like Crawford and Marsh (1989), I have made the case that although more dietary energy alone was necessary, it would still have been insufficient to initiate or support brain expansion. This is because, unlike other organs, the mammalian brain cannot use as a fuel either of the energy-dense components of the diet, that is, fat and protein (Cunnane and Crawford, 2003; Cunnane, 2005a). More importantly, increasing fat and protein intake does not address the increasing need for specific indispensable micronutrients that control and refine brain development (see Brain-Selective Nutrients section). These micronutrient deficiencies are far too prevalent today to ignore as factors that would have influenced successful evolution of the brain in hominins destined to become humans.

Current models mostly link a better quality diet acquired during human brain evolution to consumption of fresh meat, that is, organized hunting. However, these models do not adequately explain two key attributes of early human brain development that profoundly affect brain function in adults – (1) its vulnerability to nutritional deficiencies and (2) its extraordinarily high energy requirements. Both these attributes are defining features of human brain function today, so it is necessary to explain how they arose *in infants*. Among the nutrient deficiencies to which the developing brain is vulnerable, inadequate dietary availability of two nutrients in particular is prevalent worldwide in at least a billion people today – iodine and iron. Unless prevented or counteracted early, these deficiencies permanently impair functional capacity of the adult brain (see Brain-Selective Nutrients section). Whatever enabled evolving hominins to avoid this major challenge to successful brain development which, 2 million years after hominin brain expansion started, still left at least 20% of the world's population at the mercy of these major nutritional deficiencies. What circumstances would have favored the expansion of the brain yet, simultaneously, increased such developmental vulnerability before the functional payoff finally starts in late adolescence and early adulthood? Indeed, why are humans the only primate (and probably the only undomesticated mammals) with such an obvious and detrimental vulnerability to specific dietary nutrient inadequacies? In other words, if early hominins (like other primates) rarely experienced inadequate intakes of these nutrients, what enabled the hominin brain to expand while simultaneously all the while *increasing* this nutrient vulnerability? To truly explain the origins of the human brain, these present-day vulnerabilities must also be explained.

The second issue with most other models of human brain evolution is that they do not address the very high energy requirements of the *newborn and infant* brain. What feature of human metabolism evolved to ensure that the expanding and uninterrupted energy demands of the early developing brain could be met? At birth, the human brain consumes 74% of the energy needs of the baby (Holliday, 1971). Proportional to the rest of the body, the energy requirements of the infant brain exceed those of the adult brain by an astonishing threefold, yet the brain has no real survival benefit to the newborn. On the contrary, among primates, human babies are the most helpless despite their brains receiving this astonishing investment in terms of energy supply. How did such an investment evolve that not only provides no tangible survival benefit to the newborn human but, by its defenselessness, means that the human infant also puts the survival of the nursing mother at greater risk? How did this vulnerability and disproportionate commitment to supply energy to the developing brain arise in early hominins but not to other nonhuman primates?

To mask the brain's increasing developmental vulnerability as it expanded, some habitat or ecological conditions must have been fundamentally different than for other primates. Genetic modifications that favored brain expansion may be part of the answer but such mutations or changes in gene expression actually left the brain *more* developmentally vulnerable than it probably was in australopithecines (see Gene-Nutrient Interactions section). However, other near-simultaneous mutations would also have been required that increased the density of mitochondria in brain cells and supported the increasing fuel demands of the expanding brain. Assuming that the coalescence of such genetic changes favorable to brain expansion is plausible, the evolution of a brain that was increasingly vulnerable to multiple nutrient deficiencies is rather more difficult to attribute to gene mutations or to changing gene expression. However, the occupation of a particularly favorable environmental niche some time *before* hominin brain expansion started could have masked the brain's coincident increasing developmental vulnerability, thereby enabling its expansion. Occupation of other nutritionally less favorable environments would have essentially stalled or prevented human brain evolution because they did not protect against this increasing developmental vulnerability. Indeed, differential gene expression alone cannot explain why about 20% of the world's population experiences suboptimal cognitive development caused by iron and iodine deficiency, yet 80% avoids this major challenge. What environmental factor masking this persistent vulnerability of the developing brain was present 2 or more million years ago and is still present today?

As previously described (Crawford and Marsh, 1989; Crawford et al., 1992; Broadhurst et al., 1998; Crawford et al., 1999; Cunnane, 2005a), there are three main components to this new paradigm of human brain evolution: (1) an increasing energy requirement, with a concomitant increase in the dependence on ketones to back up glucose; (2) a diet providing a richer and more reliable source of brain-selective nutrients for optimum brain development and function in adult life; and (3) concomitant evolution of body fat in newborns, which supplies both the fatty acids that are substrates for ketone production and at least one vitamin-like ω3 fatty acid – *docosahexaenoic acid (DHA)* – needed for successful brain development.

This new paradigm fully recognizes that healthy brain development cannot be assumed – it depends absolutely on the right environmental circumstances and a favorable ecological niche. Human brain evolution did not occur simply in the past 50,000–100,000 years; nutritional and metabolic circumstances that favor successful development in the present-day human brain must have been met more or less continuously for about 2 million years if natural selection was to enable its evolution. The corollary is that sustained occupation of an environment, more specifically of a habitat, providing a diet that broke through the nutritional and metabolic constraints on nonhuman primate brain function, created the opportunity for sustained brain evolution in hominins. Genetic modification or mutation would also have contributed to the potential for increasing brain sophistication but not in the absence of dietary ingredients and a habitat that could sustain such a potential.

BRAIN DEVELOPMENT

Vulnerability and Critical Periods

Common to all vertebrates is the vulnerability of the brain to permanent deficit caused by acute deprivation of oxygen or nutrients, or by environmental toxicity. Despite this vulnerability, some species have developed extreme resistance to the risk of such brain damage. Brain function in goldfish appears to be unimpaired by oxygen-free water for many hours. Some carp and freshwater turtles can withstand anoxic water for months at near-freezing temperatures. Goldfish and humans are therefore examples of two extreme strategies in brain evolution; in the former, natural selection made no investment in cognitive ability but made a major investment toward protecting against adverse conditions such as extremely low oxygen and nutrient availability, or low environmental temperature. In the goldfish, the brain's minimal energy demands were part of the strategy to successful survival. If the brain is protected against environmental extremes, it is because these insults are often present and become part of the strategy of natural selection in that species. The trade-off required by that strategy is that the brain will remain small, and the range of functions will be limited so that nutrient and oxygen requirements are met even under the most difficult of circumstances.

In primate brain evolution, the opposite investment became possible; increased size and cognitive performance was favored, but the trade-off was increasing vulnerability to environmental insults and rising risk of defective development. Therefore, where vulnerability in brain development has arisen, it is because the risk of environmental insult is minimal. This does not mean that the primate brain is not at risk of developmental problems, only that the risk of exposure to such circumstances is markedly lower than in fish or turtles. Hence, evolution of increasing brain vulnerability always shadows increasing cognitive capacity and depends absolutely on a more stable environment and more secure food supply. Hominins took this strategy one step further than primates, and humans took it further still.

A key attribute of maturation in higher organisms is that normal development occurs in a series of stages or events, the success of each of which depends to some extent on the success of the preceding stage. This sequential maturation creates a series of *critical periods* in which each stage is programmed to have a certain amount of time for completion. Successful completion of subsequent stages depends in part on the successful completion of each earlier stage. Nevertheless, the next stage is usually programmed to start whether or not the previous stage is fully complete. Critical periods are not unique to the nervous system but also characterize the development of the cardiovascular, skeletal, and immune systems. Throughout human brain evolution, the *risk* of exposure to this developmental delay as the brain progresses through its critical stages must have been essentially nonexistent for this vulnerability to have been retained as such a pervasive feature of human brain development.

ENERGY REQUIREMENTS OF THE BRAIN

In adult humans, the body's functions consume on average about 1800–2000 kcal of energy/day. A disproportionately large part of that total goes to the brain, the heart, the liver, and the kidneys; at rest, they consume two-thirds of the body's energy intake, but they represent less than 6% of the body's total weight. The metabolic rate of the heart is highest at 610 kcal/kg/day, followed by the kidneys at 390, the brain at 296, and the liver at 290 kcal/kg/day. Well below these values are those of other organs, including resting skeletal muscle at about 17 kcal/kg/day. The size of the adult human brain is relatively large compared to the heart or kidneys, so it disproportionately uses more of the body's total energy needs – about 23% of the total for only 2% of body weight (Holliday, 1971). The heart and kidneys are metabolically more active than the brain, but the brain takes a higher *proportion* of the body's total energy needs. Skeletal muscle has a relatively low rate of energy consumption but, on average, still uses 25–30% of the body's energy intake because it occupies 40–50% of total body weight and represents 80% of the body's total cell mass. Most remarkably, the newborn infant brain consumes about 74% of the infant's total energy requirements (Table 3.2).

The brain's high energy requirement arises principally from the activity of the sodium-potassium pump needed for electrical signal transmission and from the biochemical processes at the synapses needed to transfer an action potential between neurons. There are significant differences in energy consumption in different brain areas, with the *auditory cortex* and *inferior colliculus* (used for hearing) having the highest energy requirement (Sokoloff, 1991). The brain normally runs about 97% on the sugar, *glucose*. The healthy adult human brain consumes ~3.5 g glucose/100 g/min, while the infant brain consumes ~5.3 g glucose/100 g/min. At the most, the brain only has a few minutes worth of energy reserve in the form of *glycogen*, which is a large molecule consisting of branched chains of glucose molecules. After about 24-h starvation, no glucose reserves remain in liver or muscle, so organs requiring glucose must either obtain it through glucose synthesis, mostly from amino acids in muscle protein, or find a different fuel to replace glucose. While this is an entirely normal process during short-term energy deficit, without a backup

TABLE 3.2 Energy requirements of the human brain from birth to adulthood (modified from Holliday, 1971)

Age (years)	Body weight (kg)	Brain weight (g)	Brain's energy consumption	
			(kcal/day)	(% of total)
Newborn	3.5	400	118	74
4–6 months	5.5	650	192	64
1–2 years	11	1045	311	53
5–6 years	19	1235	367	44
10–11 years	31	1350	400	34
14–15 years	50	1360	403	27
Adult	70	1400	414	23

fuel different from glucose, synthesis of the necessary amount of glucose for the brain alone would use up within a matter of days a significant amount of the protein present in other organs like muscle and intestine. Hence, a key challenge during human brain evolution was to develop a significant fuel reserve but one that was different from glucose and that would be easily used by the brain and compatible with the body's energy needs as a whole. Ketones are that specialized brain fuel.

Ketones – Key Players in Human Brain Function

During fasting or starvation, organs like the heart and skeletal muscle readily use fatty acids as their preferred alternative fuel to glucose. However, the brain is unable to directly use fatty acids as a fuel. The brain is technically able to burn fatty acids, but the transport of fatty acids through the blood–brain barrier is too slow to make fatty acids a useful alternative to glucose for the brain (Pardridge, 1991). During prolonged starvation, ketones can supply up to two-thirds of the adult human brain's energy needs (Cahill, 2006). Hence, there is wide consensus that ketones are by far the principal alternative fuel to glucose for the brain (Sokoloff, 1991; Cunnane, 2005a; Cahill, 2006).

In fact, the brain appears to be unique in absolutely depending on ketones to replace low availability of glucose. When food intake is interrupted and blood glucose starts to fall, ketone production gradually increases and seamlessly replaces lower availability of glucose to the brain. There are three ketones – *beta-hydroxybutyrate*, *acetoacetate*, and *acetone* – all of which easily access the brain. Saturated and monounsaturated fatty acids are easily converted (beta-oxidized) to *acetyl CoA* which, through a series of enzyme-catalyzed steps, is then condensed into ketones. Complete beta-oxidation of fatty acids without their conversion to ketones is how organs other than the brain replace low availability of glucose. The liver is capable of producing but not burning ketones, so fatty acids released from fat stores that make it to the liver without being transported into other tissues are easily converted to ketones, the main beneficiary of which is the brain.

Ketones can replace much of the human brain's requirement for glucose, but they cannot meet the brain's entire energy needs (Cahill, 2006). In addition to providing the carbon to replace *oxaloacetate* in the tricarboxylic acid cycle, glucose is essential for the brain as the precursor to *lactate*, which is exchanged when ketones are taken up by the brain's monocarboxylic acid transporter. Hence, the brain actively transports and uses glucose as well as ketones even during starvation when blood glucose is low and ketones are high.

As products of fatty acid oxidation, long-term ketone availability is linked to fat stores. Unlike glucose synthesis, ketone synthesis does not draw on amino acids so it does not threaten organ integrity by depleting tissue protein levels. Brain uptake of ketones occurs through the *monocarboxylic acid transporter* which rapidly ramps up ketone transport in proportion to rising ketone levels in the blood. The brain can perform all the steps converting ketones to fuel, including conversion of beta-hydroxybutyrate to acetoacetate, acetoacetate to *acetoacetyl CoA*, and acetoacetyl CoA to acetyl CoA (Fig. 3.1). Acetyl CoA is the final common denominator in the metabolism of all the different energy substrates for adenosine triphosphate (ATP) production. Ketone utilization is a constitutive feature of brain function because the amounts and activities of the relevant enzymes are not changed by starvation and always exceed the amount necessary to supply the brain's energy needs. Hence, the brain is always ready to burn ketones as soon as they are available; a situation consistent with their function as its main backup fuel. The factor that limits ketone use by the brain is therefore the rate of ketone production by the liver and not the brain's ability to use them. Ketone availability is principally dictated by fatty acid

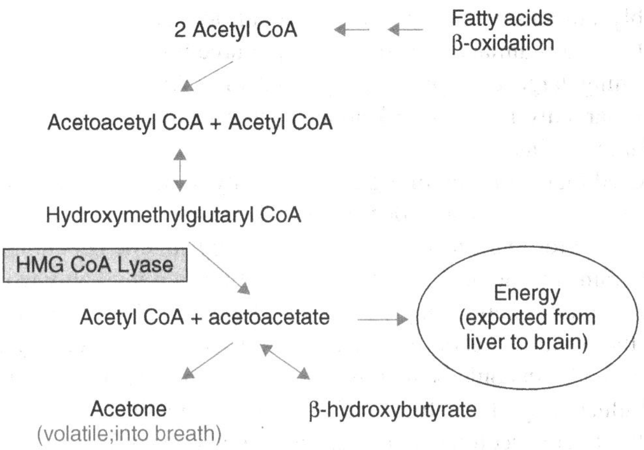

Figure 3.1 *Ketogenesis* – the process of beta-oxidizing fatty acids to the ketones – acetoacetate, beta-hydroxybutyrate, and acetone.

release from fat stores, which in turn is determined by how much insulin levels fall in response to decreasing plasma glucose; it is low insulin that permits the mobilization of fatty acids from fat stores and their conversion to ketones by the liver.

There are two key advantages to having ketone bodies as the main alternative fuel to glucose for the human brain. First, humans normally have significant body fat stores, so there is an abundant supply of fatty acids to make ketones. Second, using ketones to meet part of the brain's energy requirement when food availability is intermittent frees up some glucose for other uses and greatly reduces both the risk of detrimental muscle breakdown during glucose synthesis, as well as compromised function of other cells dependent on glucose, that is, red blood cells. One interesting attribute of ketone uptake by the brain is that it is four to five times faster in newborns and infants than in adults (Robinson and Williamson, 1980; Cremer, 1982). Hence, in a sense, the efficient use of ketones by the infant brain means that it arguably has a better fuel reserve than the adult brain. Although the role of ketones as a fuel reserve is important, in infants, they are more than just a reserve brain fuel – they are also the main substrate for brain lipid synthesis (see Baby Fat – The Reserve for Brain Lipids section).

Comparative Aspects of Ketone Metabolism

Comparing brain energy metabolism across species provides a useful perspective on potential brain fuel constraints on brain development, function, and evolution. Brain-to-body size ratios vary across species and so does ketone production in response to fasting or food deprivation. The ability of the brain to switch to ketones as the primary energy substrate is not unique to humans but appears to be better in humans than in other omnivores like the rat, pig, or monkey (Robinson and Willliamson, 1980). Dogs, sheep, and pigs are unable to achieve significant ketosis, even during prolonged fasting (Wiener et al., 1971) and, compared to their body size, they also do not have very large brains. In contrast, humans appear to have the highest ketone response to fasting or starvation. When animals with proportionally smaller brains experience food deprivation, their muscle and liver glucose stores (glycogen) last longer because their proportionally smaller brains are not using up glucose as fast as the large brain in humans. Thus, the demand for ketone production to replace low glucose availability is lower in animals that have proportionally smaller brains. Plausibly, animals with proportionally small- to medium-sized brains were not challenged to evolve a substantial alternative energy source for the brain because their brains were not becoming large enough to really need one. Alternatively, not having evolved the ability to markedly raise blood ketones during fasting may have prevented them from evolving larger brains.

I have hypothesized that evolution of a greater capacity to make ketones coevolved with human brain expansion (Cunnane, 2005a). This increasing capacity was directly linked to evolving fatty acid reserves in body fat stores *during fetal and neonatal development*. To both expand brain size and increase its sophistication so remarkably would have required a reliable and copious energy supply for a very long period of time, probably at least a million, if not two million, years. Initially, and up to a point, the energy needs of a somewhat larger hominin brain could be met by glucose and short-term glucose reserves such as glycogen and glucose synthesis from amino acids. As hominins slowly began to evolve larger brains after having acquired a more secure and abundant food supply, further brain expansion would have depended on evolving significant fat stores and having reliable and rapid access to the fuel in those fat stores. Fat stores were necessary but were still not sufficient without a coincident increase in the capacity for ketogenesis. This unique combination of outstanding fuel store in body fat as well as rapid and abundant availability of ketones as a brain fuel that could seamlessly replace glucose was the key fuel reserve for expanding the hominin brain, a reserve that was apparently not available to other land-based mammals, including nonhuman primates.

NUTRIENTS AND BRAIN FUNCTION

Adequate long-term fuel reserves are necessary but, alone, are not sufficient to guarantee the function of a large brain. Specific dietary nutrients are also needed for normal development and function of the brain and organ systems. If the mother is underfed or poorly nourished, her fetus develops more slowly, from which two possible outcomes are increasingly likely – either the baby will be born of low birth weight or it will be born prematurely. By definition, in humans, a premature birth occurs at a gestational length of 37 weeks or less, whereas normal full-term birth is at 40 weeks gestation. Low birth weight babies are skinny and small for their gestational age; they may be born at full term but may also be born prematurely. In part because of a risk of suboptimal nutrition, both these conditions – low birth weight and prematurity – impose a high risk that the developing brain will be malnourished, at least transiently.

Malnutrition and Brain Development

During starvation, brain function is affected less, and it loses less weight compared to skeletal muscle or gastrointestinal organs. During malnutrition, brain function tends to be spared more in adults than in infants or children. Because of their rapid growth and immaturity, children are the most adversely affected by hunger and malnutrition. Unlike in adults, the brains of young children are still very vulnerable to developmental delay caused by starvation. This is due to the proportionally higher energy requirements of the brain during childhood than at any other time later in life and to the incomplete maturation of the brain. Severe malnutrition rapidly dampens the exploratory behavior of children and leads to apathy. Brain activity is metabolically expensive so less behavioral responsiveness and reduced physical activity allow the developing brain to better withstand the severe metabolic impact of malnutrition. This "coping" strategy is effective in the short term but, if malnutrition is chronic, greatly increases the risk of permanent suboptimal functionality.

The main effect of nutritional deprivation on the brain is that it grows more slowly. The reduction in brain growth involves a reduction in weight, volume, number of cells, total amount of DNA, and total amount of lipid in the brain. In addition to impairing protein synthesis, malnutrition in human infants reduces the total content of cholesterol and phospholipids in the brain. Species with a prenatal brain growth spurt, for example, the guinea pig, are less affected by postnatal malnutrition or by excess competition for milk caused by a large litter than are the young of a species whose brain growth spurt is postnatal, for example, the rat or human. Different timing of the brain growth spurt in different species affects this deficit because the same nutritional deprivation induced after the brain growth spurt has much less impact than during the brain growth spurt.

Malnutrition at birth reduces brain cell number and size but malnutrition starting later in infancy (4–5 years old) may only affect brain cell size because by then the brain is closer to its maximal cell number (although malnutrition may prevent these cells from full development). If malnutrition has not impaired cell division and normal cell number in the brain, it is possible to renourish the infant and largely restore normal brain development. However, if malnutrition occurs earlier, that is, before the critical period for completing brain cell division is over, the deficit can rarely be fully corrected, and brain function will almost always be permanently impaired.

The Need for Specific Dietary Nutrients

A number of organic molecules and inorganic minerals are needed in the diet to permit normal development, maturation, and reproduction. These dietary molecules cannot be made by higher organisms and are collectively known as *nutrients*. Nutrients include certain minerals, vitamins, fatty acids, and amino acids. Other molecules such as glucose and cholesterol are important in metabolism or cell structure but are not essential dietary nutrients *per se* because adequate amounts can be made in the body irrespective of their presence in the diet. Deficient intake of nutrients causes specific symptoms that are characteristic for the nutrient in question and can only be corrected or prevented by ensuring sufficient dietary supply of that specific nutrient (e.g., the gum deterioration and bleeding that occur in scurvy as a result of vitamin C deficiency, and which can only be corrected by foods providing vitamin C). Nutrient deficiencies occur in one of two ways: either because the diet provides inadequate amounts of that nutrient, or because a disease or abnormality inhibits the ability to absorb, utilize, or retain that nutrient.

There is an important distinction between malnutrition and specific nutrient deficiency. Malnutrition is a general insufficiency of food intake. However, a nutrient deficiency can occur in the presence of sufficient intake of energy, protein, and other essential nutrients. Indeed, specific nutrient deficiencies are widespread in the world and several occur in the absence of frank malnutrition. For instance, according to the World Health Organization, inadequate intake of iodine, iron, and vitamin A impairs the health of over a billion people globally, that is, at least 20% of the world's population. Dietary deficiencies of iodine, iron, and vitamin A all adversely affect neurological development, behavior, and learning in school-aged children, that is, at an age when brain function should be maturing and securing a lifetime of opportunity. Hence, discussion of these specific nutrients is of particular relevance to the discussion of habitats and diets that enabled hominin brain evolution.

A report by the United Nations Children's Fund concludes that the "brainpower of entire nations is slipping because of a shortage of certain dietary nutrients, including iodine, iron, vitamin A and zinc" (Ramkrishnan, 2002). This report examined health and nutritional status in 80 developing countries representing about 80% of the world's population. It notes that inadequate iron intake reduces a child's IQ by five to seven points, while insufficient iodine reduces childhood IQ by 13 points. Iron deficiency is pervasive enough that it is estimated to reduce the gross domestic product of the most affected countries by up to 2%. This report also observes that "so ubiquitous is vitamin and mineral deficiency that it debilitates in some significant degree the energies, intellect and economic prospects of nations." Of note, there is no mention of protein insufficiency as a problem of comparable magnitude to these micronutrient deficiencies. Hence, inadequate dietary supply of micronutrients needed by the brain is not to be confused with generalized malnutrition, protein insufficiency, or chronic starvation.

Micronutrient deficiencies compromising brain development are clearly debilitating and widespread. Tragically, they are relatively easily preventable by supplementation with the affected nutrients. The point in relation to human brain evolution is if that about a billion people worldwide are sufficiently iodine or iron deficient to cause mild to moderate impairment in mental function (excluding tens of millions that are more severely deficient in these two nutrients), a valid concept of human brain evolution must account for this major environmental challenge to brain function in present-day humans. Clear evidence on a global scale shows that the human brain cannot develop or function normally when deficiencies of iodine, iron, zinc, and vitamin A are present. What then enabled the evolving hominin brain to confront these nutrient deficiencies so as to bypass or avoid the resulting vulnerability that so markedly impairs brain function today?

BRAIN-SELECTIVE NUTRIENTS

The term *brain-selective nutrients* was coined to identify specific nutrients that are viewed as essential for expansion and evolution of primate brain to its present-day form in humans (Cunnane et al., 1993). Brain-selective nutrients include certain minerals, vitamins, and ω3 fatty acids like *DHA*. *Brain-selective minerals* include at least five dietary minerals (or trace elements) – *iodine, iron, selenium, copper*, and *zinc*. These minerals are not more important than the vitamins (especially vitamins A and D) or DHA, but compared to the vitamins and ω3 fatty acids which are organic molecules, it is simpler to characterize the function and limiting nature of certain dietary minerals because neither plants nor animals synthesize minerals in the body.

Iodine – The Primary Brain-Selective Mineral

I have classified iodine as the primary brain-selective mineral because it is the only nutrient for which national governments have created legislation in order to enforce adequate availability in the food supply. This is a powerful testimonial both to iodine's importance for normal human health and to the exceptionally widespread extent to which suboptimal human development would otherwise continue simply because of inadequate voluntary consumption of iodine (Venturi et al., 2000; see also Chapter 6). The need to iodize salt to prevent nutritional iodine deficiency is the principal reason for proposing iodine as the primary brain-selective mineral. The richest natural dietary sources of iodine are fish, shellfish, and coastal plants, particularly algae and seaweed.

Iodine is essential for thyroid hormone production which, in effect, sets the thermostat on body metabolism. In this role, iodine is of utmost importance for normal fetal and postnatal brain development (Pharoah et al., 1971; Dussault and Ruel, 1987). Impaired ketone synthesis is one aspect of brain development adversely affected by hypothyroidism (Patel, 1979). Other roles of iodine that are of potential relevance to brain function and evolution are described in Chapter 6. In addition to causing fatigue and detrimental effects on brain development, hypothyroidism causes serious hearing impairment, probably by impeding ossification of the organ of Corti in the inner ear (Goldey et al., 1995). Even when hearing itself is not impaired, learning from the spoken word and verbally expressing ideas are impaired in hypothyroidism, indicating that this condition impairs the assimilation of auditory information. Coincidentally, the brain's hearing center has a particularly high energy requirement (Sokoloff, 1991). Expansion of the auditory region in humans undoubtedly facilitated the evolution of speech in humans, so two key aspects of human brain evolution (development of hearing and brain enlargement) both had an increasing requirement for energy and are both vulnerable to inadequate iodine status.

The need for government legislation to iodize salt so as to assure sufficient iodine intake has three important implications regarding present-day human habitats and availability of brain-selective minerals: (1) Without the availability of iodized table salt, noncoastal (inland) habitats do not provide foods rich enough to meet human iodine requirements. People eat less fish and seafood when they live far from the coasts, which is the main reason why iodine deficiency is more prevalent inland or in mountainous regions. (2) Iodine deficiency is entirely preventable by appropriate food selection, especially shellfish and other shore-based foods. Eggs can be a good source of iodine, while meat and nuts are a moderately good source (Table 3.3). (3) The retardation of cognitive development associated with iodine deficiency is of sufficient scale and severity to have globally mobilized national governments to prevent it.

Iodine deficiency is not just a problem in poor countries. Up to 10% of Europeans, especially women, are presently at risk of iodine deficiency (Wynn and Wynn, 1998; Verma and Raghuvanshi, 2001). Major national health surveys confirm that mild hypothyroidism presently goes undetected in upward of 10 million Americans and upward of 40 million Europeans (Sorcini et al., 1993; Vanderpump et al., 1995; Weetman, 1997; Cooper, 1998; Morris et al., 2001). On the basis of urinary iodine excretion decreasing by

TABLE 3.3 Approximate amount of different food groups required to meet the daily iodine requirement in nonpregnant, nonlactating adult humans expressed in comparison to fish (modified from Cunnane, 2005a).

Eggs	1.3
Shellfish	1.4
Meat	10
Nuts	10
Cereals	20
Pulses	25
Vegetables	28
Fruit	40
Milk	45

Note: Therefore, 1.3 times more eggs (total of about 190 g), 1.4 times more shellfish (total of about 210 g), and so on, would be required in comparison to fish (150 g/day). These averages are based on food composition tables (McCance and Widdowson, 2002), but clearly, there is considerable variation in each of these categories due to the different species that are included. Furthermore, variation in abundance of these foods, for example, with season or need for domestication, and any restrictions on bioavailability of iodine from these foods, has not been taken into consideration in calculated quantities required.

about 50% over the past 20 years, more than 20% of Australian women are now believed to be at risk of mild to moderate iodine deficiency (Gunton, 1999).

The human body has no mechanism to conserve iodine during iodine deficiency. The seemingly paradoxical inability of humans to conserve iodine despite its unequivocal importance could plausibly have arisen because, early in human evolution, shore-based habitats provided diets that were sufficiently rich in iodine that such a conservation mechanism was unnecessary. No other species in the wild, including primates, is known to exhibit symptoms of iodine deficiency while consuming its routine diet. The implication is that if symptoms of nutrient deficit are normally absent, then the actual intake of that nutrient is probably not limiting for reproduction, growth, or normal development. Conversely, the relatively high prevalence of mild symptoms of iodine insufficiency indicates most human diets do not provide adequate amounts of iodine. It is unreasonable to postulate that the human requirement for iodine (or other nutrients) dramatically increased only in the past few hundred years, so the prevalence of iodine deficiency seems to have arisen from increasing occupation during recent centuries by humans of geographical regions ill-suited to human iodine requirements, that is, inland or mountainous regions distant from shorelines and shore-based foods.

Goiterogens A *goiter* is a visible swelling of the thyroid caused by impaired production of thyroid hormone. It is commonly but not exclusively linked to iodine deficiency. Goiter-causing substances in food (*goiterogens*) are a major cause of iodine deficiency disorders and contribute to cretinism in many regions of the world because they bind strongly to iodine, leading to excessive iodine excretion in the urine and poor iodine absorption. Goiterogens deplete the body of iodine even in the presence of nominally adequate iodine intake. Iodine deficiency disorders are widespread where subsistence is based on vegetarian diets containing goiterogens (Remer et al., 1999).

Cassava is probably the most important goiterogenic food both because of its relatively high content of two iodine-binding goiterogens – *hydrocyanic acid* and *linamarin* – and because it is consumed very widely as a staple food. The body can detoxify hydrocyanic acid in cassava to *thiocyanate* but thiocyanate is also a strong goiterogen. Other goiterogenic foods also consumed widely by humans include maize, cabbage, rape, soybean, and mustard. These foods are frequently used in feed for farm animals where they are also goiterogenic. When fed to animals, they reduce the iodine content of meat which, in turn, reduces iodine available to humans consuming meat or dairy products from these animals.

Other Brain-Selective Minerals

Iron Iron deficiency has also become a widespread, serious nutritional problem in humans despite traditional knowledge of its health attributes and despite its abundance in the soil. Iron deficiency is a more severe problem in developing countries but is still prevalent in developed countries wherever the population is poor. Chronic iron deficiency can remain undetected for months in infants and children, and is a pernicious obstacle to normal brain and behavioral development (Lozoff and Brittenham, 1986; Pollitt and Metallinos-Katsaras, 1990; Pollitt, 1993). Unlike iron in many edible plants, iron in meat and fish is easily digested and absorbed. A major global cause of long-term iron deficiency is low meat intake and poor absorption of iron from cereal-based diets.

Iron is required for the normal production and function of several important proteins, notably *hemoglobin* for oxygen transport in blood and *myoglobin* for oxygen transport in muscle. Hence, like iodine, sufficient iron is essential for the body to efficiently use oxygen and metabolic fuels and to meet energy requirements (Erecinska and Silver, 1994). The ability of muscles to obtain oxygen determines their work capacity. Iron is needed for "activation" of oxygen by several *oxidase* and *oxygenase* enzymes, and in the *cytochrome* proteins used to transport electrons during cellular energy production in mitochondria. Hence, impaired ability to make hemoglobin due to iron deficiency is a major cause of poor muscle strength and fatigue. The brain's high energy requirement probably contributes to its vulnerability to low iron intake. Iron deficiency also impairs control of body temperature by interfering in conversion of the thyroid hormones *thyroxine* to *triiodothyronine*. This reduces the ability to conserve heat and contributes to coldness in the extremities as well as weakness and fatigability in iron deficiency.

Currently, 20% of women in the United States are sufficiently iron deficient to have no detectable plasma levels of the iron transport protein – *ferritin*. Half of these women are anemic; that is, they have low hemoglobin or red cell count in the blood. If iron deficiency persists, hemoglobin synthesis is impaired which, in turn, impairs red blood cell production. The main body pool of iron is in hemoglobin. In mild *iron deficiency anemia*, red cell iron levels may be normal, but more subtle markers of iron deficiency are usually present, that is, low serum ferritin.

On a global scale, iron deficiency appears to be a principal cause of neurodevelopmental delay and emotional fragility seen in malnourished children. Iron deficiency has a significant and specific effect on visual attention, concept acquisition, verbal scores, and school achievement; in short, several key components of healthy cognitive development (Pollitt, 1993). Part of the negative impact of iron deficiency on neurocognitive development may also relate to poor development of hearing (Roncagliolo et al., 1998). Vulnerability to iron deficiency is most apparent during early development because that is when brain growth is most rapid (Lozoff and Brittenham, 1986). Brain concentrations of iron and neurotransmitters are both reduced by dietary iron deficiency. Lower brain oxygen uptake caused by insufficient iron also appears to contribute to poorer performance on cognitive tests in iron-deficient children. Some of the decline in cognitive performance in iron-deficient children also arises via changes in brain receptors of two neurotransmitters, *dopamine* and *gamma-aminobutyric acid*.

Selenium Selenium is needed for the normal activity of an important antioxidant protein – *glutathione peroxidase*. Glutathione peroxidase defends against damage caused by oxygen attack on proteins and lipids, which releases *free radicals* that cause *lipid peroxidation* or rancidity. Glutathione peroxidase also protects the easily peroxidizable ω3 fatty acid – *DHA* – which is present in significant amounts in the brain and eye (see Chapter 4). Vitamin E is also involved in antioxidant protection, and selenium deficiency is exacerbated by vitamin E deficiency.

Selenium interacts in several important ways with two other brain-selective minerals – iron and iodine (Mitchell et al., 1997, 1998). Dietary selenium deficiency contributes significantly to the damaging effect of iodine deficiency on thyroid function (Vanderpas et al., 1990). Thus, in addition to reducing lipid peroxidation, the important role of selenium in iodine metabolism is a major reason why selenium is a brain-selective mineral. Heme production is not only dependent on iron but also on selenium, so selenium is important for efficient oxygen transport and energy metabolism. Selenium is essential for the activity of the enzyme, *iodothyronine deiodinase*, which converts one form of thyroid hormone to the other (*thyroxine* to *triiodothyronine*), a process that happens outside the thyroid itself. Iodine cannot be incorporated into triiodothyronine unless there is sufficient selenium for the deiodinase enzyme (Hotz et al., 1997; Arthur, 1999). The total amount of thyroid hormone in the blood is not necessarily affected by selenium deficiency, but the efficacy of thyroid hormone action is impaired by selenium deficiency because triiodothyronine is the form of thyroid hormone needed in some tissues (Arthur, 1999). Selenium is also involved in heme synthesis, thereby affecting the ability of iron to contribute to synthesis of hemoglobin. Dietary selenium is generally associated with proteins, so it is low in fruits and vegetables which are low in protein. Meat is a moderately good source of selenium. Fish, shellfish, and other shore-based foods are the best sources of selenium.

Copper Copper has several essential biological roles that make it a brain-selective mineral. Synthesis of lipids in the myelin sheath protecting signal transmission in nerves synthesis requires a dietary source of copper. *Hypomyelination* occurs during dietary copper deficiency, thus contributing to poor control of the skeletal muscles because the nerves do not reliably conduct signals unless they are myelinated. In humans, *Menke's disease* is a genetic abnormality in copper absorption causing progressive mental retardation, hypothermia, seizures, hypomyelination, and death during infancy. Copper deficiency in Menke's disease occurs because copper absorbed from the diet is not released from intestinal cells to the rest of the body. At least three enzymes involved in neurotransmitter production require copper as a cofactor: (1) *tyrosine hydroxylase*, which acts at one of the steps converting the amino acid, *tyrosine*, to the neurotransmitter *dopamine*; (2) *dopamine-beta-hydroxylase*, which converts dopamine to another neurotransmitter – *norepinephrine*; and (3) *monoamine oxidase*, which inactivates the monoamine neurotransmitters – *norepinephrine, serotonin*, and *dopamine*. During human evolution, expanding brain size and complexity of connections between neurons increased the need for these supporting processes, several of which depend on copper, as well as antioxidant protection of the additional membrane lipids which are easily damaged by free radicals and lipid peroxidation, additional capacity for cellular energy production, and myelination for integrity of electrical signals. The function of iron and iodine in cellular respiration also depends on copper. For instance, iron cannot prevent anemia unless there is sufficient dietary copper because copper is required for hemoglobin synthesis.

Zinc Zinc is a brain-selective mineral because zinc deficiency rapidly affects growth and development in infants and children, and because zinc is highly concentrated in the brain, particularly in the *mossy fibers* of the hippocampus where it appears to be involved in *enkephalin* binding to its receptor. Zinc is also needed for normal metabolism of the neurotransmitter – *norepinephrine*. Learning and memory are impaired in young, zinc-deficient animals, partly because zinc is also needed in the metabolism of polyunsaturated fatty acids (PUFAs), notably in the synthesis of *arachidonic acid*. Meat is a good source of zinc, but shellfish, especially oysters, is the best dietary source of zinc. *Phytate* is a significant component of cereals, but because phytate avidly binds zinc, zinc is less available from cereals than would appear to be the case from the zinc content of cereals alone. Dietary insufficiency of zinc occurs in areas where cereals are staple foods.

Brain-Selective Minerals in the Diet

The daily human requirement for these five brain-selective minerals has been estimated through research on the effect of nutrient deficiencies and the intake required preventing deficiency symptoms and optimizing development. Combining the dietary requirement

TABLE 3.4 Amount of each major food group required to meet the daily requirement for the five brain-selective minerals – iodine, iron, zinc, copper, and selenium (modified from Cunnane, 2005a).

Food group	Amount/day (g)	Most limiting nutrient
Shellfish	900	Copper
Eggs	2500	Copper
Fish	3500	Iron
Pulses	3700	Iodine
Cereals	4800	Copper
Meat	5000	Selenium
Nuts	5500	Selenium
Vegetables	8700	Zinc
Fruits	9300	Zinc
Cow's milk	47,000	Zinc

Note: This calculation is derived from food composition tables (McCance and Widdowson, 2002) and is based on the concentration of iodine, iron, selenium, zinc, and copper, and on the daily requirements for these five nutrients. Of these five minerals, the one which is least abundant in each food group becomes the most limiting nutrient (right-hand column). Hence, that nutrient determines how much of that food group has to be consumed to meet its requirement. Thus, shellfish is the best source of these five brain-selective minerals because the least amount is needed to satisfy the requirements for all five of these minerals.

values of these minerals with their concentration in various foods allows the calculation of how much of each food group would be needed by an adult human if that food group alone supplied the daily requirement for these minerals (Table 3.4). These food groupings permit distinction at a glance between broad groups of foods rich or poor in these brain-selective minerals. For instance, less than a kilogram of a selection of shellfish, (primarily mollusks and crustaceans) is needed to meet the daily requirements for all five of these five brain-selective minerals. This is much less than for any other food group, so shellfish is the best overall dietary source of these brain-selective minerals. Except for a somewhat low content of copper, eggs would be almost as good a source of these minerals as seafood. Fish is the best source of iodine and selenium but has lower levels of copper, zinc, and iron; thus, along with pulses, fish ranks below seafood and eggs, both at about 3.5 kg/day each. Pulses supply more copper, iron, and zinc than the equivalent weight of seafood, but pulses have much less iodine and selenium than seafood and contains phytate that reduces mineral absorption. They are followed by cereals, meat, and nuts at about 5 kg/day each. No one food group totally lacks all of these minerals but cow's milk is clearly the poorest source by a wide margin.

The key point of this analysis is that *any* amount of shellfish helps considerably to meet the dietary needs of humans for these important minerals. Inland areas of most continents lack sufficient dietary sources of iodine, selenium, and iron, so consumption of even small amounts of shellfish would clearly help meet the nutrient needs of normal human brain development and function. A combination of shellfish, fish, nuts, eggs, meat, and fruit seems to be more likely than exclusive consumption of one food group to the exclusion of all others. Other shore-based foods (eggs of aquatic birds, frogs, turtles, marsh plants, etc.) that also constitute "seafood" have not been analyzed in sufficient detail to state categorically that their content of brain-selective minerals would be equivalent to that of shellfish, but it is a reasonable assumption that shore-based foods in general are a good source of these nutrients (including ω3 fatty acids).

CRITICAL IMPORTANCE OF BABY FAT IN HUMANS

Among primates, baby fat is unique to humans, but its possible evolutionary importance has been given little attention (Cunnane, 2005a). We are attracted to plump babies because they appear healthy and their fatness somehow makes them innately more attractive. Perhaps because excess body fat in adults is unhealthy, physiological and evolutionary significance of body fat in babies has been overlooked. The fossil record has yielded few specimens of hominin infants, and body fat does not fossilize, so almost everything about the emergence of fatness in hominin infants is speculative. Nevertheless, like brain-selective nutrients, baby fat makes a crucial contribution to healthy brain development.

Fat Development in the Human Fetus

The near-term fetus spends considerably more energy on fat deposition than it does on protein synthesis and building lean tissue. At nearly 90% of fetal energy requirements, fat deposition is by far the major component of the increase in body weight during the final weeks of pregnancy (Battaglia and Meschia, 1973; Battaglia and Thureen, 1997). During the first two thirds of gestation, the human fetus has essentially no body fat, that is, is very lean like the fetuses of other terrestrial species. At 25–26 weeks gestation, cells under the skin start to differentiate from connective tissue and become fat cells or *adipocytes*, the function of which is to accumulate a droplet of fat or triglyceride. The differentiation of adipocytes is the start of a process during which fetal body fat doubles during the next 6–7 weeks and more than triples during the final 6–7 weeks of pregnancy. As a result, 500–600 g of fat will normally be present on the fetus by the time it is born, almost all of it accumulating under the skin and during the final 13–14 weeks of pregnancy (see Fig. 3.2). Because of the very rapid rate of fetal fat accumulation during the third trimester, babies born before term (prematurely) have less fat than do those born at term. Babies born 5 weeks early have only about half the fat they should; those born 10 weeks early have only about 15% of the fat they should.

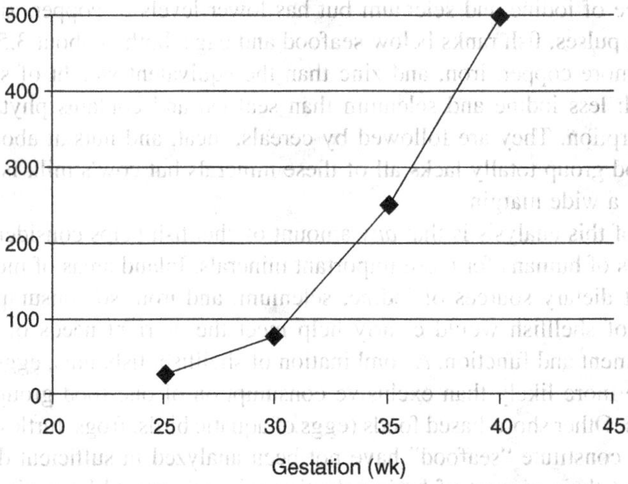

Figure 3.2 Accumulation of body fat on the human fetus during the third trimester of gestation (wk, weeks). Note that the total increase is about 10-fold what is present at week 25–26. Among all large mammals, especially primates, this amount and rate of body fat accumulation are unprecedented.

Neonatal Body Fat and Brain Development

Human baby fat is different from adult body fat in several ways:

1. Body fat in the infant is about 25% water and 70% stored fat, whereas in adults, it is about 10–12% water and 88–90% stored fat.

2. Its distribution is different; imaging studies show that baby fat is almost all directly under the skin (*subcutaneous fat*) and, unlike in adults, little or none is present as *visceral fat* around the abdominal organs (Harrington et al., 2002). In addition to the torso, baby fat occurs on the limbs and face, which is a pattern of distribution unlike that seen in adults except in extreme obesity. In adults, accumulation of subcutaneous fat poses markedly less of a risk for cardiovascular disease and heart attack than does visceral fat.

3. Of the fatty acids in newborn body fat, greater than 90% are saturated or monounsaturated fatty acids, and there are very low amounts of the two primary dietary PUFAs, the omega-6 (ω-6) fatty acid – *linoleic acid*, and the ω3 fatty acid – *alpha (α)-linolenic acid* (Hirsch et al., 1960; Sarda et al., 1987; Table 3.5). These two PUFAs are typically found at a total of 10–15% in adult body fat, an amount roughly corresponding to their consumption from vegetable oils in the diet. Hence, before birth, the placenta somehow screens out the mother's relatively high intake of these two PUFAs and sends very low amounts of them to the fetal body fat that is developing in the third trimester. This indirectly suggests that fetal body fat serves a function that differs in subtle ways from adult fat which is simply a mirror of dietary fat intake. Interestingly, unlike other organs, the brain's fatty acid composition also has very low amounts of linoleic acid and α-linolenic acid (Table 3.5).

TABLE 3.5 Average percent composition of the principal fatty acids in human infant brain as compared to body fat in adult and infant humans[a].

	Infant brain	Infant body fat	Adult body fat
Saturated fatty acids			
Palmitic acid (16:0)	28	43	20
Stearic acid (18:0)	18	4	5
Monounsaturated fatty acids			
Palmitoleic acid (16:1ω7)	3	14	4
Oleic acid (18:1ω9)	20	28	42
Polyunsaturated fatty acids			
Linoleic acid (18:2ω6)	1	2	15
Arachidonic acid (20:4ω6)	11	0.7	0.3
α-Linolenic acid (18:3ω3)	<0.1	<0.1	2
DHA (22:6ω3)	7	0.4	0.1

[a] Data are compiled and modified from Hirsch et al. (1960), Sarda et al. (1987), and Farquharson et al. (1992, 1993).

Note: Key points are that fatty acid composition of the brain differs most clearly from body fat in having a much higher content of the two polyunsaturated fatty acids, arachidonic acid, and DHA. Interestingly, infant body fat is also richer in these two fatty acids than is adult body fat. Indeed, the 0.4% DHA in infant fat represents a 1000 mg reserve of this fatty acid for the term infant (Cunnane et al., 2000). Additionally, compared to body fat in infants, adult body fat contains 7–20 times more of the two "parent" polyunsaturated fatty acids – linoleic acid and α-linolenic acid – this despite the abundant availability of these two fatty acids in maternal nutrition. These differences between the composition of infant and adult body fat suggest that the former has a subtly different function than that of adult body fat, that is, that it is better suited to supplying the two fatty acids most likely to be limiting for normal brain development – arachidonic acid and DHA.

4. Compared to adults, baby fat contains 3–4 times more *arachidonic acid* and DHA, the two principal long-chain PUFAs that are needed in the brain (Table 3.5). The two most common PUFA in adult body fat are linoleic acid and α-linolenic acid, which are vitamin-like because they have irreplaceable functions as building blocks in the body and because mammals cannot make them in their bodies; thus, they have to be supplied in the diet. Arachidonic and docosahexaenoic acids are the two PUFA most needed in the brain and, unlike other fatty acids, they are not easily burned as fuels; they function almost exclusively as membrane constituents and precursors to specialized signaling molecules called *eicosanoids* and *docosanoids*. Hence, storing nonoxidizable fatty acids such as arachidonic acid and DHA in infant body fat suggests they are there as a reserve for uses other than as fuels (see the section below on Baby Fat – The Reserve for Brain Lipids and Chapter 4).

Healthy newborns continue to gain body fat during the first few years of life so, in effect, they continue to build a fuel reserve that is particularly appropriate for (though not exclusively used by) the brain. One might well ask how this reserve could be important if it expands rather than being used up during early development. Indeed, why do fat reserves develop so rapidly during the third trimester when the fetus has no real use of its brain, at least for survival? The reason that body fat expands during late fetal development is that it provides the best possible preparation, that is, insurance, that postnatal development will proceed uneventfully. The only way it makes sense to invest so much metabolic and genetic effort in a developing brain that is so vulnerable and immature is that it happens under conditions in which the brain is highly protected by nutrient and fuel reserves.

Baby Fat – The Reserve for Brain Lipids

A key point about ketone production in the neonate is that is serves more than just as a backup fuel to glucose. Ketones are also the brain's main source of carbon to make the lipids like *cholesterol* and some fatty acids (*palmitic, stearic, oleic acids*). These lipids constitute about 55–60% of the solid matter of the human brain and are crucial structural components of brain membranes. Quantitatively, the most important brain lipid is cholesterol which is vital for brain development. Interruption of brain cholesterol synthesis has catastrophic consequences for brain development in all mammalian species. Almost no cholesterol in the diet or made elsewhere in the body gets into the brain (Jurevics and Morell, 1995). Hence, 99% of brain cholesterol is made *in situ*, almost all from ketones (Edmond, 1974; Cunnane et al., 1999). Hence, remarkably, the brain can only meet its needs for these lipids by endogenous synthesis; the presence of high or low amounts of cholesterol or saturated fatty acids in the diet has no bearing on their uptake by the brain.

This key role of ketones in brain lipid synthesis probably explains why human infants are in a continual state of mild ketonemia (Melichar et al., 1967) – not because they are underfed but because they are constantly programmed to synthesize ketones for brain lipid synthesis. Irrespective of their blood glucose level or time since last feeding, babies naturally tend to have higher blood ketones than adults. Hence, unlike in adults where they are almost exclusively an alternative fuel to glucose, in infants, ketones are constantly being produced and utilized. Indeed, mild ketonemia appears to be necessary for normal brain function in the infant (Cremer, 1982). In keeping with such a specific and important role in brain development, the brain of a human fetus or newborn is also able to extract ketones from the blood three to four times more effectively than is an adult brain (Adam et al., 1975).

It may appear inefficient to take palmitic acid stored in body fat, turn it into several ketones, transfer the ketones to the brain, and then remake palmitic acid or cholesterol within the brain. Nevertheless, that is exactly what happens. This process seems to provide better control of brain lipid composition than would otherwise be possible if dietary fats could be incorporated directly into brain membranes as they are in other organs. Since so much of human brain growth occurs postnatally, body fat stores and efficient ketogenesis seem to be an essential solution to permitting this important phase of human brain development. Other species complete their brain development much earlier than humans so this "ketone-dependent" postnatal phase that depends on ketone production from stored body fat appears peculiar to humans.

Polyunsaturates represent less than 1% of the fatty acids in body fat at birth, of which DHA is nearly half. Although seemingly a trivial amount, when multiplied by the normal 500–600 g of fat at birth, it amounts to a 1000 mg reserve of DHA (Cunnane et al., 2000). Our studies have shown that the human brain normally accumulates about 10 mg of DHA per day during the first 6 months of postnatal life. Thus, if it were supplying only the brain, this 1-g DHA reserve could theoretically last 100 days or about 3 months. We calculated that all the rest of the body needs about an equal amount of DHA as the brain so the total DHA needed by the developing infant during the first 6 months of life is about 20 mg/day. If no DHA was available from mother's milk or from a milk formula, and if all the DHA in body fat at birth was available for growth and development, this reserve would still last about 50 days (Cunnane et al., 2000). Fatty acid analysis of the human infant brain shows clearly that infants not receiving preformed dietary DHA cannot accumulate brain DHA at the same rate as those receiving a dietary source of DHA (Farquharson et al., 1992; Cunnane et al., 2000). Hence, because of slow DHA synthesis in humans (Plourde and Cunnane, 2007), DHA is an essential nutrient during infant development (Cunnane, 2005a). See the Conclusion of this chapter for further elaboration of why competing views about this point are flawed.

DHA is discussed at greater length elsewhere (see Chapters 2 and 4), so the main point here is that reserves of DHA and reserves of fatty acids that are ketone precursors are both present in infant body fat and are both important for normal brain development. In addition to the DHA reserves in body fat, mother's milk always contains some DHA so a breast-fed baby does not depend only on its own DHA reserves. However, babies fed formula milk do depend on their own DHA reserves unless they are given a special milk formula containing DHA. Most importantly, premature or low birth weight infants have much less fat and therefore have much less insurance in the form of ready-made DHA or brain fuel. In addition, the small body size but relatively larger surface area of the premature infant means it is much more likely to be in energy deficit and to be burning fatty acids as fuels.

Chimpanzee infants are like premature human infants because they have almost no body fat and therefore no reserves of brain fuel or DHA. Chimpanzee infants have highly functional brains at birth, but the lack of brain fuel and DHA reserves must curtail their postnatal brain development, just as it does in premature human infants.

Prematurity, Low Birth Weight, and Body Fat Stores

Newborn nonhuman primates are skinny and look like premature babies. Nevertheless, though lean, they are functionally well developed, while the opposite is true in lean human babies. In births occurring at <34 week gestation, pediatric intensive care is needed not only to try to maintain normal growth but also to maintain optimal body temperature and thereby reduce unnecessary energy demands so as to provide a better chance of normal brain development (Towers et al., 1997). Low body fat stores are a major component of low birth weight in human babies. Fetal body fat is almost exclusively deposited during the third trimester so premature birth is associated with a disproportionately low amount of body fat. Premature infants risk compromised organ function because of underdeveloped digestive and immune systems, inadequate fat reserves, and higher risk of delayed neurological development (Crawford et al., 1997). This is a problem similar to that faced in malnutrition when energy reserves are low to nonexistent and poor sanitation causes gastrointestinal infection and poor nutrient absorption. In effect, even the healthiest premature infant is at higher risk of malnutrition than is a healthy term infant.

Premature infants face several challenges linked to energy metabolism and brain development: (1) They have low fat stores because they are born before fat stores have been fully deposited. (2) Because they have less body fat, premature infants have proportionally larger brain size even than healthy term infants. (3) Their small body size has a proportionally larger surface area, which increases heat loss and hence energy needs to maintain normal body temperature and function. In essence, the premature infant faces the challenge of simultaneously fuelling the high cost of staying warm and developing the brain and other organs but without the fuel reserves of the term infant. Thus, there is competition between using fuel reserves to stay warm or fuel the brain. This problem is more acute the more premature the infant.

Birth weight of less than 1000 g, that is, at <27 weeks, is defined as very *low birth weight*. Although not uncommon today, such low birth weight is an extreme form of prematurity with much higher odds of delayed development. Indeed, the proportion of total births represented by very small birth weights is increasing in developed and developing countries alike (Crawford et al., 1997). The cause of this growing problem is multidimensional but includes poor maternal socioeconomic circumstances as well as poor diet selection and nutrient deficiencies in the fetus. Indeed, even before conception, poor nutrition has a negative impact on pregnancy outcome (Crawford et al., 1997). Prematurity, low birth weight, or postnatal malnutrition all increase the risk of stunted brain development and poor performance at school (Hack et al., 1991, 1994, 2002). Low fat reserves in premature infants create competition to oxidize for energy the same fatty acids that should be destined to become ketones used in making the lipids needed as components of synaptic membranes in the developing brain.

Compared to humans, low body fat reserves at birth represent less of a developmental limitation in other species, including nonhuman primates. Some are born helpless like humans; others are born "fully functional" and can feed and be transported immediately. Although fully mature, they are born without visible reserves of body fat so, in effect, other species are born with similar fuel reserves to those of premature human infants. Those born precocial have better developed brains permitting higher functionality at birth. However, unlike humans, the brains of most other species are not expanding significantly after birth, so the nutrient and energy investment in the brain is less than in humans. In other species, much of the brain's maturation is complete before birth so, in a sense, it would be superfluous to develop fetal or postnatal fat stores because the brain is not big enough and because further increases in functional capacity are not needed. The piglet is a good example of such a precocial, highly functional species at birth, but it cannot develop sufficient ketosis to save its own life if starved for more than a day or two. It has a high functioning brain at birth, but brain function does not expand much postnatally, in part because there are no fuel reserves to support it. The point is that since neonatal body fat is a crucial ally for normal human brain development and is less available to the premature infant, the evolution of the big brain in humans must have depended on simultaneous or prior evolution of neonatal body fat.

GENE – NUTRIENT INTERACTIONS

Genes and Brain Evolution

Recent advances in the technology needed to analyze the genome and manipulate gene expression have dramatically altered our understanding of gene function. This has increased research into genes that affect learning and memory and could potentially have been involved in expanding the size and functional capacity of the hominin brain (Bliss, 1999; Tang et al., 1999; Routtenberg et al., 2000). Genetic changes potentially affecting human brain evolution appear to have occurred through changes in gene expression, coding sequence, or addition or deletion of entire genes (Hill and Walsh, 2005). Though the genomes of the chimpanzee and human are now widely recognized as having very similar sequences, some gene inactivation or silencing has occurred in humans, especially for olfactory receptor genes. Several genes functioning specifically in the brain also seem to be more highly expressed or may be undergoing more rapid changes in humans compared to chimpanzees (Enard et al., 2002; Caceres et al., 2003). Though more debatable, in primates compared to rodents, more genes controlling development and function seem to have undergone positive selection, that is, nondetrimental changes in the *nonsynonymous* (noncoding) regions of DNA that encode for proteins. Significant changes in the sequence of nonsynonymous DNA frequently cause deleterious mutations but more subtle changes do occur and may therefore have been positively selected in humans.

Examples include the *FOXP2* gene, which is involved in language and speech production (Hill and Walsh, 2005), and two genes which when markedly mutated cause microcephaly (small brains) in humans but for which subtle sequence changes appear to have been positively selected in humans (Bond et al., 2002; Jackson et al., 2002;). These microcephaly genes are expressed in the upper layers of the cortex so their expression may be a constraint on further cortex development in nonhuman primates. Other genes affecting energy and lipid metabolism seem to have undergone up-regulation in the human brain (Caceres et al., 2003; Grossman et al., 2004); in both cases, the effects of these genes could well be linked to nutrients controlling brain lipid and energy metabolism (see Brain-Selective Nutrients section) or to thyroid function (see Chapter 6).

Though the vast majority of DNA is noncoding, comparative genomics between related species permits the identification of both functional elements in conserved regions of DNA not coding for proteins as well as more rapid sequence changes within these more conserved regions. *Human accelerated regions* are areas of noncoding DNA, the sequence of which has undergone more rapid change in humans than in the chimpanzee. Human accelerated regions are of interest for human brain evolution because they seem to particularly affect neurodevelopment (Pollard et al., 2006).

Several points bear consideration when attempting to establish a link between genetic events and human brain evolution (Carroll, 2003): (1) Like the evolution of many other traits, evolution of the brain probably occurred nonlinearly, with transition between rapid phases, slower phases, and frequent stasis of the overall trend toward increasing brain size (see Chapter 1 and Table 3.1). (2) Morphological change during the evolution of other species has occurred at rates similar to that of the hominin brain so human brain evolution as a process was not particularly exceptional. (3) Performance of the human brain is exceptional but does not appear to involve the appearance of totally new structures but rather the modification of existing features. In sum, the outcome of human brain evolution was unique, but the rudiments of the human brain are present in most other mammals, especially primates. This process is more compatible with a genetic event making an incremental impact on developmental expression of certain neurological traits, rather than a specific genetic event precipitating a new trait such as language. It leaves wide open the opportunity for nutrients to play a key role as mediators amplifying a genetic shift that might otherwise remain phenotypically silent and therefore not influence human brain evolution.

Several considerations are relevant to interpreting the role that specific genetic modification or mutation can have on a trait such as brain size (Carroll, 2003): (1) Continuous, quantitative traits like brain evolution usually involve changes in many genes so changes in the expression of single genes rarely have much positive impact (though they may be deleterious). (2) Genetic variation can be present even when phenotype is relatively constant so rates of change in brain size provide less information about the number of genes likely to be involved than do heritability of the trait or intensity of its selection. (3) Individual regulatory genes and those that are part of regulatory networks controlling development are commonly implicated in creating morphological change; (4) Trait variation is often due to mutations in noncoding regions of the genome. (5) Functional differences between alleles that lead to evolutionary change more commonly involve multiple nonadditive nucleotide replacements rather than those at a single site. (6) Genetic differences within a species may sometimes inform the genetic basis for interspecies divergence leading to trait evolution.

Genes and Nutrient Metabolism

Human brain evolution and the morphological and functional changes in the skull, limbs (leading to bipedalism), dentition (associated with diet), and language were most likely the product of subtle, complicated polygenic changes with modulating effects of the significant increases in availability of brain-selective nutrients and changing thyroid function (see also Chapter 6). It is instructive to see how specific gene mutations can markedly alter many aspects of nutrition and metabolism including nutrient absorption, transport, and the biological roles of nutrients. Several such mutations affect brain development or brain-selective minerals.

For example, *acrodermatitis enteropathica* is a rare disease involving a single defective gene that specifically impairs the intestinal uptake of zinc from the diet. The ensuing serious zinc deficiency impairs growth, protein synthesis, and immune function. Dietary zinc supplementation entirely corrects this abnormality which would otherwise be fatal. *Menke's disease* is a defect in copper absorption leading to copper deficiency, hypomyelination, and impaired neurodevelopment. Other specific gene mutations can cause nutrient toxicity rather than nutrient deficiency. *Wilson's disease* is a form of copper toxicity caused by a defect in copper excretion, while *phenylketonuria* impairs brain development because of accumulation of the amino acid, *phenylalanine*. These and many other inherited nutritional disorders caused by specific genetic abnormalities have only been overcome through recognition of their environmental (nutrient) component and appropriate nutritional treatment.

Gene defects causing specific anomalies in nutrient metabolism can be devastating but, with few exceptions, are still very rare and so are unlikely to influence evolutionary processes. On the other hand, some relatively common gene mutations can cause more generalized malnutrition. For instance, *cystic fibrosis* disrupts function of the pancreas and intestines, impairing absorption of fats and fat-soluble vitamins thereby causing malnutrition. In addition and perhaps more importantly, cystic fibrosis impairs lung function and the immune system. Only with aggressive antibiotic treatment and intensive nutritional management do patients now survive into adulthood. Cystic fibrosis is the most common inherited disease in Caucasian males and is therefore a good model of several important features of gene-nutrient interactions with a wide impact on human health and viability.

At least a thousand variants of the cystic fibrosis gene are now known and characterized, yet few of these variants cause symptoms of cystic fibrosis, so the presence of a defective cystic fibrosis gene is not necessarily diagnostic of the disease. Cystic fibrosis is therefore an example of why linking human brain evolution to specific genes can be challenging – which mutations will be silent? Which will be advantageous to brain function? Given that human brain evolution involved complex changes not only in the brain but in the skeleton and in the formation and function of neonatal body fat, how did a cluster of just the right gene change essentially simultaneously and uniquely in humans? Gene mutations affecting nutrient metabolism can still be nutritionally managed, which demonstrates that nutrition and genetics are not two solitudes. When it comes to brain evolution, genes are neither independent of nor more important than the environment.

CONCLUSIONS

I believe that one or more late australopithecine populations in east and southern Africa stumbled serendipitously across solutions to two major constraints on primate brain size and function, one a metabolic constraint and the other a structural constraint. Through at least 2 million years of evolution, the unique cognitive potential of the present-day adult human brain emerged as a direct consequence of evolving neonatal body fat as a reserve against the *metabolic constraint* – the voracious fuel needs yet acute vulnerability of the infant brain. More neonatal body fat stores improved the reliability of the fuel supply to the brain by guaranteeing abundant substrate for an alternative fuel to glucose in the form of ketone bodies. Ketone bodies magnified the potential for more sophisticated communication between neurons, but only in the context of a habitat and diet that permitted development of body fat and simultaneously provided a richer supply of brain-selective nutrients. These nutrients, particularly brain-selective minerals and DHA, met the need for additional membrane complexity, which released the *structural constraint* on neuronal connectivity.

Of these two constraints affecting human brain development and hence evolution, it is the first one – fuel insurance for the brain – which most clearly distinguishes humans from other primates. In my view, big human brains evolved hand in hand with fat human babies; whatever best explains the evolution of fat babies also explains big, advanced brains in humans. A lot of physiology and biochemistry has to change to improve both brain function and fetal adiposity and, for the latter, there was no previous primate blueprint.

The claim made here and in Chapters 2 and 4 that DHA was one of the critical nutrients permitting human brain evolution is not universally accepted. The two most common opposing points are (1) that humans have sufficient capacity to make DHA so did not need a pre-formed dietary source (Langdon, 2006; Carlson and Kingston 2007) and (2) that intake of pre-formed dietary DHA began too late to have influenced evolution of the human brain and was therefore a consequence of rather than a contributor to human brain evolution (Cordain et al., 2005). These arguments have been responded to elsewhere (Cunnane, 2005b; 2007; Cunnane et al., 2007), so the reasons why they are flawed will only be briefly described here.

First, it is widely understood and agreed that humans have some capacity to make DHA (Plourde and Cunnane, 2007); the key point in relation to human brain evolution and human brain development is, How much can we make *relative to the need* for DHA? Two reports and one review show that the brain of human infants not provided with dietary DHA accumulates this fatty acid at about *half the rate* it accumulates in breast-fed infants (Farquharson et al., 1992; Makrides et al., 1994; Cunnane et al., 2000). Using the breast-fed infant as a reference, the human infant's cannot make enough DHA to meet the needs of its developing brain. Hence, despite some capacity to make DHA, that capacity is insufficient when it counts most, so infants must get it in the diet. Second, as reported previously (Stewart, 1994) and detailed more extensively in Chapters 8 and 9, some early hominins were consuming fish, particularly catfish and shellfish. The claim that hominins only started consuming preformed DHA once they were arguably already human or about to become human is therefore patently misinformed. Hence, human brain development is dependent on a dietary source of pre-formed DHA, and consumption of aquatic foods containing DHA began at about the same time the brain started expanding. Furthermore and perhaps most important, this chapter attempts to make clear that DHA is but one of several brain-selective nutrients (see section on Brain-Selective Nutrients), so this important fatty acid was not alone in supporting human brain evolution. These key nutrients were also complemented by an impressive capacity in humans to use ketones to ensure brain fuel supply and unique evolution of infant fat stores to provide reserves of both DHA and fatty acids as fuels. In my opinion, DHA could not have been responsible by itself for human brain evolution, so it is overly simplistic, indeed incorrect, to focus too much attention on it alone; doing so misses the forest for the trees.

Starting at least 2 million years ago in east Africa, the fossil record shows that hominins destined to have larger brains and become humans started to occupy shore-based habitats along lakes, marshes, rivers, estuaries, and possibly some sea coasts (see Chapters 7, 8, and 10). Exploitation of such habitats permitted human brain evolution not only because of the enhanced availability of brain-selective nutrients but also because this habitat permitted a lifestyle that was less preoccupied by a nomadic search for food or a haphazard success with hunting and competition with other scavengers. In turn, this provided more free time to develop manual, social, and cognitive skills. Certain genetic changes undoubtedly accompanied or accelerated some attributes of human brain evolution but, fundamentally, the process was tied to a habitat suitably enriched in brain-selective nutrients.

The increasing vulnerability inherent in the process of enlarging the hominin brain demonstrates that brain enlargement and increasing cognitive potential were neither evolutionary objectives in themselves nor were they solutions to any problem of survival. The fact that increased developmental vulnerability of the brain was tolerable, that is, did not substantially impair reproduction, survival or, indeed, did not prevent evolution of early *Homo* species to humans, means that the organism as a whole was not put at substantially greater risk by increased vulnerability of the enlarging brain. Together with the evolution of neonatal fat stores, food supplies and habitat must have remained relatively secure and abundant as brain expansion occurred.

If hominin occupation of a shore-based habitat was crucial to prepare for human brain evolution, the prediction that follows is that moving away from a shore-based habitat would somehow imperil human brain function. A billion iodine and iron-deficient people worldwide today show that moving away from a shore-based diet decreases the intake of brain-selective minerals and has profound and widespread consequences for human brain development and function. Hence, humans are still in a phase of evolution utterly dependent on the nutrients in a shore-based diet. While healthy present day vegetarians undoubtedly have normal cognitive development, the prerequisites for optimal human brain function are more easily met by a shore-based diet and habitat providing an increased supply of brain-selective minerals as well as an accessible, abundant food supply allowing time to be committed to nascent exploratory and cultural activities including stone knapping, music, and cave art.

ACKNOWLEDGMENTS

Frequent discussions with several people have contributed significantly to the concept elaborated here, among whom Michael Crawford, Kathy Stewart, Elaine Morgan, and David Horrobin are particularly noteworthy. Tom Brenna and Frits Muskiet provided many helpful comments. The Natural Sciences and Engineering Research Council of Canada supported our research on this topic.

REFERENCES

Adam, P.A.J., Raiha, N., Rahiala, E.L. et al. 1975. Oxidation of glucose and D-beta-hydroxybuyrate by the early human fetal brain. *Acta Paediatr Scand* **64**:17–24.

Aeillo, L.C. and Dean, C. 1990. *Introduction to Human Evolutionary Anatomy*. New York: Academic.

Aeillo, L.C. and Wheeler, P. 1995. The expensive-tissue hypothesis. The brain and digestive system in human and primate evolution. *Curr Anthropol* **36**:199–221.

Allen, J.S., Damasio, H., and Grabowski, T.J. 2002. Normal neuroanatomical variation in the human brain: An MRI-volumetric study. *Am J Phys Anthropol* **118**:341–358.

Armstrong, E. 1983. Relative brain size and metabolism in mammals. *Science* **230**:1302–1304.

Arthur, J.R. 1999. Functional indicators of iodine and selenium status. *Proc Nutr Soc* **58**:507–512.

Battaglia, F.C. and Meschia, G. 1973. *Fetal and Neonatal Physiology*. Cambridge, UK: Cambridge University Press.

Battaglia, F.C. and Thureen, P.J. 1997. Nutrition of the fetus and premature infant. *Nutrition* **13**:903–906.

Bliss, T.V.P. 1999. Young receptors make smart mice. *Nature* **401**:25–27.

Blumenberg, B. 1983. The evolution of the advanced human brain. *Curr Anthropol* **24**:589–622.

Bond, J., Roberts, E., Mochida, G.H. et al. 2002. *ASPM* is a major determinant of cerebral cortical size. *Nat Genet* **32**:316–320.

Broadhurst, C.L., Cunnane, S.C., and Crawford, M.A. 1998. Rift Valley lake fish and shellfish provided brain-specific nutrition for early *Homo*. *Brit J Nutr* **79**:3–21.

Caceres, M., Lachur, J., Zapala, M.A. et al. 2003. Elevated gene expression levels distinguish human from non-human primate brains. *Proc Natl Acad Sci U S A* **100**: 13030–13055.

Cahill, G.H. Jr. 2006. Fuel metabolism in starvation. *Ann Rev Nutr* **26**:1–22.

Carlson, B.A. and Kingston, J.D. 2007. Docosahexaenoic acid, the aquatic diet, and hominin encephalization: difficulties in establishing evolutionary links. *Am J Hum Biol* **19**:132–141.

Carroll, S.B. 2003. Genetics and the making of *Homo sapiens*. *Nature* **422**:849–857.

Changeux, J.P. and Chavaillon, J. 1995. *Origins of the Human Brain*. Oxford: Clarendon Press.

Conroy, G.C. 1997. *Reconstructing Human Origins: A Modern Synthesis*. New York: WW Norton.

Cooper, D.S. 1998. Subclinical thyroid disease: A clinician's perspective. *Ann Intern Med* **129**:135–138.

Cordain, L., Eaton, S.B., Sebastian, A., Mann, N., Lindeberg, S., Watkins, B.A., O'Keefe, J.H., and Brand-Miller, J. 2005. Origins and evolution of the Western diet: health implications for the 21st century. *Am J Clin Nutr* **81**:341–354.

Crawford, M.A., Bloom, M., Broadhurst, C.L. et al. 1999. Evidence for the unique function of docosahexaenoic acid during the evolution of the modern hominin brain. *Lipids* **34**:S39–S47.

Crawford, M.A., Costeloe, K., Ghebremeskel, K. et al. 1997. Are deficits of arachidonic and docosahexaenoic acids responsible for the neural and vascular complications of preterm babies? *Am J Clin Nutr* **S66**: 1032S–1041S.

Crawford, M.A., Cunnane, S.C., and Harbige, L.S. 1992. A new theory of evolution: Quantum Theory. In *Essential Fatty Acids and Eicosanoids. Invited Papers from the Third International Congress*, ed. A. Sinclair and R. Gibson, pp. 87–95. Champaign, IL: AOCS Press.

Crawford, M.A. and Marsh, D. 1989. *The Driving Force: Food in Evolution and the Future*. London: William Heinemann.

Cremer, J.E. 1982. Substrate utilization and brain development. *J Cereb Blood Flow Metab* **2**:394–407.

Cronin, J.E., Boaz, N.T., Stringer, C.B. et al. 1981. Tempo and mode in hominin evolution. *Nature* **292**:113–122.

Cunnane, S.C. 2005a. *Survival of the Fattest*. Hackensack, NJ: World Scientific.

CUNNANE, S.C. 2005b. Origins and evolution of the Western diet: implications of iodine and seafood intake for the human brain. *Am J Clin Nutr* **82**:483.

CUNNANE, S.C. 2007. Docosahexaenoic acid and human brain evolution: missing the forest for the trees. *Br J Nutr* **97**:1021–1022.

CUNNANE, S.C. AND CRAWFORD, M.A. 2003. Survival of the fattest: Fat babies were the key to evolution of the large human brain. *Comp Biochem Physiol* **136A**: 17–26.

CUNNANE, S.C., FRANCESCUTTI, V., BRENNA, J.T. ET AL. 2000. Breast-fed infants achieve a higher rate of brain and whole body docosahexaenoate accumulation than formula-fed infants not consuming dietary docosahexaenoate. *Lipids* **35**:105–111.

CUNNANE, S.C., HARBIGE, L.S., AND CRAWFORD, M.A. 1993. The importance of energy and nutrient supply in human brain evolution. *Nutr Health* **9**:219–235.

CUNNANE, S.C., MENARD, C.R., LIKHODII, S.S. ET AL. 1999. Carbon recycling into de novo lipogenesis is a major pathway in neonatal metabolism of linoleate and alpha-linolenate. *Prostaglandins Leukot Essent Fatty Acids* **60**:387–392.

CUNNANE, S.C. PLOURDE, M., STEWART, K. AND CRAWFORD, M.A. 2007. Docosahexaenoic acid and shore-based diets in hominin encephalization: a rebuttal. *Am J Human Biol* **19**:578–581.

CUNNANE, S.C., PLOURDE, M., PIFFERI, F., BÉGIN, M., FÉART, C., AND BARBERGER-GATEAU, P. 2009. Fish, docosahexaenoic acid and Alzheimer's disease. *Progr Lipid Res* **48**:239–256.

DE SILVA, J.M. AND LESNIK, J.J. 2008. Brain size at birth throughout human evolution: A new method for estimating neonatal brain size in hominins. *J Hum Evol* **55**:1064–1074.

DUSSAULT, J.H. AND RUEL, J. 1987. Thyroid hormones and brain development. *Annu Rev Physiol* **49**:321–334.

EDMOND, J. 1974. Ketone bodies serve as important precursors of brain lipids in the developing rat. *J Biol Chem* **249**:72–80.

ENARD, W., KHAITOVITCH, P., KLOSE, J. ET AL. 2002. Intra- and interspecific variation in primate gene expression patterns. *Science* **296**:340–343.

ERECINSKA, M. AND SILVER, I.A. 1994. Iron and energy in mammalian brain. *Prog Neurobiol* **43**:37–71.

FALK, D. 1998. Hominid brain evolution. Looks can be Deceiving. *Science* **280**:1714.

FARQUHARSON, J., COCKBURN, F., PATRICK, W.A. ET AL. 1992. Infant cerebral cortex phospholipid fatty acid composition and diet. *Lancet* **340**:810–813.

FARQUHARSON, J., COCKBURN, F., PATRICK, W.A. ET AL. 1993. Effect of diet on infant subcutaneous tissue triglyceride fatty acids. *Arch Dis Child* **69**:589–593.

GOLDEY, E.S., KEHN, L.S., REHNBERG, G.L. ET AL. 1995. Effects of developmental hypothyroidism on auditory and motor function in the rat. *Toxicol Appl Pharmacol* **135**:67–76.

GROSSMAN, L.I., WILDMAN, D.E., SCHMIDT, T.R. ET AL. 2004. Accelerated evolution of the electron transport chain in anthropoid primates. *Trends Genet* **20**:578–585.

GUNTON, J.E. 1999. Iodine deficiency in ambulatory participants at a Sydney teaching hospital: Is Australia truly iodine replete? *Med J Aust* **171**:467–470.

HACK, M., BRESLAU, N., WEISSMAN, B. ET AL. 1991. Effect of very low birth weight and subnormal head size on cognitive abilities at school age. *N Engl J Med* **325**:231–237.

HACK, M., FLANNERY, D.J., SCHLUCHTER, M. ET AL. 2002. Outcomes in young adulthood for very low birth weight infants. *N Engl J Med* **346**:149–157.

HACK, M., TAYLOR, G.H., KLEIN, N. ET AL. 1994. School-age outcomes in children with birth weights under 750 g. *N Engl J Med* **331**:753–759.

HARRINGTON, T.A., THOMAS, E.L., MODI, N. ET AL. 2002. Fast and reproducible method for the direct quantitation of adipose tissue in newborn infants. *Lipids* **37**:95–100.

HILL, R.S. AND WALSH, C.A. 2005. Molecular insights into human brain evolution. *Nature* **437**:64–67.

HIRSCH, J., FARQUHAR, J.W., AHRENS, E.H. JR. ET AL. 1960. Studies of adipose tissue in man. *Am J Clin Nutr* **8**: 499–511.

HO, K.C., ROESSMANN, U., HAUSE, L. ET AL. 1981. Newborn brain weight in relation to maturity, sex, and race. *Ann Neurol* **10**:243–246.

HOLLIDAY, M. 1971. Metabolic rate and organ size during growth from infancy to maturity and during late gestation and early infancy. *Pediatrics* **47**:169–172.

HOTZ, C.S., FITZPATRICK, D.W., TRICK, K.D. ET AL. 1997. Dietary iodine and selenium interact to affect thyroid hormone metabolism of rats. *J Nutr* **127**: 1214–1218.

JACKSON, A.P., EASTWOOD, H., BELL, S.M., ET AL. 2002. Identification of microcephalin, a protein implicated in determining the size of the human brain. *Am J Hum Genet* **71**:136–142.

JERISON, H. 1973. *Evolution of the Human Brain and Intelligence*. London: Academic Press.

JUREVICS, H. AND MORELL, P. 1995. Cholesterol for synthesis of myelin is made locally, not imported into the brain. *J Neurochem* **64**:895–901.

KAPPELMAN, J. 1996. The evolution of body mass and relative brain size in fossil hominins. *J Hum Evol* **30**: 243–276.

LANGDON, J. 1997. Umbrella hypotheses and parsimony in human evolution: A critique of the Aquatic Ape Theory. *J Hum Evol* **33**:479–494.

LANGDON, J.H. 2006. Has an aquatic diet been necessary for hominin brain evolution and functional development? *Br J Nutr* **96**:7–17.

LEONARD, W.R. 2002. Food for thought. Dietary change was a driving force in human evolution. *Sci Am* **287**: 106–115.

LEONARD, W.R. AND ROBERTSON, M.L. 1994. Evolutionary perspectives on human nutrition: The influence of brain and body size on diet and metabolism. *Am J Hum Biol* **6**:77–88.

LEONARD, W.R. AND ROBERTSON, M.L. 1996. On diet, energy metabolism, and brain size in human evolution. *Curr Anthropol* **37**:125–129.

LEONARD, W.R., ROBERTSON, M.L., SNODGRASS, J.J. ET AL. 2003. Metabolic correlates of human evolution. *Comp Biochem Physiol* **136A**:5–16.

LOZOFF, B. AND BRITTENHAM, G.M. 1986. Behavioural aspects of iron deficiency. *Prog Hematol* **14**:23–53.

MACE, G.M., HARVEY, P.H., AND CLUTTON-BROCK, T.H. 1981. Brain size and ecology in small mammals. *J Zool* **193**:333–354.

MACKINNON, J. 1978. *The Ape Within Us*. New York: Holt, Rinehart and Winston.

MAKRIDES, M., NEUMANN, M.A., BYARD, R.W., SIMMER, K., AND GIBSON, R.A. 1994. Fatty acid composition of brain, retina, and erythrocytes in breast- and formula-fed infants. *Am J Clin Nutr* **60**:189–194.

MARTIN, R.D. 1981. Relative brain size and basal metabolite rate in terrestrial vertebrates. *Nature* **293**:57–60.

MARTIN, R.D. 1996. Scaling of the mammalian brain: The maternal energy hypothesis. *News Physiol Sci* **11**: 149–156.

MCCANCE, R. AND WIDDOWSON, E. 2002. *The Composition of Foods*, 6th Edition. Springer.

MELICHAR, V., DRAHOTA, Z., AND HAHN, P. 1967. Ketone bodies in the blood of full-term newborns, premature and dysmature infants and infants of diabetic mothers. *Biol Neonate* **11** (1):23–28.

MINK, J.W., BLUMENSCHINE, R.J., AND ADAMS, D.B. 1981. Ratio of central nervous system to body metabolism in vertebrates: Its constancy and functional basis. *Am J Physiol* **241**:R203–R212.

MITCHELL, J.H., NICOL, F., BECKETT, G.J. ET AL. 1997. Selenium and iodine deficiencies: Effects on brain and brown adipose tissue selenoenzyme activity and expression. *J Mol Endocrinol* **155**:255–263.

MITCHELL, J.H., NICOL, F., BECKETT, G.J. ET AL. 1998. Selenoprotein expression and brain development in preweanling selenium- and iodine-deficient rats. *J Mol Endocrinol* **20**:203–210.

MORRIS, M.S., BOSTON, A.G., JACQUES, P.F. ET AL. 2001. Hyperhomocysteinemia and hypercholesterolemia associated with hypothyroidism in the third US National Health and Nutrition Examination Survey. *Atherosclerosis* **155**:195–200.

PARDRIDGE, W.M. 1991. Blood–brain barrier transport of glucose, free fatty acids, and ketone bodies. *Adv Exp Biol Med* **291**:43–53.

PASSINGHAM, R.E. 1985. Rates of brain development in mammals including man. *Brain Behav Evol* **26**:167–175.

PATEL, M.S. 1979. Influence of neonatal hypothyroidism on the development of ketone-body-metabolizing enzymes in rat brain. *Biochem J* **184**:169–172.

PHAROAH, P.O.D., BUTTFIELD, I.H., AND HETZEL, B.S. 1971. Neurological damage to the fetus resulting from severe iodine deficiency during pregnancy. *Lancet* **13**: 308–310.

PINKER, S. 2002. *The Blank Slate*. New York: Penguin.

PLOURDE, M. AND CUNNANE, S.C. 2007. Extremely limited conversion of long chain omega 3 polyunsaturates in adults: Implications for their dietary essentiality and use as supplements. *Appl Physiol Nutr Metab* **32**:619–634.

POLLARD, K.S., SALAMA, S.R., LAMBERT, N. ET AL. 2006. An RNA gene expressed during cortical development evolved rapidly in humans. *Nature* **443**:167–172.

POLLITT, E. 1993. Iron deficiency and cognitive function. *Annu Rev Nutr* **13**:521–537.

POLLITT, E. AND METALLINOS-KATSARAS, E. 1990. Iron deficiency and behaviour. Constructs, methods, and validity of the findings. *Nutr Brain* **8**:101–148.

RAMKRISHNAN, U. 2002. Prevalence of micronutrient malnutrition worldwide. *Nutr Rev* **60**:S45–S52.

REMER, T., NEUBERT, A., AND MANX, F. 1999. Increased risk of iodine deficiency with vegetarian nutrition. *Brit J Nutr* **8**:45–49.

ROBINSON, A.M. AND WILLIAMSON, D.H. 1980. Biological roles of ketone bodies as substrates and signals in mammalian tissues. *Physiol Rev* **60**:143–187.

RONCAGLIOLO, M., GARRIDO, M., WALTER, T. ET AL. 1998. Evidence of altered central nervous system development in infants with iron deficiency anemia at six months. Delayed maturation of auditory brainstem responses. *Am J Clin Nutr* **68**:683–690.

ROUTTENBERG, A., CANTALLOPS, I., ZAFFUTO, S. ET AL. 2000. Enhanced learning after overexpression of a brain growth protein. *Proc Nat Acad Sci U S A* **97**:7657–7662.

RUFF, C.B., TRINKAUS, E., AND HOLIDAY, T.W. 1997. Body mass and encephalization in Pleistocene *Homo*. *Nature* **387**:173–176.

SARDA, P., LEPAGE, G., ROY, C.C. ET AL. 1987. Storage of medium-chain triglyceride in adipose tissue of orally fed infants. *Am J Clin Nutr* **45**:399–405.

SOKOLOFF, L. 1991. Measurement of local cerebral glucose utilization and its relation to local functional activity in the brain. *Adv Exp Biol Med* **291**:21–42.

SORCINI, M., BALESTRAZZI, P., GRANDOLFO, M.E. ET AL. 1993. The national register of infants with congenital hypothyroidism detected by neonatal screening in Italy. *J Endocrinol Invest* **16**:573–577.

STEWART, K.M. 1994. Early hominid utilisation of fish resources and implications for seasonality and behaviour. *J Human Evol* **27**:229–245.

TANG, Y.P., SHIMIZU, E., DUBE, G.R. ET AL. 1999. Genetic enhancement of learning and memory in mice. *Nature* **401**:63–68.

TATTERSALL, I. 1998. *Becoming Human: Evolution and Human Uniqueness*. New York: Harcourt Brace.

TEAFORD, M.F. AND UNGAR, P.S. 2000. Diet and the evolution of the earliest human ancestors. *Proc Natl Acad Sci U S A* **97**:13506–13511.

TOWERS, H.M., SCHULZE, K.F., RAMAKRISHNAN, R. ET AL. 1997. Energy expended by low birth weight infants in the deposition of protein and fat. *Pediatr Res* **41**:584–589.

VANDERPAS, J.B., CONTEMPRÉ, B., DUALE, N.L. ET AL. 1990. Iodine and selenium deficiency associated with cretinism in northern Zaire. *Am J Clin Nutr* **52**:1087–1093.

VANDERPUMP, M.P.J., TUNBRIDGE, W.M.G., FRENCH, J.M. ET AL. 1995. The incidence of thyroid disorders in the community: A twenty-year follow-up of the Wickham Survey. *Clin Endocrinol* **43**:55–68.

VENTURI, S., DONATI, F.M., AND VENTURI, M. 2000. Environmental iodine deficiency: A challenge to the evolution of terrestrial life. *Thyroid* **10**:727–729.

VERMA, M. AND RAGHUVANSHI, R.S. 2001. Dietary iodine intake and prevalence of iodine deficiency disorders in adults. *J Nutr Environ Med* **11**:175–180.

WEETMAN, A.P. 1997. Hypothyroidism: Screening and subclinical disease. *BMJ* **314**:1175–1178.

WIENER, R., HIRSCH, H.J., AND SPITZER, J.J. 1971. Cerebral extraction of ketones and their penetration into CSF in the dog. *Am J Physiol* **220**:1542–1546.

WOOD, B. AND COLLARD, M. 1999. The human genus. *Science* **284**:65–72.

WYNN, M. AND WYNN, A. 1998. Human reproduction and iodine deficiency: Is it a problem in the UK? *J Nutr Environ Med* **8**:53–64.

CHAPTER 4

METABOLIC AND MOLECULAR ASPECTS OF THE CRITICAL ROLE OF DOCOSAHEXAENOIC ACID IN HUMAN BRAIN FUNCTION

J. Thomas Brenna

DOCOSAHEXAENOIC ACID (DHA) MOLECULAR STRUCTURE

DHA is a long-chain polyunsaturated fatty acid (PUFA) that is the most unsaturated fatty acid found in mammals. DHA is commonly designated "22:6n-3" or "22:6ω3" to indicate that it has 22 carbon atoms and 6 double bonds. Overwhelmingly, the most abundant PUFA in mammalian tissue have their double bonds configured in a specific arrangement known as "homoallylic," in which double bonds and methylene units (–CH2–) alternate along the carbon chain. Because of this regularity, designation of the number of double bonds in the molecule and the omega-3 (ω3) position fully specifies the position of every double bond in the molecule, as shown in Fig. 4.1. Chemical structure also dictates that double bonds must be configured with the *cis* or *trans* geometry. DHA is consistent with the vast majority of mammalian unsaturated fatty acids, and PUFA, with respect to this structural feature with the double bonds in the *cis* configuration. Both of these molecular structural features are of importance to DHA's unique biophysical properties.

The structural characteristics of DHA impart substantial instability to chemical attack by active agents, for instance, various forms of activated oxygen. Specifically, the methylene group (–CH2–) between the double bonds is considered to be doubly activated by the adjacent double bonds, and is referred to as being "doubly allylic." The *cis* configuration of the double bonds further activates this position along the carbon chain. These features of DHA make it particularly vulnerable to chemical reaction by activated molecular oxygen ("singlet oxygen," $^1O_2$) or other chemical agents, which destroy DHA immediately. The resulting DHA hydroperoxides are also unstable and further react with other biomolecules, typically other fats, proteins, or DNA. The damaged products of these secondary reactions no longer function properly, and must be removed by metabolic repair mechanisms. Another product is a relatively stable oxygenated form of DHA that can activate receptors and modulate physiological responses of cells.

All PUFAs are subject to attack by oxygen, but DHA is uniquely vulnerable among mammalian PUFA because its five doubly allylic sites make it highly prone to oxidation. When extracted from tissue as a component of fish oil, for instance, DHA is rapidly

Human Brain Evolution, Edited by Stephen C. Cunnane and Kathlyn M. Stewart

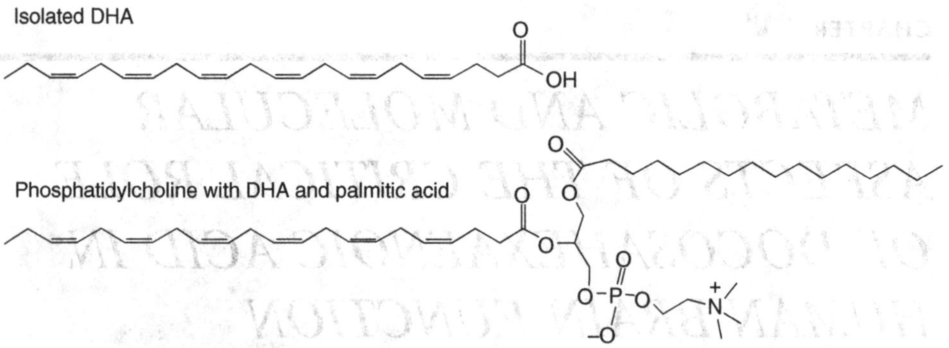

Figure 4.1 Chemical structures of docosahexaenoic acid (DHA). Top: Isolated DHA as a non-esterified fatty acid obtained from the diet. Double bonds are in the *cis* geometry and alternate with –CH2– groups, both of which are chemically less stable than alternative structures. Bottom: DHA as found in a membrane phospholipid, phosphatidyl choline, along with the nonessential saturated fatty acid palmitic acid (16:0). DHA can be exchanged among various phospholipid types and is found paired with many different fatty acids within phospholipids; however, its structural role is always as a component of similar molecules.

degraded in air (Sinclair, 1986). Within tissue, it is much more stable, indicating that biological systems have developed extraordinary means to protect DHA from oxidation and to scavenge and replace damaged DHA. Some mechanisms for protection are understood, for instance, the presence of fat soluble compounds such as vitamin E that are known as *antioxidants* because they are oxidized preferentially to DHA and form stable products that can be excreted, thereby sparing more oxidizable molecules like DHA. Similar considerations apply to antioxidant molecules synthesized endogenously, such as glutathione peroxidase. The efficiency of antioxidants in protecting DHA is impressive, considering that DHA concentrations are highest in tissues that have high metabolic rates and, necessarily, the highest levels of activated oxygen (Brenna and Diau, 2007). Hence, tissues that need DHA devote considerable energy toward its acquisition and preservation.

DHA AND NEURAL FUNCTION

DHA has long been known to be a major component of neural tissue, where it is found in relatively high concentrations, particularly in the brain (Rathbone, 1965) and retina (Anderson, 1970). Brain DHA concentrations are remarkably constant across many terrestrial species irrespective of the diversity of their natural diets (Crawford et al., 1976). The same is true in many fish species living in environments over a broad temperature range (Farkas et al., 2000). These data strongly imply that DHA has a specific molecular role, and many controlled studies show that this role cannot be filled by ω-6 PUFA that are prevalent in the seed-oil-based modern human diet. More specifically, this role of DHA cannot be filled by the closest ω-6 PUFA molecular homologue – docosapentaenoic acid (DPAω6) (22:5n-6) – which, except for having only one less double bond, is identical to DHA.

Also present in tissues is another structural near-homologue to DHA and DPAω6, the ω3 PUFA referred to as DPAω3. By depriving laboratory animals of dietary ω3 PUFA,

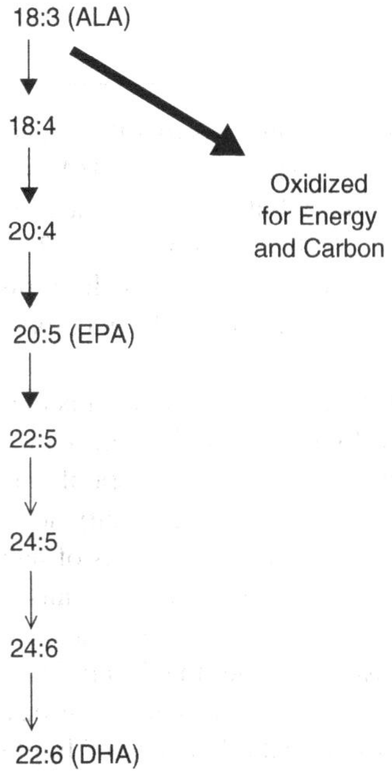

Figure 4.2 Accepted pathway for docosahexaenoic acid (DHA) synthesis. α-linolenic acid (ALA) is primarily oxidized and used as a source of energy, and for carbon to synthesize nonessential compounds such as saturated fatty acids and cholesterol. This pathway leads to biosynthesis of ω3 long-chain PUFA, but in humans, yields a very low degree of conversion to DHA. The sole ω3 PUFA in terrestrial plants is ALA, while all ω3 are found in seafood, the richest source of DHA other than consumption of animal brain. Docosapentaenoic acid (DPAω3; 22:5) is an intermediate in DHA biosynthesis, and must undergo three biochemical steps to be converted to DHA. All biochemical steps take place in cells in the endoplasmic reticulum, apart from the last step that requires translocation of 24:6 to the peroxisomes for one round of beta-oxidation.

DPAω6 can be experimentally substituted for DHA into animal tissues. Such deprivation of dietary ω3 PUFA causes tissue DPAω6 concentrations to rise and replaces both DPAω3 and DHA (Salem et al., 2001). Many studies report impairment in tissue function when DPAω6 replaces DHA. There is, however, no dietary means to experimentally replace DHA with DPAω3, so there are no direct studies of the functional consequences of replacing DHA with DPAω3.

There are, however, several good indications that DHA is required for a role that neither DPAω3 nor DPAω6 can fulfill: First, DHA concentrations are substantially greater than those of DPAω3 in breast milk and all tissues. This is perhaps not entirely coincidental because breast milk has a similar profile of long-chain PUFA as is found in neural tissue, the most rapidly expanding tissue during much of the human nursing period. Second, according to the widely accepted pathway of DHA biosynthesis, DPAω3 is subject to an additional three biochemical steps requiring translocation from one cellular organelle to another (the endoplasmic reticulum to the peroxisomes) to be converted to DHA, as shown in Fig. 4.2. Recent data indicate that neurons can perform synthesis of DHA (Kaduce et al., 2008), in contrast to previous data (Williard et al., 2001). However, in adult humans, the DHA synthesis pathway appears to be very inefficient and essentially stops at DPAω3 (see Dietary Need for Preformed DHA section).

DPAω3 can and does accumulate in the brain but, compared to DHA, this accumulation is relatively low (Kaduce et al., 2008). DPAω3 has one fewer double bond than DHA and thus is somewhat less vulnerable to chemical attack, so if DPAω3 were able to perform the same molecular functions as DHA, there would be no advantage for the system to expend the considerable energy required to synthesize DHA. Finally, DPAω3 has no other known metabolic role other than as an intermediate in the biosynthesis of DHA from DPAω3. By this reasoning, it seems likely that DHA performs some molecular role uniquely well compared to DPAω3.

METABOLIC AND BIOPHYSICAL CONSIDERATIONS

The very high concentration of DHA in neural membranes suggests that it is indispensable for some as yet ill-defined role in the central nervous system (CNS). DHA is found in tissues with high metabolic rates (Brenna and Diau, 2007), but it is not burned for energy. Hence, once acquired, it is avidly retained. Although the brain is only about 2% of body weight in adult humans, it consumes 20% of whole body energy, and much of this energy expense appears to be devoted to phospholipid recycling, specifically for membrane maintenance (Purdon and Rapoport, 2007).

Among the many perplexing questions in biophysics and metabolism is the extraordinary need of neural cell function for a long-chain PUFA containing six double bonds. The biochemical literature is replete with discussions invoking the concept of "membrane fluidity," to explain what PUFA and long-chain PUFA do. Membrane fluidity derives from several observations. For given average molecular weights, melting points of natural fats decrease as the number of *cis* double bonds increase. As a general trend, plants grown in lower temperatures have a higher content of unsaturated lipids. This is achieved by a variety of mechanisms, including alteration of expression of genes that insert double bonds into fatty acids, and by altering the stability of the functional proteins, the desaturases (Penfield, 2008). This concept has been extended to animals, but it is of limited value because the addition of double bonds beyond three has minimal effects on melting points. For instance, the melting points of two long-chain PUFA differing in only one double bond, arachidonic acid (AA; 20:4ω6) and eicosapentaenoic acid (EPA; 20:5ω3), are −50°C and −54°C, respectively. In fact, the melting point of DHA is −44°C (Nishijima, 2007), which predicts that fluidity would actually decrease if DHA substitutes for arachidonic or EPA in a membrane. These trends are not evident in the neural tissue of any species regardless of the temperature (Farkas et al., 2000).

Figure 4.3 presents the chemical structures for the three 22 carbon PUFAs found at highest concentration in the CNS. DHA has six double bonds arranged in a regular methylene-interrupted pattern along the carbon chain. The two DPA have five double bonds also arranged in the same methylene-interrupted pattern and located in the same positions along the chain as is found in DHA. Compared to DHA, the two DPAs can be thought of as missing a single double bond at one or the other end of the molecule. From a physical chemical point of view, the presence or absence of just one more or less double bond would not be expected to impart dramatic differences to a molecule. Indeed, it is difficult to show any biophysical difference between these molecules when they are modeled theoretically or experimentally as embedded within a membrane. In fact, a difference in the way that the chains behave in membranes has been shown for DHA compared to DPAω6, but not for DPAω3 (Eldho et al., 2003). However, biochemical measurements of the interaction of *G-protein-coupled receptors*, such as the light-sensing protein – *rhodopsin* – show that the presence of the one extra bond markedly changes their interaction with the membrane in which they are embedded (Feller et al., 2003), and alters the function of rhodopsin (Gawrisch et al., 2008; Soubias et al., 2008). These findings are fully consistent with the unique compromise of retinal response observed in human infants deprived specifically of DHA but not of any other PUFA.

G-protein-coupled receptors are ubiquitous in neural tissue, and the possible role of DHA-containing phospholipids in their function has been studied for rhodopsin, and is illustrated in Fig. 4.4. The conventional cartoon of the fluid mosaic model of membrane phospholipid bilayers with an embedded transmembrane protein is shown at the top.

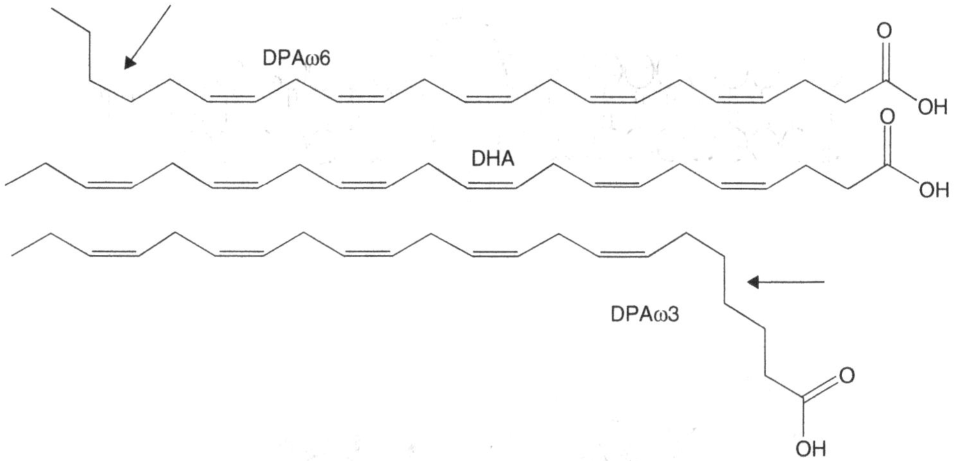

Figure 4.3 Comparison of the chemical structures for docosahexaenoic acid (DHA), and the two docosapentaenoic acids (DPA) in the CNS. DPAn-6 is missing a double bond near the methyl end of the molecule, whereas DPAω3 is missing a double bond near the carboxyl end of the molecule, where it bonds to the phospholipid. All three structures have a series of methylene-interrupted cis double bonds in the center of the molecule. DPAω6 cannot be converted to DHA, and accumulates when ω3 PUFA are absent in the diet. DPAω3 is present at one-tenth the concentration of DHA.

The phospholipids are represented as circles as the hydrophilic head groups that interact with water at the inner and outer boundaries of the membrane, while the two straight lines represent the fatty acids. A more detailed view from three-dimensional modeling is also presented. This representation shows rhodopsin as a series of seven ribbons that span the width of a membrane bilayer; in the vertebrate retina this bilayer is the membrane of the disks in the rod outer segments. *Phosphatidylcholine* surrounds rhodopsin in two dimensions though the details of their intimate points of contact are only beginning to be understood on a molecular basis. Figure 4.4 shows eight identical molecules of phosphatidylcholine with DHA and *palmitic acid* (16:0) arranged as a bilayer. The saturated palmitic acid is configured as a straight carbon chain that projects directly into the membrane, while DHA is represented as a meandering chain. Palmitic acid, like all saturates, tends to assume a three-dimensional geometry that enables close interactions among chains in adjacent molecules. Were this to be the dominant interaction, as would be expected in a membrane dominated by saturated fatty acids, the membrane would be rigid and inflexible.

In a membrane with a high concentration of long-chain PUFA, the situation is different. The methylene-interrupted *cis* double bonds of DHA cannot form straight or regular chains. Rather, the ends of these DHA chains are now known to rapidly wander from the interior to the exterior of the membrane when they are far from protein. In membrane regions close to protein, the molecular interactions may diverge radically from their behavior far away from protein. There is evidence that DHA and saturated fatty acids interact with rhodopsin differently, and that DHA interactions are more favorable for solubilizing the protein within the membrane (Grossfield et al., 2006a,b). Although this work is still in progress, it is now clear that phospholipids enriched in long-chain PUFA that cannot be biosynthesized de novo by humans, that is, DHA, has a very specific role in neural membranes.

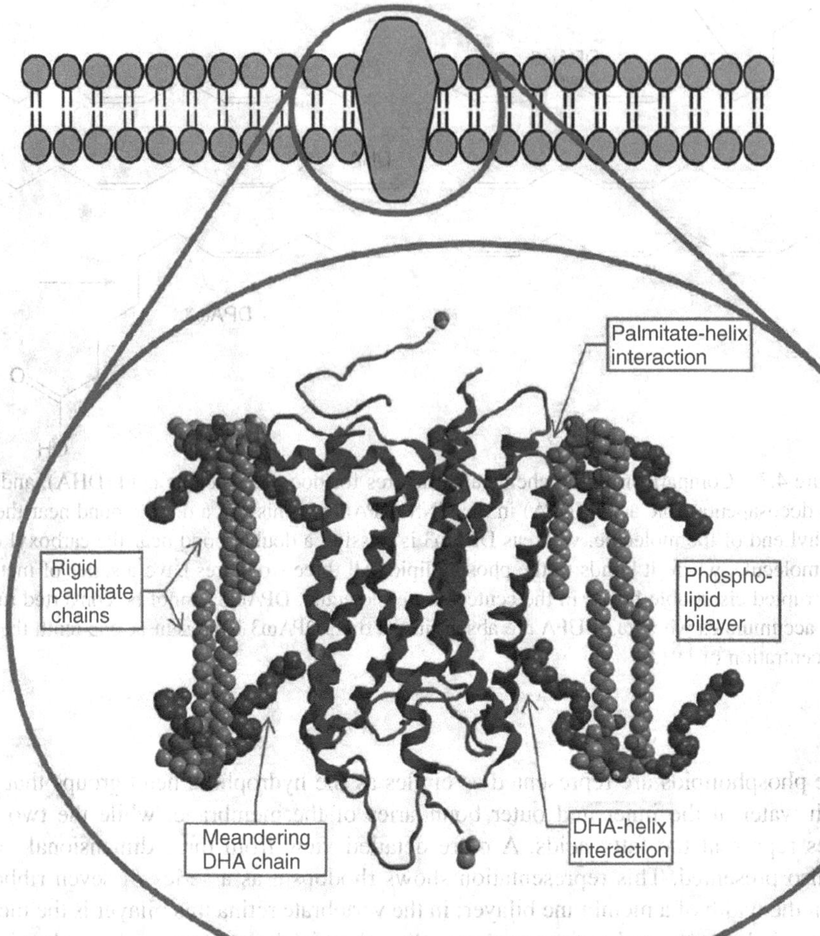

Figure 4.4 Top: Classical representation of a membrane bilayer embedded with a transmembrane protein. *Inset*: Schematic representation of rhodopsin (blue ribbon) embedded in a membrane. *Eight* identical phosphatidylcholine-docosahexaenoic acid (DHA)-palmitic acid molecules are arranged around the transmembrane rhodopsin protein. Palmitic acid (gray) forms straight chains that are rigid and interaction with one another between molecules. DHA (green) chains meander from the interior to the exterior surface of the membrane. Next to rhodopsin, the fatty chains interact very differently based on their shapes, with DHA thought to solubilize rhodopsin more strongly than palmitate or other saturates. (See color insert.)

FUNCTIONAL IMPORTANCE OF DHA IN RETINAL AND NEURAL MEMBRANES

In rats (Benolken et al., 1973) and primates (Neuringer et al., 1986), deprivation of dietary ω3 PUFA leads to a failure of the visual system to respond properly to light. In many of these studies, ω6 PUFA were provided, but all ω3 PUFA were deleted from the diet. Reproduction was apparently normal, as was growth. The absence of ω3 PUFA led to the expected marked decrease in DHA in all tissues, including retina and brain. These studies, and many since, show that ω3 PUFA deficiency leads to an accumulation of DPAω6, synthesized from the parent ω6 PUFA – linoleic acid (LA)– or possibly other ω-6 PUFA that are available.

Tissue DPAω6 is now known to rise in tissues as DHA falls during dietary ω3 PUFA deficiency (Greiner et al., 2003). The rise in DPAω6 is not specific, however, since another long-chain ω-6 PUFA, adrenic acid (22:4ω6) with four double bonds, also rises during ω3 PUFA deficiency. Nevertheless, the observation that tissue levels of DPAω6 at 90% of that of DHA cannot support visual and brain function similar to DHA is direct evidence for a critical and highly specific role of DHA. Further evidence is available from repletion studies in which animals on an ω3 PUFA-deficient diet are changed to a diet containing DHA. Tissue composition recovers within a few weeks in most (Connor et al., 1990) but not all tissues (Anderson et al., 2005). In spite of >85% recovery of retina DHA, most but not all functional parameters associated with the retinal response to light recover in the long term when deficiency is *in utero* (Anderson et al., 2005). These studies with acute ω3 PUFA deficiency, and/or repletion, are unequivocal evidence that the high concentrations of DHA found in the retina and brain under normal dietary conditions are not simply dictated by the particular food consumed. Clearly, DHA is shunted to the retina and brain, and it plays a role that cannot be replicated by any other fatty acid, ω3 or ω6 PUFA included. Further, DHA deprivation *in utero* leads to programmed suboptimal function that is not reversible with repletion even up to 3 years postnatally (Anderson et al., 2005).

Neural tissue is vulnerable to low dietary ω3 PUFA supply early in life when the brain places severe demands on ω3 for structural lipid. However, in adults placed on low ω3 PUFA oils, the brain is more effective at maintaining its concentration of ω3 long-chain PUFA even though other tissues become depleted (Bourre et al., 1992). In adults, ω3 deficiency does not affect brain composition, at least in the medium term, and thus any functional deficiencies are unlikely to be found with the same tests used for developing animals. Functional deficits are thus most likely to be observed in the next generation, in animals deprived perinatally of ω3 PUFA. Strikingly, both primate (Diau et al., 2003) and human infants (Gibson et al., 2001) consuming dietary preformed DHA show improved retinal responses compared with those with diets containing only the parent ω3 PUFA, α-linolenic acid (ALA).

DIETARY NEED FOR PREFORMED DHA

Modern methods of food production have led to the widespread use of foods that were not a part of the food chain before the industrial revolution of the nineteenth century. With respect to the fat component, mechanical crushers and organic solvent extraction have enabled production of vast quantities of seed oils that now constitute the majority of visible liquid fat in the modern human diet. Soy oil, for instance, now accounts for a remarkable 20% of the calories consumed in America. Prior to these developments, a much higher proportion of dietary oils were derived from fruits such as the olive, which yields an oil with a different mix of phytochemicals than seed oils. More importantly, fats and oils visible in the diet constituted a much smaller fraction of dietary energy than they do now. The composition of these oils has thus become far more important to human health than it once was.

Whether from fruit or seeds, plant-based oils differ from animal fats in several ways. The most important for essential fatty acid nutrition is that plant oils contain only LA and ALA, as well as small amounts of shorter-chain PUFA, primarily 16:2ω6 and 16:3ω3, mostly in the leaves. However, they contain no long-chain PUFA such as DHA. Vegans, who consume no animal fats, get all of their PUFA in this form, and then metabolically must biosynthesize all long-chain PUFA from these parent PUFA.

The earliest papers that characterized long-chain PUFA synthesis in rats showed that the dietary mix of LA and ALA caused a reciprocal change in the proportion of long-chain PUFA found in tissue depending on the predominance of one or the other parent PUFA (Mohrhauer and Holman, 1963). The biochemical mechanism is now known to be competition for the same set of enzymes that biosynthesize long-chain PUFA of both ω3 and ω6 families (Nakamura and Nara, 2004). Foods generally support a diet ratio of LA/ALA = 10, and current recommendations for modern societies are to maintain this ratio nearer to 4. These observations collectively inspired studies on ω3 PUFA deficiency in the 1980s in primates and rats. These studies induced ω3 PUFA deficiency with diets containing safflower oil or sunflower oil, both seed oils with about 55–75% of fatty acids as LA and less than 0.3% as ALA, thus representing an ω3 PUFA-deficient oil with a ratio well above 100. As expected, rhesus monkey ω3 PUFA tissue levels dropped precipitously, and their visual acuity was significantly lower than in controls (Neuringer et al., 1984, 1986). Numerous studies in rats showed deficits in the function of the visual system (Stinson et al., 1991, Weisinger et al., 1996) and in learning/cognition (Yamamoto et al., 1987; Bourre et al., 1989; Okaniwa et al., 1996; Salem et al., 2001) that induce dramatic changes in long-chain PUFA composition and in neural structure (Homayoun et al., 1988; Yoshida et al., 1997).

Whether the dietary ω3 PUFA that optimizes retinal function must be DHA or could be one of its precursors was largely answered by studies in humans and nonhuman primates in the 1980s and 1990s. Numerous studies show that the most vulnerable infants are those born very premature. The study of human premature infants ("premies") became a priority in the early 1980s when the introduction of neonatal respirators increased the survival of very low birth weight infants (<1200 g) from the low single digits to 90% by 1990 (Shiono and Behrman, 1995). It is now routine for the newborn to survive birth after a gestational period of 6 months. The last third of gestation is the critical period for expansion of the two tissues that are uniquely large at birth in humans relative to other primates: the brain and adipose (see also Chapter 3). Rather than acquiring nutrients required for rapid expansion of these tissues by placental transfer, infants born prematurely must obtain them from the diet. Importantly, however, the natural food for mammals – milk – did not evolve to meet the nutritional needs of an infant born with little body fat and a brain whose growth is about to accelerate to its highest rate of growth. It was thus recognized that breast milk may not necessarily be optimal for very premature infants.

More importantly, the composition of infant formulas of the time was dictated by nutrient requirements worked out primarily with studies in laboratory animal models and then refined by human observations. During the 1980s, the only PUFA in infant formula were LA and ALA, that is, the precursors to the ω6 and ω3 long-chain PUFA, respectively. It was assumed that since rats could synthesize all the long-chain PUFA they needed from these parent PUFA, so could human infants. Normally, humans, including the smallest infants, can synthesize some long-chain PUFA from LA and ALA. This has been established in metabolic experiments in humans with stable isotopes and controlled feeding trials (Brenna, 2002). However, a recent analysis of 19 studies that reported on adult human nutrition supplementation of the plant-derived ω3 precursor ALA (18:3ω3) show that DHA in the blood or breast milk does not increase in response to additional dietary ALA. Hence, nearly all available evidence points to such a limited DHA biosynthetic *capacity* that, in adults, statistically, plasma DHA levels do not increase when DHA is not provided (Plourde and Cunnane, 2007). Incidentally, supplementation of EPA in adults also does not increase blood DHA.

The strongest evidence for a need for DHA for retinal and neural development is in infants. The modest ability of infants to biosynthesize DHA from ALA (Carnielli et al., 1996; Salem et al., 1996) is, however, insufficient to support optimal brain DHA accumulation (Farquharson et al., 1992; Makrides et al., 1994; Cunnane et al., 2000). To improve DHA status, DHA itself must be consumed (Brenna et al., 2009). This situation contrasts to the most prominent other brain PUFA, AA, the concentration of which is not substantially compromised in infants when it is absent from the diet. Thus, in infants, brain DHA is significantly reduced when dietary DHA is not provided. The data are now sufficiently compelling for industrialized countries that, in North America for instance, availability of formulas containing DHA will go from 0% in 2001 to over 90% of formula produced for term infants in 2009. DHA is expected to be a routine component of all formulas in the near future.

DHA INTAKE DURING PREGNANCY AND LACTATION: EFFECTS ON HIGHER CNS FUNCTIONS OF THE MOTHER AND INFANT

Based on studies of tissue composition and a consideration of biosynthetic pathways, it is now clear that dietary supply of PUFA more seriously affects tissue composition of DHA than AA. Since this vulnerability is known to be perinatal, the composition of human milk provides a guide to appropriate intakes of DHA and AA. All human milk samples studied to date contain DHA (Brenna et al., 2007). Breast milk DHA concentration is very sensitive to dietary supply of DHA (Gibson et al., 1997), but is unaffected by higher dietary levels of ALA (Francois et al., 2003).

Data addressing the question of whether dietary DHA is needed for optimal development were generated by studies in human and primate neonates who were provided with formulas with acceptable levels of ALA, but with or without long-chain PUFA, specifically DHA and AA. A series of studies first appearing around 1990 demonstrated convincingly that formula with DHA provided an advantage to preterm infant vision and cognition (Gibson et al., 2001). More specifically, the development of visual acuity, a function related to nerve development, is accelerated by dietary DHA. Problem solving (Willatts et al., 1998) and cognitive maturity was also improved with dietary DHA. Crucially, studies also showed that infants consuming formulas without long-chain PUFA had lower brain DHA than infants consuming human milk (Farquharson et al., 1992; Makrides et al., 1994; Cunnane et al., 2000).

More recent work has focused more on the possible advantage to both mother and infant of breast milk containing DHA at the higher end of the human milk distribution and less on the absolute minimum amount of DHA needed. This work was in part prompted by studies showing an association between higher DHA intake and more mature CNS at birth (Cheruku et al., 2002). A number of studies on supplementation of lactating women with DHA show advantages to the nursing infant of higher DHA in breast milk for cognition (Helland et al., 2001; Dunstan et al., 2008; Innis and Friesen, 2008) and immune function (Dunstan et al., 2003; Krauss-Etschmann et al., 2008), with follow-up measurements as late as 16 years of age.

There are also data suggesting that consumption of DHA and the intermediate ω3 PUFA, EPA, are related to development of affective disorders, primarily depression and other issues related to mood. In two separate meta-analyses of several studies in adults, daily multi-gram doses of DHA and EPA significantly decrease symptoms of various affective disorders (Freeman et al., 2006; Lin and Su, 2007). A very recent study shows improvement in depressive symptoms for pregnant women consuming DHA and EPA (Su et al., 2008). Combined with knowledge about the changes in tissue composition that accompany low ω3 PUFA or low DHA consumption, these data are compelling evidence that optimal cognitive function and behavior depends on a supply of preformed DHA in the diet.

SUMMARY

Many animals find ecological niches by evolving organs that are unique compared to other mammals. The elephant's trunk, the giraffe's neck, the hummingbird's breast muscles, and the claw of the many fiddler crab species are familiar examples. In most cases, the dietary needs for these structures are not substantially unlike those for the rest of the body, and the biological niche supplies any added nutrients required for growth of the runaway organ.

Humans are the species of the runaway brain (Wills, 1993). Neural tissue composition is unlike that of any large organ, and thus requires efficient acquisition and use of highly specific components. It is very clear that among the 50 or so nutrients required for mammalian health, those derived from dietary PUFA are a disproportionate fraction of neural tissue in every species yet studied. In humans, disproportionately large brain size means that a disproportionate intake of the nutrients required for optimal brain development, including DHA, must be acquired through the diet. In this regard, the dietary necessity for DHA is no different than that of vitamin A, which can be acquired either preformed from animal sources, or as an inefficient precursor. Moreover, the human would be no different from the many species of cats, which can technically biosynthesize DHA (Pawlosky et al., 1994) but at level much too low to support their requirements. Along with other nutrients, the need to acquire preformed dietary long-chain PUFA leads cats to be obligate carnivores, a niche they have filled with prey of all but the largest animals. As elaborated elsewhere in this volume, expansion of the hominin brain in an environment providing a ready-made dietary source of DHA is consistent with the developmental needs and hence, evolution of our species' runaway organ.

REFERENCES

ANDERSON, G.J., NEURINGER, M., LIN, D.S. ET AL. 2005. Can prenatal n-3 fatty acid deficiency be completely reversed after birth? Effects on retinal and brain biochemistry and visual function in rhesus monkeys. *Pediatr Res* **58**:865–872.

ANDERSON, R.E. 1970. Lipids of ocular tissues. IV. A comparison of the phospholipids from the retina of six mammalian species. *Exp Eye Res* **10**:339–344.

BENOLKEN, R.M., ANDERSON, R.E., AND WHEELER, T.G. 1973. Membrane fatty acids associated with the electrical response in visual excitation. *Science* **182**:1253–1254.

BOURRE, J.M., DUMONT, O.S., PICIOTTI, M.J. ET AL. 1992. Dietary alpha-linolenic acid deficiency in adult rats for 7 months does not alter brain docosahexaenoic acid content, in contrast to liver, heart and testes. *Biochim Biophys Acta* **1124**:119–122.

BOURRE, J.M., FRANCOIS, M., YOUYOU, A. ET AL. 1989. The effects of dietary alpha-linolenic acid on the composition of nerve membranes, enzymatic activity, amplitude of electrophysiological parameters, resistance to poisons and performance of learning tasks in rats. *J Nutr* **119**: 1880–1892.

BRENNA, J.T. 2002. Efficiency of conversion of alpha-linolenic acid to long chain n-3 fatty acids in man. *Curr Opin Clin Nutr Metab Care* **5**:127–132.

BRENNA, J.T., AND DIAU, G.Y. 2007. The influence of dietary docosahexaenoic acid and arachidonic acid on central nervous system polyunsaturated fatty acid composition. *Prostaglandins Leukot Essent Fatty Acids* **77**:247–250.

BRENNA, J.T., SALEM, N. JR., SINCLAIR, A.J. ET AL. 2009. Alpha-linolenic acid supplementation and conversion to n-3 long-chain polyunsaturated fatty acids in humans. *Prostaglandins Leukot Essent Fatty Acids* **80**:85–91.

BRENNA, J.T., VARAMINI, B., JENSEN, R.G. ET AL. 2007. Docosahexaenoic and arachidonic acid concentrations in human breast milk worldwide. *Am J Clin Nutr* **85**: 1457–1464.

Carnielli, V.P., Wattimena, D.J., Luijendijk, I.H. et al. 1996. The very low birth weight premature infant is capable of synthesizing arachidonic and docosahexaenoic acids from linoleic and linolenic acids. *Pediatr Res* **40**:169–174.

Cheruku, S.R., Montgomery-Downs, H.E., Farkas, S.L. et al. 2002. Higher maternal plasma docosahexaenoic acid during pregnancy is associated with more mature neonatal sleep-state patterning. *Am J Clin Nutr* **76**:608–613.

Connor, W.E., Neuringer, M., and Lin, D.S. 1990. Dietary effects on brain fatty acid composition: The reversibility of n-3 fatty acid deficiency and turnover of docosahexaenoic acid in the brain, erythrocytes, and plasma of rhesus monkeys. *J Lipid Res* **31**:237–247.

Crawford, M.A., Casperd, N.M., and Sinclair, A.J. 1976. The long chain metabolites of linoleic avid linolenic acids in liver and brain in herbivores and carnivores. *Comp Biochem Physiol B* **54**:395–401.

Cunnane, S.C., Francescutti, V., Brenna, J.T. et al. 2000. Breast-fed infants achieve a higher rate of brain and whole body docosahexaenoate accumulation than formula-fed infants not consuming dietary docosahexaenoate. *Lipids* **35**:105–111.

Diau, G.Y., Loew, E.R., Wijendran, V. et al. 2003. Docosahexaenoic and arachidonic acid influence on preterm baboon retinal composition and function. *Invest Ophthalmol Vis Sci* **44**:4559–4566.

Dunstan, J.A., Mori, T.A., Barden, A. et al. 2003. Fish oil supplementation in pregnancy modifies neonatal allergen-specific immune responses and clinical outcomes in infants at high risk of atopy: A randomized, controlled trial. *J Allergy Clin Immunol* **112**:1178–1184.

Dunstan, J.A., Simmer, K., Dixon, G. et al. 2008. Cognitive assessment of children at age 2(1/2) years after maternal fish oil supplementation in pregnancy: A randomised controlled trial. *Arch Dis Child Fetal Neonatal Ed* **93**:F45–F50.

Eldho, N.V., Feller, S.E., Tristram-Nagle, S. et al. 2003. Polyunsaturated docosahexaenoic vs docosapentaenoic acid-differences in lipid matrix properties from the loss of one double bond. *J Am Chem Soc* **125**:6409–6421.

Farkas, T., Kitajka, K., Fodor, E. et al. 2000. Docosahexaenoic acid-containing phospholipid molecular species in brains of vertebrates. *Proc Natl Acad Sci U S A* **97**:6362–6366.

Farquharson, J., Cockburn, F., Patrick, W.A. et al. 1992. Infant cerebral cortex phospholipid fatty-acid composition and diet. *Lancet* **340**:810–813.

Feller, S.E., Gawrisch, K., and Woolf, T.B. 2003. Rhodopsin exhibits a preference for solvation by polyunsaturated docosohexaenoic acid. *J Am Chem Soc* **125**: 4434–4435.

Francois, C.A., Connor, S.L., Bolewicz, L.C. et al. 2003. Supplementing lactating women with flaxseed oil does not increase docosahexaenoic acid in their milk. *Am J Clin Nutr* **77**:226–233.

Freeman, M.P., Hibbeln, J.R., Wisner, K.L. et al. 2006. Omega-3 fatty acids: Evidence basis for treatment and future research in psychiatry. *J Clin Psychiatry* **67**: 1154–1967.

Gawrisch, K., Soubias, O., and Mihailescu, M. 2008. Insights from biophysical studies on the role of polyunsaturated fatty acids for function of G-protein coupled membrane receptors. *Prostaglandins Leukot Essent Fatty Acids* **79**:131–134.

Gibson, R.A., Chen, W., and Makrides, M. 2001. Randomized trials with polyunsaturated fatty acid interventions in preterm and term infants: Functional and clinical outcomes. *Lipids* **36**:873–883.

Gibson, R.A., Neumann, M.A., and Makrides, M. 1997. Effect of increasing breast milk docosahexaenoic acid on plasma and erythrocyte phospholipid fatty acids and neural indices of exclusively breast fed infants. *Eur J Clin Nutr* **51**:578–584.

Greiner, R.S., Catalan, J.N., Moriguchi, T. et al. 2003. Docosapentaenoic acid does not completely replace DHA in n-3 FA-deficient rats during early development. *Lipids* **38**:431–435.

Grossfield, A., Feller, S.E., and Pitman, M.C. 2006a. Contribution of omega-3 fatty acids to the thermodynamics of membrane protein solvation. *J Phys Chem B* **110**:8907–8909.

Grossfield, A., Feller, S.E., and Pitman, M.C. 2006b. A role for direct interactions in the modulation of rhodopsin by omega-3 polyunsaturated lipids. *Proc Natl Acad Sci U S A* **103**:4888–4893.

Helland, I.B., Saugstad, O.D., Smith, L. et al. 2001. Similar effects on infants of n-3 and n-6 fatty acids supplementation to pregnant and lactating women. *Pediatrics* **108**:E82.

Homayoun, P., Durand, G., Pascal, G. et al. 1988. Alteration in fatty acid composition of adult rat brain capillaries and choroid plexus induced by a diet deficient in n-3 fatty acids: Slow recovery after substitution with a nondeficient diet. *J Neurochem* **51**:45–48.

Innis, S.M. and Friesen, R.W. 2008. Essential n-3 fatty acids in pregnant women and early visual acuity maturation in term infants. *Am J Clin Nutr* **87**:548–557.

Kaduce, T.L., Chen, Y., Hell, J.W. et al. 2008. Docosahexaenoic acid synthesis from n-3 fatty acid precursors in rat hippocampal neurons. *J Neurochem* **105**: 1525–1535.

Krauss-Etschmann, S., Hartl, D., Rzehak, P. et al. 2008. Decreased cord blood IL-4, IL-13, and CCR4 and increased TGF-beta levels after fish oil supplementation of pregnant women. *J Allergy Clin Immunol* **121**: 464–470 e6.

Lin, P.Y. and Su, K.P. 2007. A meta-analytic review of double-blind, placebo-controlled trials of antidepressant efficacy of omega-3 fatty acids. *J Clin Psychiatry* **68**: 1056–1061.

Makrides, M., Neumann, M.A., Byard, R.W. et al. 1994. Fatty acid composition of brain, retina, and erythrocytes in breast- and formula-fed infants. *Am J Clin Nutr* **60**:189–194.

Mohrhauer, H. and Holman, R.T. 1963. Effect of linolenic acid upon the metabolism of linoleic acid. *J Nutr* **81**:67–74.

NAKAMURA, M.T. AND NARA, T.Y. 2004. Structure, function, and dietary regulation of delta6, delta5, and delta9 desaturases. *Annu Rev Nutr* **24**:345–376.

NEURINGER, M., CONNOR, W.E., LIN, D.S. ET AL. 1986. Biochemical and functional effects of prenatal and postnatal omega 3 fatty acid deficiency on retina and brain in rhesus monkeys. *Proc Natl Acad Sci U S A* **83**: 4021–4025.

NEURINGER, M., CONNOR, W.E., VAN PETTEN, C. ET AL. 1984. Dietary omega-3 fatty acid deficiency and visual loss in infant rhesus monkeys. *J Clin Invest* **73**: 272–276.

NISHIJIMA, M. 2007. Lipidbank. Japanese Conference on the Biochemistry of Lipids, Tokyo, Japan.

OKANIWA, Y., YUASA, S., YAMAMOTO, N. ET AL. 1996. A high linoleate and a high alpha-linolenate diet induced changes in learning behavior of rats. Effects of a shift in diets and reversal of training stimuli. *Biol Pharm Bull* **19**:536–540.

PAWLOSKY, R., BARNES, A., AND SALEM, N. JR. 1994. Essential fatty acid metabolism in the feline: Relationship between liver and brain production of long-chain polyunsaturated fatty acids. *J Lipid Res* **35**:2032–2040.

PENFIELD, S. 2008. Temperature perception and signal transduction in plants. *New Phytol* **179**:615–628.

PLOURDE, M. AND CUNNANE, S.C. 2007. Extremely limited conversion of long chain omega 3 polyunsaturates in adults: Implications for their dietary essentiality and use as supplements. *Appl Physiol Nutr Metab* **32**:619–634.

PURDON, A.D. AND RAPOPORT, S.I. 2007. Energy consumption by phospholipid metabolism in the mammalian brain. In *Neural Energy Utilization: Handbook of Neurochemistry and Molecular Biology*. New York: Springer.

RATHBONE, L. 1965. The effect of diet on the fatty acid compositions of serum, brain, brain mitochondria and myelin in the rat. *Biochem J* **97**:620–628.

SALEM, N. JR., MORIGUCHI, T., GREINER, R.S. ET AL. 2001. Alterations in brain function after loss of docosahexaenoate due to dietary restriction of n-3 fatty acids. *J Mol Neurosci* **16**:299–307; discussion 317–321.

SALEM, N. JR., WEGHER, B., MENA, P. ET AL. 1996. Arachidonic and docosahexaenoic acids are biosynthesized from their 18-carbon precursors in human infants. *Proc Natl Acad Sci U S A* **93**:49–54.

SHIONO, P.H. AND BEHRMAN, R.E. 1995. Low birth weight: Analysis and recommendations. *The Future of Children* **5**. http://futureofchildren.org/futureofchildren/publications/journals/article/index.xml?journalid=60&articleid=369.

SINCLAIR, H. 1986. History of EFA and their prostanoids: Some personal reminiscences. *Prog Lipid Res* **25**: 667–672.

SOUBIAS, O., NIU, S.L., MITCHELL, D.C. ET AL. 2008. Lipid-rhodopsin hydrophobic mismatch alters rhodopsin helical content. *J Am Chem Soc* **130**:12465–12471.

STINSON, A.M., WIEGAND, R.D., AND ANDERSON, R.E. 1991. Recycling of docosahexaenoic acid in rat retinas during n-3 fatty acid deficiency. *J Lipid Res* **32**: 2009–2017.

SU, K.P., HUANG, S.Y., CHIU, T.H. ET AL. 2008. Omega-3 fatty acids for major depressive disorder during pregnancy: Results from a randomized, double-blind, placebo-controlled trial. *J Clin Psychiatry* **69**:644–651.

WEISINGER, H.S., VINGRYS, A.J., AND SINCLAIR, A.J. 1996. Effect of dietary n-3 deficiency on the electroretinogram in the guinea pig. *Ann Nutr Metab* **40**:91–98.

WILLATTS, P., FORSYTH, J.S., DIMODUGNO, M.K. ET AL. 1998. Effect of long-chain polyunsaturated fatty acids in infant formula on problem solving at 10 months of age. *Lancet* **352**:688–691.

WILLIARD, D.E., HARMON, S.D., KADUCE, T.L. ET AL. 2001. Docosahexaenoic acid synthesis from n-3 polyunsaturated fatty acids in differentiated rat brain astrocytes. *J Lipid Res* **42**:1368–1376.

WILLS, Christopher. 1993. *The Runaway Brain: The Evolution of Human Uniqueness*. New York: Basic Books.

YAMAMOTO, N., SAITOH, M., MORIUCHI, A. ET AL. 1987. Effect of dietary alpha-linolenate/linoleate balance on brain lipid compositions and learning ability of rats. *J Lipid Res* **28**:144–151.

YOSHIDA, S., YASUDA, A., KAWAZATO, H. ET AL. 1997. Synaptic vesicle ultrastructural changes in the rat hippocampus induced by a combination of alpha-linolenate deficiency and a learning task. *J Neurochem* **68**:1261–1268.

CHAPTER 5

LESSONS FROM SHORE-BASED HUNTER-GATHERER DIETS IN EAST AFRICA

Frits A.J. Muskiet and Remko S. Kuipers

INTRODUCTION

There is increasing awareness that the chronic degenerative "Western" diseases are caused more by lifestyle choices and man-made changes in the environment than by imperfections in the human genome. This awareness touches the very basics of "evolutionary medicine" (Williams and Nesse, 1991; Nesse and Williams, 1994; Nesse and Williams, 1998), which is a young discipline that acknowledges that the genome is the result of millions of years of evolution with constant shaping by the environment. Consequently, a particular environment supports optimal homeostasis, which may be defined as the integration and balance of physiological functions supported by optimal interaction between environment and genome. In this view, the so called *disease susceptibility genes* that are frequently described in this era of the genome should be considered as misnomers that might rather be named "alleles with proven sensitivity to a faulty environment" or "alleles with poorly recognized evolutionary advantages" (Harris and Malyango, 2005). Examples of such disease susceptibility genes include the sensitivity of the *methylenetetrahydrofolate reductase* (MTHFR) C677T polymorphism to dietary folate (Ozturk et al., 2005) and the malaria heterozygote advantage that is conferred by the sickle cell gene (Aidoo et al., 2002). Even the relatively widespread autosomal dominant *LDL-receptor* mutations that are known to cause familial hypercholesterolemia and confer a highly elevated risk of coronary artery disease may have been conserved because, in heterozygotes, they confer an advantage against *Gram negative infection* (Sijbrands et al., 2001).

Affluent countries have in the recent past been able to diminish the influence of several unfavorable conditions of existence, notably famine and some major infections. Effective control of infections during infancy and childhood greatly reduces mortality, which increases the chances of reaching reproductive age, augments average life expectancy, and also improves overall health. Meanwhile, due to the rising incidence of diseases and conditions such as obesity, coronary artery disease, type 2 diabetes mellitus, osteoporosis, and certain cancers, including those of the colon, breast and prostate, we have unfortunately also introduced new challenges that have increased the number of years in gradually deteriorating health toward the end of our life cycle (Eaton et al., 2002). It is becoming increasingly clear that major pregnancy complications, and some psychiatric and neurodegenerative diseases may need to be added to this list (Cordain et al., 2003; Reaven, 2005; Moran et al., 2006; McIntyre et al., 2007; Pasinetti and Eberstein 2008). These dis-

Human Brain Evolution, Edited by Stephen C. Cunnane and Kathlyn M. Stewart

eases of affluence are clearly attributable in large measure to "Western" lifestyles (Willett, 2002). They affect us mostly after reproductive age, so consequently exert little selection pressure and also have not prevented the world's population from rising. Eliminating the underlying, voluntarily introduced, unfavorable conditions of existence and returning to the dietary balance and life style on which our genome has evolved is likely to restore homeostasis, may increase the number of years in health, will certainly reduce the costs of health care, but may not add much to life expectancy (Eaton et al., 2002).

Knowledge of the environment in which our genome evolved is of utmost importance to the understanding of homeostasis and may provide information on present-day nutrient requirements. This requires reconstruction of our evolutionary roots by using data from many disciplines, including anthropology, archeology, comparative anatomy, (patho) physiology, and genetics. A complementary approach is the study of populations with traditional dietary habits and lifestyles in general. After briefly discussing our genetic background, the way we adapt to environment, some of the conflicts between Western diets and the human genome, and the influence of "brain selective nutrients" (notably long-chain polyunsaturated fatty acids [long-chain PUFA]), we present here some of our own data on the diets of various populations living in Tanzania. Some of these populations have diets that might still be close to those of the hunter-gatherers that once lived in the East African cradle of anatomically modern humans. We focus on their long-chain PUFA status and consumption of saturated fatty acids. For this we used the fatty acid compositions of their breast milk as a proxy, since it reflects to a large extent human dietary habits (Francois et al., 1998; Jensen, 1999). Since it is reasonable to consider the land–water interface as a plausible ecological niche in which hominins have evolved (Broadhurst et al., 1998; Crawford et al., 1999; Crawford et al., 2001; Broadhurst et al., 2002; Gibbons, 2002a; Cunnane and Crawford, 2003; Crawford, 2006), we also modeled the composition of the diet that might have been consumed by early *Homo sapiens* living in the East African locations where the land meets with water.

OUR GENETIC BACKGROUND

Until about 6 million years ago, humans shared a common African ancestor with the present-day chimpanzee and bonobo. Hominins[1] have since then developed a large brain, bipedalism, language, social interaction, control over fire, and toolmaking skills. They also changed from a vegetarian to a hunting-gathering omnivore-carnivore diet. These formidable changes needed sizeable compensations in our physiology (Colagiuri and Brand, 2002; Cunnane and Crawford, 2003; Leonard et al., 2003). A large brain consumes a lot of glucose (see Chapter 3), which has rendered us sensitive to the development of insulin resistance, notably at times of increased competition for the available glucose, such as during pregnancy when the fetus competes metabolically with the mother (Homko et al., 1999), and during infection and inflammation, when the brain's energy demands may compete with those of the immune system (Calder et al., 2007). Many of the current typically Western diseases center around this weak spot which, together with the relative loss of muscle, might be regarded as one of the many trade-offs accompanying human brain evolution.

The oldest *H. sapiens* found to date originates from what is currently Ethiopia and is about 160,000 years old (Stringer, 2003; White et al., 2003). About 100,000 years ago, early humans started to spread across the world to become the only member of the genus *homo* to currently inhabit the earth. This third "Out-of-Africa" diaspora (Stringer, 2003) is well supported by genetic analyses (Rosenberg et al., 2002; Zhivotovsky et al., 2003; Ramachandran et al., 2005; Schroeder et al., 2007), and necessitated adaptation to new conditions of existence. Changes in physical characteristics gave rise to the concept of *race*, a term which is better described as "geographic place of origination," as shown by studies of human genetic variation (Weber, 1999; Rosenberg et al., 2002; Zhivotovsky et al., 2003; Manica et al., 2005; Ramachandran et al., 2005). Nevertheless, the number of new alleles that have been added to *H. sapiens'* gene pool since the "Out-of Africa" movement is small compared with the genetic variation that was already present in its founder population. If the total genetic variance is set at 100%, three points become apparent: first, that 93–95% of the variance can be ascribed to differences between individuals belonging to a single population within a single race; second, that the differences between populations within a single race account for only 2%, and third, that the five races do not differ genetically by more than 3–5% (Rosenberg et al., 2002).

The domination of famine during hominin evolution has shaped our genome (Chakravarthy and Booth, 2004; Prentice et al., 2005) to what has been named by Neel in 1962 – the *thrifty genotype* (Neel, 1999). The thrifty genotype hypothesizes that limited food resources in the past favored alleles that promote efficient bodily storage of energy reserves as fat. In conditions of plentiful nutrition, this genotype predisposes to obesity and the associated degenerative diseases. Genetic analyses increasingly suggest that we are all carriers of the thrifty genotype: the alleles with demonstrated risk of obesity in meta analyses exhibit high frequency and confer low relative obesity risk (Van den Berg, 2008), a point predicted from the phenomena of high frequency-low penetrance and low-frequency-high penetrance (Willett, 2002; The Wellcome Trust Case Control Consortium, 2007).

The vast majority of functional alleles that have been added to the current genetic variance results from selection pressure that was exerted by a changing environment over the past 160,000 years. They include alleles not only for protection against sunlight (skin [de]pigmentation; Jablonski and Chaplin, 2000; Harris and Meyer, 2006), and infectious diseases such as malaria (Cserti and Dzik, 2007), small pox, yellow fever, typhus, cholera (Balter, 2005), but also for adaptation by adults to new components in the diet such as lactose in milk (persistent expression of the enzyme – *lactase* [Harris and Meyer, 2006]), amylase to digest starchy carbohydrates (Perry et al., 2007), and gluten in grains (Cordain, 1999).

The molecular clock hypothesis predicts that our nuclear DNA changes at a rate of about 0.5% per million years or 1.7×10^{-9} substitutions/site/year (Ingman et al., 2000). The genuine rate of genomic change is obviously dependent of the actual mutation rate, the magnitude of selection pressure, and other features of population genetics, including genetic drift. There is some evidence that the rapidity of these changes increased after the declining influence of famine, infection, and genetic drift, and the concomitant growth of the human population starting some 40,000 years ago (Hawks et al., 2007). Nevertheless, genetically, present-day humans remain almost the same as the first *H. sapiens*, some 160,000 years ago; that is, we are still largely adapted to the East African ecosystem in which our genome evolved, with some adaptations since the Out-of-Africa diaspora. Subsequent changes in the environment are clearly capable of introducing new selection pressures, with outcomes ranging not only from no effect to extinction, but also to intermediate outcomes such as altered susceptibility to chronic degenerative diseases that diminish health in adult years without having a major influence on the chance of reaching reproductive age.

ADAPTATION TO THE CONDITIONS OF EXISTENCE

Resolution of conflicts between our current environment and our ancient genome would benefit from a better understanding of the environmental factors that are involved, notably the mechanisms by which organisms adapt to the "conditions of existence."

Adaptation of our genome underlies evolution and occurs by natural selection, as conceptualized by Darwin in 1859. It occurs by spontaneous mutation and subsequent selection by the environment, sexual (partner) selection, competition within and between species, and by two neo-Darwinian forces in population genetics – *gene flow* (nonrandom mixing caused by migration) and *genetic drift* (random loss of alleles in small populations). Adaptation of the human DNA base pair sequence is, however, a slow process that cannot support adjustment to environmental changes occurring at intermediate or rapid pace. The needed flexibility for rapid adaptation is provided by genetically encoded mechanisms that allow adjustment of phenotype by epigenetics and by the interaction of the environment with sensors, such as those of the sensory organs, but also the many sensors that remain unnoticed (Muskiet, 2006). Epigenetics is about the regulation of the translation of genotype to phenotype (Jiang et al., 2004; Feil, 2006; Van Vliet et al., 2007). It operates via meta-stable epigenetic *marks*, notably DNA methylation, *histone* modification, and non-coding RNA. These have many functions, including the conveying of environmental instructions to our genome. Animal studies have shown that epigenetic marks may become transmitted to the next generation, in which case they persist during meiosis (Morgan and Whitelaw, 2008; Youngson and Whitelaw, 2008). Serving as a form of genetic memory, this type of meta-stable inheritance of phenotypic characteristics has the clear evolutionary advantage to pass information on the predicted environment to the next generation and is most probably at the basis of the developmental origins of health and disease (Godfrey et al., 2007; Waterland and Michels, 2007; Gluckman et al., 2008), that is, the *thrifty phenotype*. The thrifty phenotype hypothesizes that limited food resources cause physical and metabolic adaptations of the fetus that, when mismatched, predispose to disease in later life (Gluckman et al., 2005; Godfrey and Barker, 2000). In a more general sense, this phenomenon conveys instructions on the predicted environment to the fetus, allowing phenotypic adjustments that increase the chance to reach reproductive age.

Nutrients and other environmental factors are increasingly recognized to influence epigenetic marks (Waterland and Jirtle, 2004; Lillycrop et al., 2005, 2008; Cooney, 2006; Burdge et al., 2007), either directly or indirectly via bodily sensors. For instance, *peroxisomal proliferator-activated receptors* (PPARs; [Bensinger and Tontonoz, 2008; Castrillo and Tontonoz, 2004; Clarke, 2004]), the vitamin D receptors (Bouillon et al., 2008), the *retinoid X receptors* (RXR), and the *retinoic acid receptors* (RAR; Francis et al., 2003; McGrane, 2007) are examples of nuclear transcription factors that serve as ligand-driven sensors. PPARs are at the crossroads of metabolism and inflammation. They are promiscuous "lipid sensors" that can bind many unsaturated fatty acids and their *eicosanoid* metabolites, notably ω3 PUFA and their derivatives (Deckelbaum et al., 2006; Bragt and Popeijus, 2008; Gani, 2008; Gani and Sylte, 2008; Itoh and Yamamoto, 2008). The vitamin D receptor and RXR bind *1,25-dihydroxyvitamin D* and *9-cis retinoic acid*, respectively. As much as 3% of the human genome is either directly or indirectly regulated by the vitamin D endocrine system, illustrating its widespread function in the translation of environmental signals into an adapted phenotype (Bouillon et al., 2008). The fish PUFA – *docosahexaenoic acid* (DHA) – is also a ligand of RXR, while RAR binds *all-trans retinoic acid, retinal, retinyl acetate,* and *retinol* (McGrane, 2007). DHA, vitamin A, and vitamin D are "dinosaurs" among the many environmental signal molecules. DHA and vitamin A both have 600 million year track records in neuronal and visual function, of which the latter is a modification of the photosynthesis occurring in blue algae from some 3 billion years ago (Crawford et al., 2001; see Chapter 2 in this book). DHA has numerous functions, ranging from the physicochemical properties of membranes and thereby the functionality of their embedded proteins, to modulation of neurotransmission, gene expression, receptor activities, and immunity and inflammation (Farooqui et al., 2007). Vitamin

D has probably been a functional molecule for more than 750 million years, and might initially have been used as a natural sunscreen to protect lower life forms from the damaging effects of high energy UV radiation and as a mediator for UV signal transduction (Holick 1992, 2003). Thus, it seems that complexity in evolution is driven more by an increase of the number of receptors than by an increase of the arsenal of environmental signal molecules that influence them (Venter et al., 1988). Especially the "dinosaurs" among these signaling molecules have gained multiple functions that are not necessarily related.

Nuclear transcription factor-ligand complexes usually do not support transcription by themselves, but need to *heterodimerize* to facilitate gene transcription, notably with RXR (e.g., PPAR/RXR, VDR/RXR, RAR/RXR). Their modes of action illustrate the need for balance between such signaling systems, that is, balance between dietary ω3 PUFA from fish, dietary vitamins A and D, and the synthesis of vitamin D_3 in skin by UV B exposure. Given a certain genome and its many polymorphisms, it is clear that maintenance of homeostasis in such systems is the result of many environmental factors that operate in concert, and that it will be virtually impossible to define evolution-optimized homeostasis by studying the dose–response relationships of each of the environmental factors in isolation, let alone their numerous interactions.

WESTERN DIETS AND THE HUMAN GENOME

Since the agricultural revolution some 10,000 years ago, humans have gradually changed their diets, changes which accelerated after the beginning of the industrial revolution nearly 200 years ago. The seven major dietary changes as recognized by Cordain et al. (2005) are summarized in Fig. 5.1 and include a shift of dietary macronutrient composition toward carbohydrates at the expense of protein; increased intake of ω-6 PUFA, notably *linoleic acid* from refined seed oils; increased intake of saturated fatty acids and industrially produced *trans-fatty acids*; decreased intake of ω3 PUFA (both *alpha-linolenic acid* and those from fish); a shift toward a high intake of refined carbohydrates with a high *glycemic index* (e.g., highly processed grains, sucrose, fructose syrups); a decreased intake of various micronutrients (e.g., folate, vitamin D, magnesium, zinc); a shift toward acid-producing foods like meat and grains at the expense of base-producing fruits and

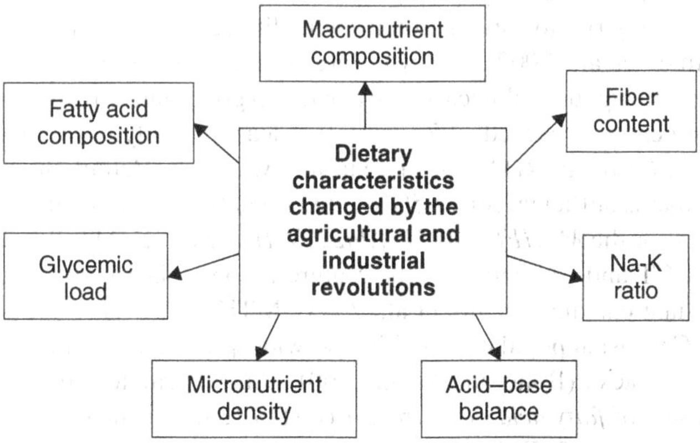

Figure 5.1 Seven crucial nutritional characteristics that have changed since the agricultural and industrial revolutions (Cordain et al., 2005). Reproduced from Muskiet et al. (2006) with permission of Elsevier.

vegetables; an increase of sodium (salt) intake and a reduction of potassium intake; and, finally, decreased intake of fiber.

These dietary changes together with an energy intake that does not match the actual energy requirements of more sedentary lifestyles have caused a conflict with our genome that appears to underlie chronic, degenerative, typically Western diseases. The resulting human phenotype centers around the *metabolic syndrome* and chronic low-grade inflammation, which are risk factors for many associated diseases and conditions, including obesity, coronary artery disease, type 2 diabetes mellitus, osteoporosis, and certain types of cancer (notably colon, breast, prostate), fertility problems (polycystic ovary syndrome), pregnancy complications (gestational diabetes, preeclampsia), some psychiatric diseases (major and postpartum depression, schizophrenia, autism), and neurodegenerative diseases (Alzheimer's disease, Parkinson's disease) (Cordain et al., 2003; Reaven, 2005; WCR/AICR, 2007). The metabolic syndrome, also referred to as the insulin resistance syndrome, is a combination of four frequently occurring conditions and symptoms – overweight, disturbed glucose homeostasis, dyslipidemia, and high blood pressure, with microalbuminuria also being considered as part of the metabolic syndrome by the World Health Organization (Haffner et al., 2006).

Up to now, many if not all of the polymorphisms[2] implicated in Western disease refer to perfectly normal genes that do not cause disease by themselves and are likely to have been present in the first *H. sapiens*. With a generation time of 20–25 years, the nucleotide sequence of our genome has simply not become adapted to these new disadvantageous conditions of existence. It seems that the original genetically determined "survival strategy" based on the thrifty genotype has become counterproductive now that many in Westernized countries can eat whatever and whenever they want and need little physical activity for food procurement. The current *obesogenic environment* did not exist in the past, was consequently not part of selection pressure, and genetic adaptations to it would therefore also not be expected. Consequently, apart from some rare mutations (Farooqi and O'Rahilly, 2006), obesity is not a genetic disease. The opening sentence of many of the current papers on Western diseases stating that the condition in question is caused by "interaction between genes and environment" distracts from its principal causation by faulty diet and lifestyle and therefore does not carry useful information from a public health perspective. The identification of the underlying genes is nevertheless important for health care, since it may help us target treatments in those who are more likely to develop disease related to the metabolic syndrome due to under- or overexposure to the underlying environmental, often dietary, factor(s).

Polymorphisms giving rise to variant proteins with different sensitivities to nutrients, such as vitamins (Ames et al., 2002), are widespread and may cause higher nutrient requirements in the homozygous and occasionally heterozygous states. If the prevalence of a nutrient-sensitive genotype exceeds 2.5%, it will at least partially be included in the *recommended dietary allowance* (RDA) for that nutrient which, by definition, equals the recommended intake that is sufficient for 97.5% of the population (Yates et al., 1998). For example, homozygotes for the *MTHFR C677T (Ala222Val)* allele (i.e., MTHFR TT) who, compared to MTHFR CT and CC counterparts, require higher folate intake for optimal functioning of the variant enzyme (Ozturk et al., 2005). MTHFR TT has a prevalence of about 10–20% in the Caucasian population and is also widespread among the other races, excluding Sub-Saharan blacks (Botto and Yang, 2000). Other examples are the recently identified polymorphisms of *fatty acid desaturase 1* (FADS 1, also named *delta-5 desaturase*) and *fatty acid desaturase 2* (FADS 2; *delta-6 desaturase*) which both exhibit lower activities in the shared biosynthetic pathways toward *arachidonic acid* (AA) and DHA (Schaeffer et al., 2006). Normally, these polymorphisms are unlikely to cause deficiencies by themselves, but may do so under unfavorable environmental conditions, that is, low long-chain PUFA intake. DHA is at least conditionally essential (Cunnane, 2003; Muskiet et al., 2004) and, compared to α-linolenic acid, preformed DHA is clearly preferred for the developing brain (Burdge, 2006). Absence of DHA in the diet of young rats causes organ DHA depletion to which the brain, retina, testes, skeletal muscle, and brown adipose tissue are most resistant (DeMar et al., 2008). Consequently, the common occurrence of FADS polymorphisms may indicate that long-chain PUFA intakes have been sufficient to confer Darwinian fitness to their carriers. The Inuit might be even more dependent on a dietary source of long-chain PUFA, because their multigeneration high intake of marine foods may have led to selection for low FADS1 activity, which would render them semi-obligate carnivores who, similar to cats, are in need of a preformed source of long-chain PUFA (Gibson and Sinclair, 1981). Since, by definition, RDAs are meant to be adequate for 97.5% of all humans, sub-Saharan blacks with low MTHFR TT prevalence may not set the RDA for folate, while the Inuit with their low FADS1 activity may define optimal long-chain PUFA intakes.

Regarding the incompatibility of the human genome and current Western diets, it is important to note that present RDAs and dietary advice are based on relatively conservative scientific evidence. For instance, the current "adequate intake" of vitamin D is based solely on its role in the skeleton with insufficient recognition of its numerous nonskeletal functions (Vieth et al., 2007). Also the higher vitamin D status of primates in the wild and subjects exposed to more traditional dosages of ultraviolet B radiation have not been taken into account (Vieth, 2006). Our hunting-gathering ancestors probably consumed 600 mg vitamin C per day, while the current population in the United States consumes 77–109 mg/day (Eaton et al., 1997). Setting the vitamin C RDA at 60 mg/day because it prevents deficiency disease conveys the suggestion that higher intakes are redundant and implies that supplements are superfluous, and possibly have adverse effects. By analogy, the complexity and interdependence of the antioxidant and pro-oxidant network (Vertuani et al., 2004) may not become apparent from large scale intervention studies with one or two antioxidants showing no effects (Sesso et al., 2008; Gaziano et al., 2009; Lippman et al., 2009) or unfavorable outcomes (Bjelakovic et al., 2007). In contrast to the strict attitude in developing recommendations of nutrients that have more abundantly been consumed in the past, it has proven difficult to discontinue or limit the consumption of some nutrients that were eaten by our ancient ancestors in moderation, or not at all, including salt, industrially produced trans-fatty acids, and the recently introduced high fructose syrups. Thus, the currently reigning paradigm for establishing RDAs and dietary recommendations will not facilitate correcting the many dietary changes introduced since the agricultural and industrial revolutions, while poorly researched novel food products will continue to be introduced.

BRAIN-SELECTIVE NUTRIENTS IN HEALTH AND DISEASE

The sites at which the fossil remains of human ancestors have been discovered suggest that the evolution of *H. sapiens* took place on an ω3 PUFA-rich diet in East or South African ecosystems that were notably located in places where the land meets with (fresh) water (Broadhurst et al., 1998, 2002; Crawford et al., 1999; Gibbons, 2002b; Marean et al., 2007). The Out-of-Africa Diaspora largely took place via coastlines (Stringer, 2000), including to the Americas after crossing the Bering Strait (Wang et al., 2007). Compared with hunting on the savanna, food from land–water ecosystems is relatively easy to obtain and rich in heme-iron, iodine, vitamins A and D, and ω3 PUFA from both vegetables and fish (see also Chapter 3). Each of these nutrients (Crawford et al., 1999, 2001; Crawford 2006), vitamin D included (McGrath et al., 2004; Kalueff et al., 2006; Kiraly et al., 2006; Almeras et al., 2007; McCann and Ames, 2008), has important functions for human brain growth, development, and function. Exploitation of this ecosystem with its abundant *brain-selective nutrients*, may well be at the base of human brain evolution (Cunnane and Crawford, 2003). Nevertheless, this advantageous dietary composition seems somewhat abandoned since the Out-of-Africa diaspora, because deficiencies of many of these particular nutrients are among the most widely encountered in the world today (Broadhurst et al., 1998, 2002; Crawford et al., 2001). Iodine is added to table salt in many countries, and margarines and milk have become popular food products for fortification with vitamins A and D.

The dietary composition of human ancestors is also apparent from the current human (patho)physiology. The nonskeletal functions of vitamin D are increasingly recognized, and the extent of the vitamin D deficiency/hypovitaminosis D pandemic is gaining acceptance (Holick, 2008; Holick and Chen, 2008). Epidemiological data demonstrate a negative association of fish consumption with coronary artery disease and (postpartum) depression (Hibbeln, 1998, 2002), while landmark trials with α-linolenic acid (de Lorgeril et al., 1999) and ω3 PUFA from fish (GISSI-Prevenzione trial, 1999; Yokoyama et al., 2007; Lee et al., 2008) in coronary artery disease, as well as meta analyses of randomized controlled trials with fish oil and notably *eicosapentaenoic acid* (EPA) in depression (Freeman et al., 2006; Lin and Su, 2007; Ross et al., 2007; Sinclair et al., 2007) support the causality of these relations. Although more research is needed, it has become clear that the intake of both the parent PUFA and long-chain PUFA are important for our health across the entire life cycle, but that their intakes have been subject to tremendous change (Simopoulos, 2008).

DIETARY FATTY ACIDS AT THE LAND–WATER INTERFACE

Our interest in the reconstruction of the diet of hominins is primarily focused on the people living in Chole, which is a small island in the Indian Ocean, close to the Tanzanian mainland, located south of the much larger Mafia Island. The Chole population is a mixture of

Figure 5.2 Selected fatty acids in the milk of Tanzanian tribes and populations, as compared with countries in our world data set.
Data (in g%) are indicated as box plots, showing medians (black bar), interquartile range (box length), minimum and maximum of the sample or smallest and largest values inside a "reasonable" distance from the end of the box (crossbars at the far ends of whiskers), values between 1.5–3 box lengths from either end of the box (outliers, indicated by 1) and values >3 box lengths from either end of the box (extremes, indicated by *). Countries were denoted according to ISO 3166-1 and the corresponding ISO 3166-1-alpha-2 code elements for countries. Investigated tribes and locations in Tanzania are denoted as TZ, followed by the first three letters of the tribe or alternatively by the first three letters of the sampling location in case the local population has become mixed due to nineteenth-century gathering of people from all over East-Africa, such as notably has occurred in coastal regions. AG, Antigua; BZ, Belize; DM, Dominica; IL, Jerusalem-Israel; LC, St. Lucia; AN, Curaçao-Netherlands Antilles; NL, The Netherlands; PK, Pakistan-Islamabad; PS, Jerusalem-Palestine; SR, Surinam; TZ-CHO, Chole; TZ-DAR, Dar-es-Salaam; TZ-HAD, Hadzabe; TZ-IRA, Iraqw; TZ-KER, Kerewe; TZ-MAA, Maasai; TZ-NYA, Nyakius; TZ-NYI, Nyarimba; TZ-SON, Sonjo; VC, St. Vincent. Reproduced from Kuipers et al. (2007), with permission of Elsevier.

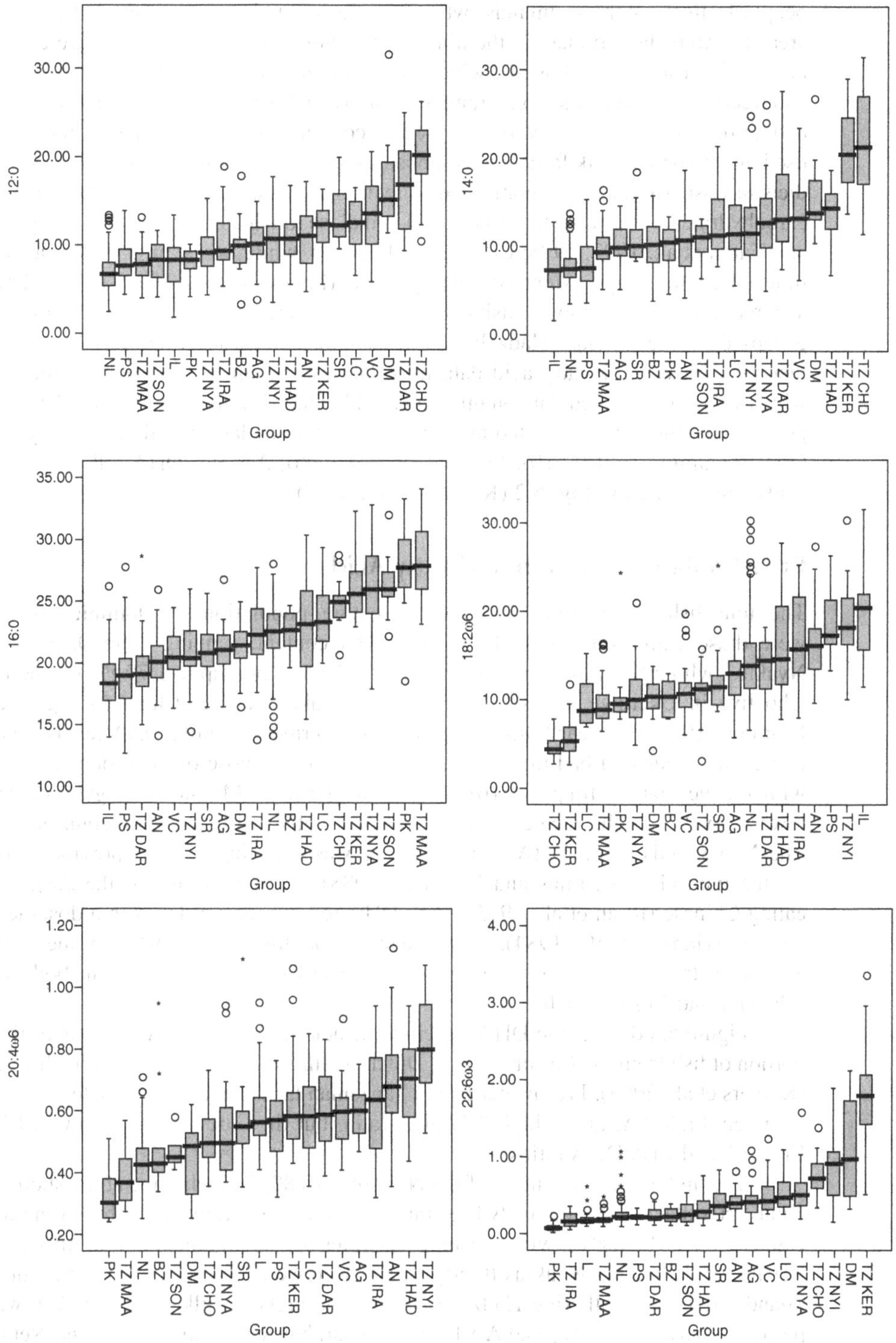
12:0
30.00
20.00
10.00
0.00
NL PS TZ MAA TZ SON IL PK TZ NYA TZ IRA BZ AG TZ NYI TZ HAD AN TZ KER SR LC VC DM TZ DAR TZ CHD
Group
14:0
30.00
20.00
10.00
0.00
IL NL PS TZ MAA AG SR BZ PK AN TZ SON TZ IRA LC TZ NYI TZ NYA TZ DAR VC DM TZ HAD TZ KER TZ CHD
Group
16:0
35.00
30.00
25.00
20.00
15.00
10.00
IL PS TZ DAR AN VC TZ NYI SR AG DM TZ IRA NL BZ TZ HAD LC TZ CHD TZ KER TZ NYA TZ SON PK TZ MAA
Group
18:2ω6
30.00
20.00
10.00
0.00
TZ CHO TZ KER LC TZ MAA PK TZ NYA DM BZ VC TZ SON SR AG NL TZ DAR TZ HAD TZ IRA AN PS TZ NYI IL
Group
20:4ω6
1.20
1.00
0.80
0.60
0.40
0.20
PK TZ MAA NL BZ TZ SON DM TZ CHO TZ NYA SR L PS TZ KER LC TZ DAR VC AG TZ IRA AN TZ HAD TZ NYI
Group
22:6ω3
4.00
3.00
2.00
1.00
0.00
PK TZ IRA L TZ MAA NL PS TZ DAR BZ TZ SON TZ HAD SR AN AG VC LC TZ NYA TZ CHO TZ NYI DM TZ KER
Group

people from the African inlands, who were transported and concentrated in the coastal areas by Arab slave traders in the nineteenth century. Their staple foods are coconut and marine fish (preferably boiled), which are combined with plentiful intake of vegetables and freely available fruits (e.g., oranges, mangos, and bananas) growing all over the island and an occasional flying fox, one of the last colonies of which resides on this island. We used mature breast milk fatty acid composition in Chole as a proxy to reconstruct hominin diets in East Africa. A diet composed of lean meat, fish, vegetables, and fruits with little carbohydrates from staple foods is considered close to what our hunter-gathering ancestors ate (Eaton et al., 1997, 1998; Cordain et al., 2000), while breast milk fatty acid composition reflects to a large extent our dietary habits (Francois et al., 1998; Jensen, 1999). We also recruited three groups of fish-eating controls (Kerewe, Nyakius, and Nyiramba), four groups from the inlands (Hadzabe, Maasai, Sonjo, and Iraqw) and also compared with historical breast milk fatty acid data from Dar-es-Salaam and the many countries from which we have collected data on milk fatty acid composition during the past 25 years. The percent composition of selected milk fatty acids, that is, lauric acid (12 : 0), myristic acid (14 : 0), palmitic acid (16 : 0), linoleic acid (18:2ω6), AA and DHA, all in g/100 g fatty acids, are depicted in Fig. 5.2 (Kuipers et al., 2007).

Breast Milk Polyunsaturated Fatty Acids

The four fish-eating tribes and the fish-eating population in Dominica (Caribbean) were those exhibiting the highest milk DHA content, ranging from 0.53 g% for the Nyakius, who inhabit the northern shores of Lake Malawi, up to 1.79 g% for the Kerewe, who live in Ukerewe, an island in southeast Lake Victoria. Chole (0.73 g%) and the Nyiramba (0.91 g%; near Lake Kitangiri) had intermediate milk DHA levels. These five fish-eating groups all had milk DHA levels well above those of the groups living inland, with the vegetarian Iraqw (0.16 g%) and the pastoral Maasai (0.20 g%) showing the lowest levels which, incidentally, were comparable with Western countries such as The Netherlands. Milk DHA of the Kerewe was even higher than previously reported for the Inuit (1.4 g%; Innis and Kuhnlein, 1988), but not as high as the 2.8 g% in fish-eating Chinese (Ruan et al., 1995). Milk DHA reflects dietary DHA in a dose-dependent manner (Harris et al., 1984). Consequently, the high milk DHA in the fish-eating Tanzanian tribes was clearly derived from the DHA that is abundant in both the local saltwater and freshwater fishes.

Figure 5.3 depicts the DHA, arachidonic acid (AA), and EPA contents of the edible portion of fish from the Caribbean Sea, Dutch North Sea and East African freshwater lakes (Kuipers et al., 2005). Freshwater fish in Tanzanian lakes have intermediate DHA content, relatively high AA, and a high AA/DHA ratio, but they are relatively low in EPA, and EPA/AA and EPA/DHA ratios.

The high milk AA content of the Nyiramba (0.80 g%), who live at the shore of Lake Kitangiri, reflects their relatively high intake of AA from freshwater fish. AA in milk from Chole women (0.50 g%), who consume saltwater fish was comparable to milk AA of the freshwater fish-eating Nyakius (0.50 g%) and Kerewe (0.58 g%). From the groups living inland, the hunter-gathering Hadzabe showed the highest milk AA (0.71 g%), while the pastoral Maasai (0.37 g%) had AA levels comparable with Pakistan and The Netherlands (0.43 g%). A recent study in which AA was supplemented together with DHA during pregnancy and lactation showed that, like DHA, milk AA is also responsive to dietary AA (van Goor et al., 2009).

Within each of the four fish-eating East-African populations, milk DHA/AA correlated positively with the DHA/AA ratio in fish obtained locally, but between them, the

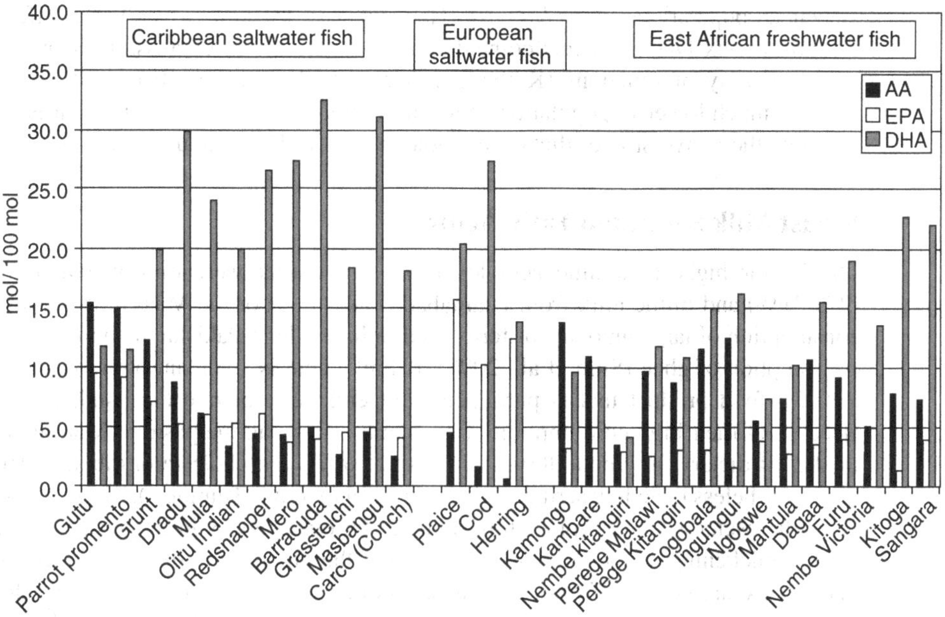

Figure 5.3 AA, EPA, and DHA contents in fish from the Caribbean Sea, North Sea and Tanzanian freshwater lakes.
AA, arachidonic acid (closed bars); EPA, eicosapentaenoic acid (open bars); DHA, docosahexaenoic acid (gray bars). Gutu and Parrot promento are parrot fish that live from coral, and grunt is a reef fish. These fish are not abundantly eaten in the Caribbean. Carco is the local name for a conch. Note the relatively high AA contents of the Tanzanian lake fish, the relatively high EPA content of the North Sea fish, and the relatively high DHA content of the Caribbean fish. Reproduced from Kuipers et al. (2005), with permission of Elsevier.

milk DHA/AA ratios differed (Kuipers et al., 2007). Thus, depending on source of long-chain PUFA, the dietary DHA/AA ratio is subject to variation, a variation to which early humans living at the land–water interface would also have been subjected.

The very low linoleic acid in the breast milk in Chole (4.23 g%) and of the Kerewe (5.20 g%) contrasts with the very high values in Jerusalem (20.31 g%) and in the Nyiramba (18.11 g%) and the vegetarian Iraqw (15.52 g%). The latter figures are typical of populations using large amounts of vegetable oils for cooking and other dietary purposes, and seem to coincide with westernization of the diet. For comparison, the linoleic acid content of breast milk in the United States was 5–7 g% in 1945 and has gradually increased to about 16 g% in 2000 (Ailhaud et al., 2006).

Taken together, it is clear that breast milk DHA and AA increase in response to the consumption of both freshwater and saltwater Tanzanian fish and that the levels reached, notably those of DHA, are well above those encountered in populations living in the inland or in countries with typically Western dietary habits. Milk DHA levels in Tanzanian shore-dwelling populations may well be as high as those previously noted in the populations living in Arctic regions, who have an estimated daily intake of 14-g long-chain ω3 PUFA (Feskens and Kromhout, 1993). In this context, it is of interest to note that a milk formula with a DHA content in the high normal range of human milk (i.e., about 1.00 g%), for example, as encountered in the Tanzanian fish-eating populations, raises DHA in the newborn baboon cerebral cortex, as compared to a formula with a DHA content in the low normal human milk range (i.e., about 0.33 g%), that is, in The Netherlands and

Tanzanian populations eating less fish (Hsieh et al., 2007). Moreover, these differences in cerebral cortex DHA contents coincide with changes in the expression of proteins with a wide variety of functions (Kothapalli et al., 2007). On the other hand, linoleic acid status is much lower in populations consuming traditional diets, a factor that is known to facilitate the conversion of dietary α-linolenic acid to EPA (Liou et al., 2007).

Breast Milk Saturated Fatty Acids

The by far highest palmitic acid levels in breast milk were observed in the Maasai (27.90 g%) and in the milk from Islamabad (Pakistan). For the Maasai, this is due to the consumption of large amounts of fermented milk, while in Pakistan it can be traced to the consumption of ghee (Smit et al., 2000). Palmitic acid is abundant both in milk and butterfat, a food product that is particularly hypercholesterolemic (Mensink et al., 2003). Because of their high consumption of both milk and meat, the pastoral Maasai have high intakes of dietary fat (about 300 g/day), saturated fat, and cholesterol (about 600 mg/day) but nevertheless have low serum total-cholesterol (about 3.3 mmol/L). They exhibit extensive atherosclerosis with lipid infiltration and fibrous changes in the aorta, together with intimal thickening of the coronary arteries. On the other hand, they also have low body mass index, are virtually free of clinical signs of coronary artery disease, exhibit low coronary artery disease mortality, have smaller hearts, are remarkably fit, and their blood pressure shows only a slight tendency to increase with age (Mann et al., 1964, 1965, 1972). Hence, despite their high saturated fat intake, paradoxically, the Maasai have low cholesterol and almost nonexistent coronary artery disease mortality, probably because of their low body mass index, low carbohydrate consumption, and their remarkable fitness.

Another paradox comes from the exceptionally high levels of lauric acid (20.17 g%) and myristic acid in the breast milk of Chole women (Kuipers et al., 2007). These high levels of lauric acid were only approached by the women of Dar-es-Salaam (16.89 g%) and the Kerewe (12.28 g%). Chole and Dar-es-Salaam are both located in the Tanzanian coastal region, where palm trees are abundant and coconuts freely available. We also reported high levels of lauric acid in breast milk the Caribbean islands of Dominica, St. Vincent, and St. Lucia, and in Surinam, where coconuts or coconut oil are part of the diets as well (Kuipers et al., 2007). A similar picture emerges for myristic acid, which is less abundant in coconuts compared with lauric acid. Here again, Chole women scored highest (21.19 g%), followed by the Kerewe (20.26 g%). The Kerewe do not consume coconuts and do not use coconut oil for cooking, but have high carbohydrate intake from "ugali" (corn wheat porridge) and "muhogo" (cassava root). The lauric to myristic acid ratio in coconuts is about 2.61 and in breast milk from Chole, Dar-es-Salaam, and the Kerewe it is 0.92, 1.15, and 0.62, respectively, making this ratio the most distinctive in the Kerewe.

Consuming lauric and myristic acids raises their content in breast milk, while the consumption of carbohydrate-rich diets raises the milk content of medium chain fatty acids (6:0–14:0) by *de novo* synthesis from glucose, especially that of myristic acid and to a lesser extent lauric acid (Francois et al., 1998; Insull et al., 1959; Jensen, 1999). In addition, both lauric acid and myristic acid not only raise LDL-cholesterol, but they also raise HDL-cholesterol. In fact, lauric acid has a more favorable effect on the total cholesterol/HDL-cholesterol ratio than any other fatty acid, either saturated or unsaturated. The corresponding effect of myristic acid is less pronounced (Mensink et al., 2003). In contrast, for equivalent amounts of dietary energy, carbohydrates cause the highest increase in total cholesterol/HDL-cholesterol (Mensink et al., 2003), suggesting an atherosclerosis-promoting effect of the carbohydrate-rich diet of the Kerewe.

These observations linking the intake of coconuts to the particular breast milk composition of coastal women in East Africa may inform the discussion about the diets of hominins. We consider it unlikely that abundant dietary carbohydrate intake would have in the past served as a major substrate for medium chain fatty acid production in the human mammary gland, especially not to the extent as presently observed in the Kerewe and Nyiramba. Prior to the start of the agricultural revolution some 10,000 years ago, carbohydrate-rich diets were not part of our dietary habits (Cordain, 1999), especially not those with a high glycemic index, which are increasingly implicated as a causal factor in the epidemics of chronic degenerative Western diseases (Cordain et al., 2003; Weinberg, 2004; Westman et al., 2007; Barclay et al., 2008; Forsythe et al., 2008; Kushner and Doerfler, 2008; Volek et al., 2008a,b). A moderately high content of 6-14 carbon (medium) chain fatty acids in breast milk, that is, as seen in East-African women eating coconuts, appears to confer favorable properties to the newborn. The content of these fatty acids increases with advancing lactation (Gibson and Kneebone, 1981; Lammi-Keefe and Jensen, 1984), they serve as easily absorbable energy sources, and they exhibit broad-spectrum antiviral and antimicrobial properties (Kabara et al., 1972; Sun et al., 2002).

The anti-microbial effects seem to be due especially to *caprylic acid* (8 : 0) and lauric acid, and to a much lesser extent to *caproic acid* (6 : 0), *capric acid* (10 : 0), and myristic acids (Sun et al., 2002). Interestingly, the breast milk from coconut-eating women of Chole and Dar-es-Salaam exhibit the highest caprylic acid (not shown) and lauric acid contents, which theoretically offers an enhanced protection against infectious diseases. In addition, in our entire human milk data set, lauric acid in breast milk correlates inversely with its content of linoleic acid and positively with AA and DHA (Smit et al., 2003), suggesting both that the typically high linoleic acid intakes in Western countries adversely affect the antimicrobial properties of breast milk conveyed by lauric acid, and that lauric acid, AA, and DHA do not compete among themselves for incorporation into breast milk lipids. Finally, coconut water is isotonic (Campbell-Falck et al., 2000) and provides an excellent source of sterile fluid to compensate for fluid losses in a hot tropical climate with little freshwater in the immediate surroundings.

Whether coconuts could have been part of the diet of early *H. sapiens* is unknown, since no coconut fossils have been found in East-Africa. However, not many fossilized coconuts have been found anywhere, which is probably because eventually they decompose on the seashores of the hot and humid tropics (Harries, 1979). One of the exceptions is a silicified coconut fruit from the Chinchilla sands in southern Queensland (Australia), which was dated to the late Pliocene, about 2 million years ago, suggesting an already wide distribution range at that time (Rigby, 1995). Coconuts are capable of sprouting after having floated in the sea for up to 110 days and their theoretical range of dissemination by water is estimated to be anywhere between the African East coast and the American West coast, wherever currents are favorable (Harries, 1978). It seems therefore plausible that coconuts would have been available to hominins and early humans and that they would have served as easy and nutritious meals and drinks for opportunistic hunter-gatherers (Harries, 1979) living along the African coast, while they could also have contributed to the Out-of-Africa diaspora which occurred along the shores of the Indian Ocean into Asia (Macaulay et al., 2005; Stringer, 2000).

TANZANIAN BREAST MILK FATTY ACIDS VERSUS WESTERN RECOMMENDATIONS

Comparison of the breast milk fatty acid composition of Tanzanian populations reveals that many of the Tanzanian human milk samples do not comply with current Western recommendations for infant milk formulae (Table 5.1) (Kuipers et al., 2007), a fact that

TABLE 5.1 Percentage human milk samples that do not fulfill the recommendations as issued by indicated organizations

Committee	Recommendation	Chole (n = 20)	Kerewe (n = 30)	Nyakius (n = 30)	Nyiramba (n = 35)	Hadzabe (n = 28)	Maasai (n = 27)	Sonjo (n = 9)	Iraqw (n = 18)	All (n = 197)
CD91, EU	12:0 ≤ 15 g%	90	10	3	3	4	0	0	11	13
CD91, EU	14:0 ≤ 15 g%	85	87	27	17	32	7	0	28	37
EU, ESP05	12:0 : 14:0 ≤ 20 g%	100	100	63	63	86	26	33	50	68
ESP05	18:2ω6 ≥ 5 g%	65	47	3	0	0	0	11	0	15
LSRO, EU	18:2ω6 ≥ 8 g%	100	87	27	0	4	33	22	6	34
CD91	18:2ω6 ≥ 9 g%	100	90	37	0	4	56	22	11	40
GR, WS	18:2ω6 ≥ 10 g%	100	93	50	0	14	67	33	22	47
FAO, ESP91	18:2ω6 ≥ 11 g%	100	93	63	3	21	81	44	22	53
ESP, CD91	18:2ω6 ≤ 20 g%	0	0	3	46	32	0	0	33	16
ESP, EU	18:2ω6 ≤ 27 g%	0	0	0	3	4	0	0	0	1
LSRO	18:2ω6 ≤ 35 g%	0	0	0	0	0	0	0	0	0
FAO, ESP05, EU	18:3ω3 ≥ 1.0 g%	100	93	73	37	96	96	78	78	80
GR	18:3ω3 ≥ 1.4 g%	100	97	97	66	100	100	89	89	91
SI, WS	18:3ω3 ≥ 1.5 g%	100	97	100	69	100	100	89	89	92
LSRO	18:3ω3 ≥ 1.75 g%	100	97	100	83	100	100	89	89	95

LSRO	18:3ω3 ≤ 4.0 g%	0	0	0	0	0	0	0	0	0
CHF, EU	20:4ω6 ≥ 0.35 g%	0	0	0	0	0	41	0	6	7
WS	20:4ω6 ≥ 0.5 g%	50	20	53	0	7	85	89	33	37
GR	20:4ω6 ≥ 0.6 g%	85	50	73	14	36	100	100	44	58
FAO	20:4ω6 ≥ 0.8 g%	100	87	93	49	75	100	100	78	82
SI	20:4ω6 ≤ 1.0 g%	0	3	0	17	0	0	0	0	4
WS	20:5ω3 ≤ 0.10 g%	70	97	20	71	14	11	22	11	43
CHF, EU	22:6ω3 ≥ 0.2 g%	0	0	3	6	25	56	44	67	22
WS	22:6ω3 ≥ 0.35 g%	0	0	23	20	57	93	56	94	40
FAO, GR	22:6ω3 ≥ 0.4 g%	5	0	27	20	68	93	78	100	44
CD96, SI, EU	20:5ω3 ≤ 22:6ω3	0	0	0	0	0	0	0	0	0
EU	LCPω6 ≤ 22:6ω3	5	93	3	0	0	0	0	0	15
SI, ESP, CD96, EU	LCPω6 ≤ 2 g%	0	3	3	83	46	0	0	28	25
SI, ESP, CD96, EU	LCPω3 ≤ 1 g%	60	97	23	71	4	0	11	0	38

CD91, Commission Directive 1991; CD96, Commission Directive 1996; CHF, Child Health Foundation, 2001; ESP, ESPHGAN 1991 and 2005; EU, European Commission, Scientific Committee on Food 2003; FAO, Food and Agriculture Organization of the United Nations, 1994; GR, Health Council of the Netherlands (Gezondheidsraad), 2001; LSRO, Life Science Research Office, 1998; SI, Statutory Instruments, 2001; WS, Workshop Statement, 2000. LCP, long-chain polyunsaturated fatty acids (≥C20 and ≥3 methylene-interrupted cis-double bond). Reproduced from Kuipers et al. (2007), with permission of Elsevier.

holds true for many other non-Western populations (Smit et al., 2003). Most of the current recommendations for infant milk formulae composition are based on observations on human milk in Western countries; that is, they are derived from mothers with inherently high risk of Western degenerative diseases, who themselves commonly do not comply with dietary guidelines issued by their own governmental public health agencies. For instance, in 2003, the mean intake of long-chain ω3 PUFA by Dutch women was 84 mg/day (Kruizinga et al., 2007), yet the current recommendation is 450 mg/day (Health Council of the Netherlands, 2006), and there is good evidence to recommend even higher intakes (Mozaffarian, 2008). Hence, the basis for dietary recommendations can be tenuous (Muskiet et al., 2007; van Goor et al., 2008).

Recent recommendations for the long-chain PUFA content of infant formulas and baby foods, as endorsed by the World Association of Perinatal Medicine, the Early Nutrition Academy and the Child Health Foundation, advise a DHA level of 0.2–0.5%. (Koletzko et al., 2008). The guidelines recommend AA also be present in an amount at least equal to DHA, while EPA is not to exceed the added DHA. The basis of this recommendation is that at least 0.2% DHA in milk formula is necessary to achieve benefits on functional endpoints, while the 0.5% upper limit is set because "systematic evaluation of higher levels of intake have not been published." However, there are currently at least four meta analyses of randomized controlled trials showing that there are no benefits to the fortification of infant formula with long-chain PUFA (Makrides et al., 2005; Simmer, 2008a,b; Smithers et al., 2008). In other words, the recommended addition of long-chain PUFA, notably DHA and AA, to infant formula seems to be based more on common sense than on the strength of the currently available scientific evidence (McCann and Ames, 2005). This raises a question of whether the widely established paradigm of randomized controlled trials and, preferably, subsequent publication of meta-analyses of many randomized controlled trials in high impact journals, was the intention of Sackett et al. (2000) when they first issued their booklet on "evidence-based medicine" in 2000.

It is also clear that there are many paradoxes regarding the influence of dietary saturated fatty acids on cholesterol and their relationships with coronary artery disease. Each 1 energy % increase in dietary saturated fatty acids not only increases serum LDL-cholesterol by 0.033–0.045 mmol/L, but also increases HDL-cholesterol with 0.011–0.013 mmol/L (Institute of Medicine, 2005). Moreover, dietary saturated fatty acids are more likely to induce an unfavorable lipid profile in the context of a carbohydrate-rich diet (Forsythe et al., 2008; Volek et al., 2008a,b). There is no doubt that statins lower LDL-cholesterol, and they are effective in both primary and secondary coronary artery disease prevention. However, their LDL-cholesterol lowering coincides with a reduction of circulating C-reactive protein (O'Keefe et al., 2006), which has been attributed to their pleiotropic effects (Liao and Laufs, 2005). It is conceivable that the genuine lesson here is in the prevention of the combination of chronic low-grade inflammation, insulin resistance, and compensatory hyperinsulinemia, which are hallmarks of the metabolic syndrome. Chronic low-grade inflammation may be triggered, or insufficiently prevented, by a combination of the currently low dietary ω3/ω6 PUFA ratio (Rao et al., 2007; Esposito et al., 2008; Mayer and Seeger, 2008; Serhan and Chiang, 2008; Serhan et al., 2008; Simopoulos, 2008), high dietary intakes of saturated fatty acids (Wendel et al., 2007) and industrially produced *trans*-fatty acids (Mozaffarian and Willett, 2007), an unbalanced antioxidant network (Vertuani et al., 2004), insufficient physical exercise (Roubenoff, 2007), and probably other dietary and lifestyle changes adopted since the industrial revolution.

ESTIMATED FATTY ACID INTAKES FROM SHORE-BASED PALEOLITHIC DIETS

Using our data on the fatty acid composition of freshwater and saltwater African fish, we have estimated the fatty acid intakes of hominins living in an East-African land–water ecosystem, and consuming a 3,000 kcal diet (Cordain et al., 1998). We have employed several models that vary in the ratio of animal/plant foods, meat/fish ratio and the percentages fat in fish, meat, and plants. The content of an individual fatty acid in grams per 100 g meat was calculated by (organ or tissue as % of its total edible portion) × (% total fatty acids in that organ or tissue) × (individual fatty acid as % of total fatty acids in that organ or tissue). The plant fatty acid composition was as previously described (Eaton et al., 1998; Guil et al., 1996). The models are modifications of the original work of Eaton et al. and Cordain et al. They assume a savanna-like diet with an animal[meat/fish]/plant composition of 35[100/0]/65 energy% (Table 5.2, model 1; Eaton et al., 1997, 1998) and a 55[100/0]/45 composition with 5 g% meat fat and 5 g% plant fat (Table 5.2, model 2; Cordain et al., 2000). Model 3 is the median outcome of all models that were investigated by us and in which the animal/plant ratios and meat/fish ratios were varied. Model 3 (ranges) provides the ranges of intakes, as derived from all models. The meat fat contents varied from 5–10 g%, the plant fat contents from 2.5–5 g%, and the fish fat content from 2.5–10 g%. The fish fatty acid composition of Tanzanian lakes and ocean was derived from Kuipers et al. (2005).

The median model had an animal[meat/fish]/plant composition of 35[30/70]/65 energy% and its ranges were 70-30[0-40/100-60]/30-70 energy%. Each model complied with the following three constraints: protein below 35 energy%, linoleic acid intake above 1.7 energy%, and (EPA + DHA)/AA above 2.25 g/g. The protein constraint is derived from the initiation of "rabbit starvation" at protein intakes above 35 energy% and is probably caused by approaching the maximum capacity of the liver to convert the excess nitrogen into urea (Cordain et al., 2000). The linoleic acid constraint is derived from the original data of Barr et al. (1981), who concluded that "at least 1.7% of calories as linoleic acid is needed to prevent biochemical essential fatty acid deficiency from either intravenous or oral sources during total parenteral nutrition," but are widely misquoted as having established a need of 2 energy% linoleic acid (see, e.g., Institute of Medicine, 2005). Finally, we also introduced a constraint of 2.25 for the (EPA + DHA)/AA ratio, which is derived from the currently recommended intake of 450 mg EPA + DHA per day to lower coronary artery disease risk (Health Council of the Netherlands, 2006; Mozaffarian and Rimm, 2006; Psota et al., 2006) and a 200 mg/day AA intake from a typical Western diet (Astorg et al., 2004). Given the role of AA in the initiation of inflammation, and the roles of EPA and DHA in its resolution, it seems that the (EPA + DHA)/AA ratio might describe the optimal dietary long-chain ω3/ω6 PUFA balance to prevent progression to chronic inflammation, scarring, and fibrosis (Serhan and Chiang, 2008; Serhan et al., 2008; Simopoulos, 2008).

As would be expected, the median intakes (model 3) and their ranges (model 3, ranges), indicate that the addition of fish to an originally savanna-like Paleolithic diet, as described by Eaton and Cordain (Eaton et al., 1997, 1998; Cordain et al., 2000), greatly increased the intakes of the sum of EPA and DHA to a median of 12.8 (range: 4.90–28.3) g/day. The intakes of alpha-linolenic acid (12.7; 6.57–17.3) and AA (2.69; 1.73–3.55) were also high, while that of linoleic acid (8.58; 5.53–11.3) was low in all models. The calculated intake of long-chain ω3 PUFA compares well with those of the Greenland Inuit who have lifetime consumption of long-chain ω3 PUFA of about 14 g/d (Feskens and Kromhout, 1993).

TABLE 5.2 Estimated energy, macronutrient, and fatty acids intakes from a Paleolithic diet, as compared with intakes from a current typically Western diet, recommended dietary allowances (RDAs), adequate intakes (AIs), and acceptable macronutrient distribution ranges (AMDRs)

Nutrient	Unit	Paleolithic diet Model 1 savanna	Paleolithic diet Model 2 savanna	Paleolithic diet Model 3 meat/fish mixed	Paleolithic diet Model 3 (range) meat/fish mixed	2003 NL diet (19–30 years)	USA recommendations RDA, AI of AMDR
			Animal/Plant ratio				
Animal{meat/fish}/plant	en%{en%/en%}/en%	35{100/0}/65	55{100/0}/45	35{30/70}/65	70–30{0–40/60–100}/30–70		
			Macronutrients				
Energy	kcal/day	3000	3000	3000		M 2760 and F 1921	M 2842 and F 2477 (a)
Protein	en%	37	35	29	(22–35)	14.3	AMDR 10–35
Carbohydrate	en%	41	28	37	(19–48)	48.2	AMDR 45–65
Fat	en%	22	37	33	(20–46)	34.4	AMDR 20–35
			Essential fatty acids				
SAFA	en%	10.0	12.2	12.0	(6.9–17)	M 12.9 and F 13.1	as low as possible; NL < 10
MUFA	en%	7.4	9.2	4.15	(3.1–10)	M 11.4 and F 11.2	
PUFA	en%	9.4	10.8	14.70	(9.3–21)	M 6.8 and F 6.5	NL 12
ω3	en%	5.5	6.4	10.20	(6.9–14)	M 0.7 and F 0.66	
ALA (alpha-linolenic acid; 18:3ω3)	en%	4.0	4.7	4.0	(2.1–5.4)	M 0.63 F 0.59	AMDR 0.6–1.2
LCPω3	en%	0.3	0.6	4.9	(1.8–10.7)		AMDR 0.06–0.12
ω6	en%	4.7	5.6	5.60	(3.4–7.6)	M 5.85 and F 5.52	
LA (linoleic acid; 18:2ω6)	en%	2.8	4.5	2.7	(1.7–3.6)	M 5.8 F 5.5	AMDR 5–10
LCPw6	en%	0.9	0.8	2.6	(1.0–5.5)		AMDR 0.5–1.0
Trans fatty acids	en%					1.1	as low as possible; NL < 1

SAFA	g/day	31.7	38.8	38.0	(21.8–53.9)	33.4	as low as possible
MUFA	g/day	23.4	29.2	13.2	(9.9–32.2)		
PUFA	g/day	29.8	34.4	46.7	(29.4–66.6)		
ω3	g/day	13.7	20.3	32.3	(21.8–44.4)		
ALA (alpha-linolenic acid; 18:3ω3)	g/day	12.6	15.0	12.7	(6.57–17.3)		AI M 1.6 and F 1.1
EPA (eicosapentaenoic acid; 20:5ω3)	g/day	0.39	0.71	3.15	(1.33–6.61)		
DPA (docosapentaenoic acid; 22:5ω3)	g/day	0.42	0.96	2.24	(0.86–4.71)		
DHA (docosahexaenoic acid; 22:6ω3)	g/day	0.27	0.41	9.61	(3.57–21.7)		
EPA+DHA	g/day	0.66	1.12	12.8	(4.90–28.3)	M 0.103 and F 0.84	USA, UK, NL 0.45
LCPω3	g/day	1.08	2.01	15.4	(5.85–33.9)		
ω6	g/day	17.9	13.4	17.8	(10.9–24.2)		
LA (linoleic acid; 18:2ω6)	g/day	8.84	14.3	8.58	(5.53–11.3)		AI M 17 and F 12
AA (arachidonic acid; 20:4ω6)	g/day	1.81	2.41	2.69	(1.73–3.55)	M 0.195 and F 0.160	
LCPω6	g/day	2.81	2.54	8.23	(3.27–17.4)		
LCPUFA	g/day	3.75	4.70	23.70	(9.1–51.3)		
Trans fatty acids	g/day					2.8	as low as possible
PUFA/SAFA	g/g	1.40	1.10	1.2	(1.05–1.37)		
ω3/ω6	g/g	1.19	1.13	1.84	(1.45–2.05)		
LA/ALA	g/g	0.70	1.04	1.52	(0.93–1.79)		
(EPA+DHA)/AA	g/g	0.49	0.47	2.47	(2.25–2.66)		
LCPω3/LCPω6	g/g	0.72	0.72	1.87	(1.75–1.96)		
Cholesterol	mg/day	480	830	514	(321–748)		as low as possible

Data from the Western diet are derived from the 2003 Dutch National Food Survey (Ocke et al., 2004; Kruizinga et al., 2007) and the U.S. Food and Nutrition Board 1989 (Eaton et al., 1997). AA intake is taken from Astorg
et al. (2004). U.S. recommendations (RDA, AI, AMDR, UL) are from the U.S. Dietary Reference Intakes (DRI USA, 2008) and EPA + DHA intake recommendations are from Psota et al. (Gezondheidsraad, 2006; Psota et al., 2006). LCPω-6 is overestimated since it was calculated as "the gap" between PUFA and all ω3 and ω6 fatty acids. For fatty acid abbreviations, see Appendix; NL, The Netherlands; UK, United Kingdom; M, male; F, female; RDA, recommended dietary allowance; AI, adequate intake; AMDR, acceptable macronutrient distribution range; (a); 30 years, length 1.65 m, body mass index 24,99 kg/m^2, active (PAL, physical activity level 1.6–1.9).

The similarity between the breast milk long-chain ω3 PUFA content of the Kerewe (Lake Victoria) and Canadian Inuit supports these calculations. The data are also in line with estimates of Broadhurst et al. (2002), who calculated a daily intake of 29 g fish from a 2196 kcal diet by native Africans living on the shore of Lake Malawi. With an average of 15 g% EPA plus DHA of the local fish, this would imply an intake of at least 6 g ω3 long-chain PUFA/day. Crawford et al. (1999) estimated the daily intake of contemporary populations living near East African lakes (Nyasa and Turkana) at 1–4 g ω3 long-chain PUFA and 0.5–1.0 g AA, which further emphasizes the higher intake of ω3 long-chain PUFA compared with AA.

Taken together, we suggest that East-African hominins living at the land–water interface had access to food that was relatively easy to obtain, was rich in α-linolenic acid, ω3 long-chain PUFA, and AA, low in linoleic acid and saturated fatty acids, and had a relatively high (EPA + DHA)/AA ratio.

CONCLUSIONS

Hominins probably evolved in a land–water ecosystem where they consumed a diet that was rich in meat, fish, vegetables, and fruits, and low in starchy carbohydrates. They were also relatively fit. Their diet would have been high in α-linolenic acid from edible green plants, rich in AA, EPA, and DHA from meat and fish, low in linoleic acid, saturated fatty acids, and starchy carbohydrates and contained no *trans*-fatty acids, sucrose, or fructose from industrial sources. This hunter-gatherer diet was gradually abandoned starting at the beginning of the agricultural revolution, some 10,000 years ago, and the changes have been accelerating since the start of the industrial revolution, starting nearly 200 years ago. Together with low physical activity and other lifestyle changes, these and other dietary changes have created a conflict with the human genome that was largely configured during the Paleolithic period. The pathophysiological effects of these changes are diverse, but center on the metabolic syndrome, and are increasingly ascribed to the induction of a pro-inflammatory state that progresses silently but inevitably toward chronic low-grade inflammation. This condition affects virtually all organs and systems and may possibly begin at conception, and possibly even prior to gametogenesis through, potentially heritable, epigenetic alterations. At first sight, the outcome is a seemingly unrelated spectrum of diseases, including coronary artery disease, type 2 diabetes mellitus, some types of cancer, pregnancy complications, some psychiatric diseases, and neurodegenerative diseases. Data from evolutionary medicine and contemporary traditional people, and perhaps common sense, encourage rethinking of "homeostasis" and a return to the dietary habits and lifestyle on which our genes have evolved. Of special importance, it seems essential to avoid the vicious cycle initiated by basing dietary and general lifestyle recommendations on observations from Western societies as a whole. Due to the multiple dietary and lifestyle changes involved, it is questionable whether clear dietary recommendations will ever come from expensive randomized controlled trials of single nutrients, especially since these require large study numbers, long observation periods, elucidation of many dose–response relationships, and investigation of numerous interactions with other nutrients.

Contemporary societies living by traditional standards provide us with clues to optimize genome-environment homeostasis. Unfortunately, many traditional peoples have meanwhile become dependent on food programs with typically Western approaches, such as the use of carbohydrate-rich staple food and linoleic acid-rich seed oils, or have already adopted a (quasi) Western lifestyle. A combination of data from traditional living societies, (patho)physiological evidence, anthropometrics, archeology, genetics, nutrigenomics, and classical trials will indicate the diet on which human genes evolved and, consequently, what diet supports optimal human health. Only then will it be clear that we have really understood Darwin's message.

NOTES

1. Members of the Tribe *hominini*, which includes all extinct and extant human species that ever evolved, that is, chimpanzees and gorillas excluded.

2. A difference in DNA sequence occurring with ≥1% population frequency, as opposed to the <1% frequency of a mutation.

REFERENCES

Aidoo, M., Terlouw, D.J., Kolczak, M.S., McElroy, P.D., ter Kuile, F.O., Kariuki, S., Nahlen, B.L., Lal, A.A., and Udhayakumar, V. 2002. Protective effects of the sickle cell gene against malaria morbidity and mortality. *Lancet* **359** (9314):1311–1312.

Ailhaud, G., Massiera, F., Weill, P., Legrand, P., Alessandri, J.M., and Guesnet, P. 2006. Temporal changes in dietary fats: Role of n-6 polyunsaturated fatty acids in excessive adipose tissue development and relationship to obesity. *Prog Lipid Res* **45** (3):203–236.

Almeras, L., Eyles, D., Benech, P., Laffite, D., Villard, C., Patatian, A., Boucraut, J., kay-Sim, A., McGrath, J., and Feron, F. 2007. Developmental vitamin D deficiency alters brain protein expression in the adult rat: Implications for neuropsychiatric disorders. *Proteomics* **7** (5):769–780.

Ames, B.N., Elson-Schwab, I., and Silver, E.A. 2002. High-dose vitamin therapy stimulates variant enzymes with decreased coenzyme binding affinity (increased K(m)): Relevance to genetic disease and polymorphisms. *Am J Clin Nutr* **75** (4):616–658.

Astorg, P., Arnault, N., Czernichow, S., Noisette, N., Galan, P., and Hercberg, S. 2004. Dietary intakes and food sources of n-6 and n-3 PUFA in French adult men and women. *Lipids* **39** (6):527–535.

Balter, M. 2005. Evolutionary genetics: Are humans still evolving? *Science* **309** (5732):234–237.

Barclay, A.W., Petocz, P., Millan-Price, J., Flood, V.M., Prvan, T., Mitchell, P., and Brand-Miller, J.C. 2008. Glycemic index, glycemic load, and chronic disease risk – A meta-analysis of observational studies. *Am J Clin Nutr* **87** (3):627–637.

Barr, L.H., Dunn, G.D., and Brennan, M.F. 1981. Essential fatty acid deficiency during total parenteral nutrition. *Ann Surg* **193** (3):304–311.

Bensinger, S.J. and Tontonoz, P. 2008. Integration of metabolism and inflammation by lipid-activated nuclear receptors. *Nature* **454** (7203):470–477.

Bjelakovic, G., Nikolova, D., Gluud, L.L., Simonetti, R.G., and Gluud, C. 2007. Mortality in randomized trials of antioxidant supplements for primary and secondary prevention: Systematic review and meta-analysis. *JAMA* **297** (8):842–857.

Botto, L.D. and Yang, Q. 2000. 5,10-Methylenetetrahydrofolate reductase gene variants and congenital anomalies: A HuGE review. *Am J Epidemiol* **151** (9):862–877.

Bouillon, R., Carmeliet, G., Verlinden, L., van Etten, E., Verstuyf, A., Luderer, H.F., Lieben, L., Mathieu, C., and Demay, M. 2008. Vitamin D and human health: Lessons from vitamin D receptor null mice. *Endocr Rev* **29** (6):726–776.

Bragt, M.C. and Popeijus, H.E. 2008. Peroxisome proliferator-activated receptors and the metabolic syndrome. *Physiol Behav* **94** (2):187–197.

Broadhurst, C.L., Cunnane, S.C., and Crawford, M.A. 1998. Rift Valley lake fish and shellfish provided brain-specific nutrition for early H. *Br J Nutr* **79** (1): 3–21.

Broadhurst, C.L., Wang, Y., Crawford, M.A., Cunnane, S.C., Parkington, J.E., and Schmidt, W.F. 2002. Brain-specific lipids from marine, lacustrine, or terrestrial food resources: Potential impact on early African *H. sapiens*. *Comp Biochem Physiol B Biochem Mol Biol* **131** (4):653–673.

Burdge, G.C. 2006. Metabolism of alpha-linolenic acid in humans. *Prostaglandins Leukot Essent Fatty Acids* **75** (3): 161–168.

Burdge, G.C., Hanson, M.A., Slater-Jefferies, J.L., and Lillycrop, K.A. 2007. Epigenetic regulation of transcription: A mechanism for inducing variations in phenotype (fetal programming) by differences in nutrition during early life? *Br J Nutr* **97** (6):1036–1046.

Calder, P.C., Dimitriadis, G., and Newsholme, P. 2007. Glucose metabolism in lymphoid and inflammatory cells and tissues. *Curr Opin Clin Nutr Metab Care* **10** (4): 531–540.

Campbell-Falck, D., Thomas, T., Falck, T.M., Tutuo, N., and Clem, K. 2000. The intravenous use of coconut water. *Am J Emerg Med* **18** (1):108–111.

Castrillo, A. and Tontonoz, P. 2004. Nuclear receptors in macrophage biology: At the crossroads of lipid metabolism and inflammation. *Annu Rev Cell Dev Biol* **20**: 455–480.

Chakravarthy, M.V. and Booth, F.W. 2004. Eating, exercise, and "thrifty" genotypes: Connecting the dots toward an evolutionary understanding of modern chronic diseases. *J Appl Physiol* **96** (1):3–10.

Clarke, S.D. 2004. The multi-dimensional regulation of gene expression by fatty acids: Polyunsaturated fats as nutrient sensors. *Curr Opin Lipidol* **15** (1):13–18.

Colagiuri, S. and Brand, M.J. 2002. The "carnivore connection" – Evolutionary aspects of insulin resistance. *Eur J Clin Nutr* **56** (Suppl. 1.):S30–S35.

Cooney, C.A. 2006. Germ cells carry the epigenetic benefits of grandmother's diet. *Proc Natl Acad Sci U S A* **103** (46):17071–17072.

Cordain, L. 1999. Cereal grains: Humanity's double-edged sword. *World Rev Nutr Diet* **84**:19–73.

Cordain, L., Eades, M.R., and Eades, M.D. 2003. Hyperinsulinemic diseases of civilization: More than just Syndrome X. *Comp Biochem Physiol A Mol Integr Physiol* **136** (1):95–112.

Cordain, L., Eaton, S.B., Sebastian, A., Mann, N., Lindeberg, S., Watkins, B.A., O'Keefe, J.H., and Brand-Miller, J. 2005. Origins and evolution of the Western diet: Health implications for the 21st century. *Am J Clin Nutr* **81** (2):341–354.

Cordain, L., Gotshall, R.W., Eaton, S.B., and Eaton, S.B. III. 1998. Physical activity, energy expenditure and fitness: An evolutionary perspective. *Int J Sports Med* **19** (5):328–335.

Cordain, L., Miller, J.B., Eaton, S.B., Mann, N., Holt, S.H., and Speth, J.D. 2000. Plant-animal subsistence ratios and macronutrient energy estimations in worldwide hunter-gatherer diets. *Am J Clin Nutr* **71** (3):682–692.

Crawford, M.A. 2006. Docosahexaenoic acid in neural signaling systems. *Nutr Health* **18** (3):263–276.

Crawford, M.A., Bloom, M., Broadhurst, C.L., Schmidt, W.F., Cunnane, S.C., Galli, C., Gehbremeskel, K., Linseisen, F., Lloyd-Smith, J., and Parkington, J. 1999. Evidence for the unique function of docosahexaenoic acid during the evolution of the modern hominid brain. *Lipids* **34**:S39–S47.

Crawford, M.A., Bloom, M., Cunnane, S., Holmsen, H., Ghebremeskel, K., Parkington, J., Schmidt, W., Sinclair, A.J., and Broadhurst, C.L. 2001. Docosahexaenoic acid and cerebral evolution. *World Rev Nutr Diet* **88**:6–17.

Cserti, C.M. and Dzik, W.H. 2007. The ABO blood group system and *Plasmodium falciparum* malaria. *Blood* **110** (7):2250–2258.

Cunnane, S.C. 2003. Problems with essential fatty acids: Time for a new paradigm? *Prog Lipid Res* **42** (6): 544–568.

Cunnane, S.C. and Crawford, M.A. 2003. Survival of the fattest: Fat babies were the key to evolution of the large human brain. *Comp Biochem Physiol A Mol Integr Physiol* **136** (1):17–26.

de Lorgeril, M., Salen, P., Martin, J.L., Monjaud, I., Delaye, J., and Mamelle, N. 1999. Mediterranean diet, traditional risk factors, and the rate of cardiovascular complications after myocardial infarction: Final report of the Lyon Diet Heart Study. *Circulation* **99** (6): 779–785.

Deckelbaum, R.J., Worgall, T.S., and Seo, T. 2006. N-3 fatty acids and gene expression. *Am J Clin Nutr* **83** (6): 1520S–1525S.

DeMar, J.C. Jr., Dimartino, C., Baca, A.W., Lefkowitz, W., and Salem, N. Jr. 2008. Effect of dietary docosahexaenoic acid on biosynthesis of docosahexaenoic acid from alpha-linolenic acid in young rats. *J Lipid Res* **49** (9):1963–1980.

DRI USA. Dietary Reference Intakes USA. http://fnic.nal.usda.gov/nal_display/index.php?info_center=4&tax_level=2&tax_subject=256&topic_id=1342.17-3-2008.24-6-2008. Ref Type: Electronic Citation.

Eaton, S.B., Cordain, L., and Lindeberg, S. 2002. Evolutionary health promotion: A consideration of common counterarguments. *Prev Med* **34** (2):119–123.

Eaton, S.B., Eaton, S.B. III, and Konner, M.J. 1997. Paleolithic nutrition revisited: A twelve-year retrospective on its nature and implications. *Eur J Clin Nutr* **51** (4): 207–216.

Eaton, S.B., Eaton, S.B. III, Sinclair, A.J., Cordain, L., and Mann, N.J. 1998. Dietary intake of long-chain polyunsaturated fatty acids during the Paleolithic. *World Rev Nutr Diet* **83**:12–23.

Esposito, G., Giovacchini, G., Liow, J.S., Bhattacharjee, A.K., Greenstein, D., Schapiro, M., Hallett, M., Herscovitch, P., Eckelman, W.C., Carson, R.E., and Rapoport, S.I. 2008. Imaging neuroinflammation in Alzheimer's disease with radiolabeled arachidonic acid and PET 2. *J Nucl Med* **49** (9):1414–1421.

Farooqi, S. and O'Rahilly, S. 2006. Genetics of obesity in humans. *Endocr Rev* **27** (7):710–718.

Farooqui, A.A., Horrocks, L.A., and Farooqui, T. 2007. Modulation of inflammation in brain: A matter of fat. *J Neurochem* **101** (3):577–599.

Feil, R. 2006. Environmental and nutritional effects on the epigenetic regulation of genes. *Mutat Res* **600** (1–2): 46–57.

Feskens, E.J. and Kromhout, D. 1993. Epidemiologic studies on Eskimos and fish intake. *Ann N Y Acad Sci* **683**:9–15.

Forsythe, C.E., Phinney, S.D., Fernandez, M.L., Quann, E.E., Wood, R.J., Bibus, D.M., Kraemer, W.J., Feinman, R.D., and Volek, J.S. 2008. Comparison of low fat and low carbohydrate diets on circulating fatty acid composition and markers of inflammation. *Lipids* **43** (1):65–77.

Francis, G.A., Fayard, E., Picard, F., and Auwerx, J. 2003. Nuclear receptors and the control of metabolism. *Annu Rev Physiol* **65**:261–311.

Francois, C.A., Connor, S.L., Wander, R.C., and Connor, W.E. 1998. Acute effects of dietary fatty acids on the fatty acids of human milk. *Am J Clin Nutr* **67** (2): 301–308.

Freeman, M.P., Hibbeln, J.R., Wisner, K.L., Davis, J.M., Mischoulon, D., Peet, M., Keck, P. E. Jr., Marangell, L.B., Richardson, A.J., Lake, J., and Stoll, A.L. 2006. Omega-3 fatty acids: Evidence basis for treatment and future research in psychiatry. *J Clin Psychiatry* **67** (12): 1954–1967.

GANI, O.A. 2008. Are fish oil omega-3 long-chain fatty acids and their derivatives peroxisome proliferator-activated receptor agonists? *Cardiovasc Diabetol* **7**:6.

GANI, O.A. AND SYLTE, I. 2008. Molecular recognition of docosahexaenoic acid by peroxisome proliferator-activated receptors and retinoid-X receptor alpha. *J Mol Graph Model* **27** (2):217–224.

GAZIANO, J.M., GLYNN, R.J., CHRISTEN, W.G., KURTH, T., BELANGER, C., MACFADYEN, J., BUBES, V., MANSON, J.E., SESSO, H.D., AND BURING, J.E. 2009. Vitamins E and C in the prevention of prostate and total cancer in men: The Physicians' Health Study II randomized controlled trial 1. *JAMA* **301** (1):52–62.

Gezondheidsraad. 2006. *Richtlijnen goede voeding 2006-achtergronddocument. Den Haag, publicatie nr A06/08.*

GIBBONS, A. 2002a. American Association of Physical Anthropologists meeting. Humans' head start: New views of brain evolution. *Science* **296** (5569):835–837.

GIBBONS, A. 2002b. Becoming human: In search of the first hominids. *Science* **295** (5558):1214–1219.

GIBSON, R.A. AND KNEEBONE, G.M. 1981. Fatty acid composition of human colostrum and mature breast milk. *Am J Clin Nutr* **34** (2):252–257.

GIBSON, R.A. AND SINCLAIR, A.J. 1981. Are Eskimos obligate carnivores? *Lancet* **1** (8229):1100.

GISSI-PREVENZIONE TRIAL. 1999. Dietary supplementation with n-3 polyunsaturated fatty acids and vitamin E after myocardial infarction: Results of the GISSI-Prevenzione trial. Gruppo Italiano per lo Studio della Sopravvivenza nell'Infarto miocardico. *Lancet* **354** (9177):447–455.

GLUCKMAN, P.D., HANSON, M.A., COOPER, C., AND THORNBURG, K.L. 2008. Effect of in utero and early-life conditions on adult health and disease. *N Engl J Med* **359** (1): pp. 61–73.

GLUCKMAN, P.D., HANSON, M.A., MORTON, S.M., AND PINAL, C.S. 2005. Life-long echoes – A critical analysis of the developmental origins of adult disease model. *Biol Neonate* **87** (2):127–139.

GODFREY, K.M. AND BARKER, D.J. 2000. Fetal nutrition and adult disease. *Am J Clin Nutr* **71** (5):S1344–S1352.

GODFREY, K.M., LILLYCROP, K.A., BURDGE, G.C., GLUCKMAN, P.D., AND HANSON, M.A. 2007. Epigenetic mechanisms and the mismatch concept of the developmental origins of health and disease. *Pediatr Res* **61** (5 Pt 2):5R–10R.

GUIL, J.L., TORIJA, M.E., GIMENEZ, J.J., AND RODRIGUEZ, I. 1996. Identification of fatty acids in edible wild plants by gas chromatography. *J Chromatogr A* **719** (1):229–235.

HAFFNER, S.M., RUILOPE, L., DAHLOF, B., ABADIE, E., KUPFER, S., AND ZANNAD, F. 2006. Metabolic syndrome, new onset diabetes, and new end points in cardiovascular trials. *J Cardiovasc Pharmacol* **47** (3):469–475.

HARRIES, H.C. 1978. The evolution, dissemination and classification of *Cocos nuciferia* L. *The Botan Rev* **44**:265–320.

HARRIES, H.C. 1979. Nuts to the Garden of Eden. *Principes* **23**:143–148.

HARRIS, E.E. AND MALYANGO, A.A. 2005. Evolutionary explanations in medical and health profession courses: Are you answering your students' "why" questions? *BMC Med Educ* **5** (1):16.

HARRIS, E.E. AND MEYER, D. 2006. The molecular signature of selection underlying human adaptations. *Am J Phys Anthropol Suppl* **43**:89–130.

HARRIS, W.S., CONNOR, W.E., AND LINDSEY, S. 1984. Will dietary omega-3 fatty acids change the composition of human milk? *Am J Clin Nutr* **40** (4):780–785.

HAWKS, J., WANG, E.T., COCHRAN, G.M., HARPENDING, H.C., AND MOYZIS, R.K. 2007. Recent acceleration of human adaptive evolution. *Proc Natl Acad Sci U S A* **104** (52):20753–20758.

HEALTH COUNCIL OF THE NETHERLANDS. 2006. *Guidelines for a Healthy Diet 2006. The Hague: Health Council of the Netherlands, 2006; publication no. 2006/21.*

HIBBELN, J.R. 1998. Fish consumption and major depression. *Lancet* **351** (9110):1213.

HIBBELN, J.R. 2002. Seafood consumption, the DHA content of mothers' milk and prevalence rates of postpartum depression: A cross-national, ecological analysis. *J Affect Disord* **69** (1–3):15–29.

HOLICK, M.F. 1992. Evolutionary biology and pathology of vitamin D. *J Nutr Sci Vitaminol (Tokyo)* Spec No: 79–83.

HOLICK, M.F. 2003. Vitamin D: A millenium perspective. *J Cell Biochem* **88** (2):296–307.

HOLICK, M.F. 2008. The vitamin D deficiency pandemic and consequences for nonskeletal health: Mechanisms of action. *Mol Aspects Med* **29** (6):361–368.

HOLICK, M.F. AND CHEN, T.C. 2008. Vitamin D deficiency: A worldwide problem with health consequences. *Am J Clin Nutr* **87** (4):S1080–S1086.

HOMKO, C.J., SIVAN, E., REECE, E.A., AND BODEN, G. 1999. Fuel metabolism during pregnancy. *Semin Reprod Endocrinol* **17** (2):119–125.

HSIEH, A.T., ANTHONY, J.C., ERSEN-SCHADE, D.A., RUMSEY, S.C., LAWRENCE, P., LI, C., NATHANIELSZ, P.W., AND BRENNA, J.T. 2007. The influence of moderate and high dietary long-chain polyunsaturated fatty acids (LCPUFA) on baboon neonate tissue fatty acids. *Pediatr Res* **61** (5 Pt 1):537–545.

INGMAN, M., KAESSMANN, H., PAABO, S., AND GYLLENSTEN, U. 2000. Mitochondrial genome variation and the origin of modern humans. *Nature* **408** (6813):708–713.

INNIS, S.M. AND KUHNLEIN, H.V. 1988. Long-chain n-3 fatty acids in breast milk of Inuit women consuming traditional foods. *Early Hum Dev* **18** (2–3):185–189.

INSTITUTE OF MEDICINE, FOOD AND NUTRITION BOARD. 2005. *Dietary Reference Intakes for Energy, Carbohydrate, Fiber, Fat, Fatty Acids, Cholesterol, Protein and Amino Acids*. Washington, DC: The National Academies Press.

INSULL, W. JR., HIRSCH, J., JAMES, T., AND AHRENS, E.H. JR. 1959. The fatty acids of human milk. II. Alterations produced by manipulation of caloric balance and exchange of dietary fats. *J Clin Invest* **38** (2):443–450.

ITOH, T. AND YAMAMOTO, K. 2008. Peroxisome proliferator activated receptor gamma and oxidized docosahexaenoic acids as new class of ligand. *Naunyn Schmiedebergs Arch Pharmacol* **377** (4–6):541–547.

Jablonski, N.G. and Chaplin, G. 2000. The evolution of human skin coloration. *J Hum Evol* **39** (1):57–106.

Jensen, R.G. 1999. Lipids in human milk. *Lipids* **34** (12): 1243–1271.

Jiang, Y.H., Bressler, J., and Beaudet, A.L. 2004. Epigenetics and human disease. *Annu Rev Genomics Hum Genet* **5**:479–510.

Kabara, J.J., Swieczkowski, D.M., Conley, A.J. and Truant, J.P. 1972. Fatty acids and derivatives as antimicrobial agents. *Antimicrob Agents Chemother* **2** (1): 23–28.

Kalueff, A.V., Minasyan, A., Keisala, T., Kuuslahti, M., Miettinen, S., and Tuohimaa, P. 2006. The vitamin D neuroendocrine system as a target for novel neurotropic drugs. *CNS Neurol Disord Drug Targets* **5** (3):363–371.

Kiraly, S.J., Kiraly, M.A., Hawe, R.D., and Makhani, N. 2006. Vitamin D as a neuroactive substance: Review. *ScientificWorldJournal* **6**:125–139.

Koletzko, B., Lien, E., Agostoni, C., Bohles, H., Campoy, C., Cetin, I., Decsi, T., Dudenhausen, J.W., Dupont, C., Forsyth, S., Hoesli, I., Holzgreve, W., Lapillonne, A., Putet, G., Secher, N.J., Symonds, M., Szajewska, H., Willatts, P., and Uauy, R. 2008. The roles of long-chain polyunsaturated fatty acids in pregnancy, lactation and infancy: Review of current knowledge and consensus recommendations. *J Perinat Med* **36** (1):5–14.

Kothapalli, K.S., Anthony, J.C., Pan, B.S., Hsieh, A.T., Nathanielsz, P.W., and Brenna, J.T. 2007. Differential cerebral cortex transcriptomes of baboon neonates consuming moderate and high docosahexaenoic acid formulas. *PLoS ONE* **2** (4):e370.

Kruizinga, A.G., Westenbrink, S., van Bosch, L.M.C., and Jansen, M.C.J.F. 2007. *TNO Kwaliteit van leven. De inneming van omega-3 en -6 vetzuren, van vitamines A en E, bij jong volwassenen. Aanvullende berekeningen op basis van voedselconsumptiepeiling 2003. V7451.*

Kuipers, R.S., Fokkema, M.R., Smit, E.N., van der Meulen, J., Boersma, E.R., and Muskiet, F. A. 2005. High contents of both docosahexaenoic and arachidonic acids in milk of women consuming fish from lake Kitangiri (Tanzania): Targets for infant formulae close to our ancient diet? *Prostaglandins Leukot Essent Fatty Acids* **72** (4):279–288.

Kuipers, R.S., Smit, E.N., van der Meulen, J., Dijck-Brouwer, D.A.J., Boersma, E.R., and Muskiet, F.A. 2007. Milk in the island of Chole (Tanzania) is high in lauric, myristic, arachidonic and docosahexaenoic acids, and low in linoleic acid reconstructed diet of infants born to our ancestors living in tropical coastal regions. *Prostaglandins Leukot Essent Fatty Acids* **76** (4):221–233.

Kushner, R.F. and Doerfler, B. 2008. Low-carbohydrate, high-protein diets revisited 1. *Curr Opin Gastroenterol* **24** (2):198–203.

Lammi-Keefe, C.J. and Jensen, R.G. 1984. Lipids in human milk: A review. 2: Composition and fat-soluble vitamins. *J Pediatr Gastroenterol Nutr* **3** (2):172–198.

Lee, J.H., O'Keefe, J.H., Lavie, C.J., Marchioli, R., and Harris, W.S. 2008. Omega-3 fatty acids for cardioprotection. *Mayo Clin Proc* **83** (3):324–332.

Leonard, W.R., Robertson, M.L., Snodgrass, J.J., and Kuzawa, C.W. 2003. Metabolic correlates of hominid brain evolution. *Comp Biochem Physiol A Mol Integr Physiol* **136** (1):5–15.

Liao, J.K. and Laufs, U. 2005. Pleiotropic effects of statins. *Annu Rev Pharmacol Toxicol* **45**:89–118.

Lillycrop, K.A., Phillips, E.S., Jackson, A.A., Hanson, M.A., and Burdge, G.C. 2005. Dietary protein restriction of pregnant rats induces and folic acid supplementation prevents epigenetic modification of hepatic gene expression in the offspring. *J Nutr* **135** (6):1382–1386.

Lillycrop, K.A., Phillips, E.S., Torrens, C., Hanson, M.A., Jackson, A.A., and Burdge, G.C. 2008. Feeding pregnant rats a protein-restricted diet persistently alters the methylation of specific cytosines in the hepatic PPARalpha promoter of the offspring. *Br J Nutr* **100** (2): 278–282.

Lin, P.Y. and Su, K.P. 2007. A meta-analytic review of double-blind, placebo-controlled trials of antidepressant efficacy of omega-3 fatty acids. *J Clin Psychiatry* **68** (7): 1056–1061.

Liou, Y.A., King, D.J., Zibrik, D., and Innis, S.M. 2007. Decreasing linoleic acid with constant alpha-linolenic acid in dietary fats increases (n-3) eicosapentaenoic acid in plasma phospholipids in healthy men. *J Nutr* **137** (4): 945–952.

Lippman, S.M., Klein, E.A., Goodman, P.J., Lucia, M.S., Thompson, I.M., Ford, L.G., Parnes, H.L., Minasian, L.M., Gaziano, J.M., Hartline, J.A., Parsons, J.K., Bearden, J.D. III, Crawford, E.D., Goodman, G.E., Claudio, J., Winquist, E., Cook, E.D., Karp, D.D., Walther, P., Lieber, M.M., Kristal, A.R., Darke, A.K., Arnold, K.B., Ganz, P.A., Santella, R.M., Albanes, D., Taylor, P.R., Probstfield, J.L., Jagpal, T.J., Crowley, J.J., Meyskens, F.L. Jr., Baker, L.H., and Coltman, C.A. Jr. 2009. Effect of selenium and vitamin E on risk of prostate cancer and other cancers: the Selenium and Vitamin E Cancer Prevention Trial (SELECT) 1. *JAMA* **301** (1):39–51.

Macaulay, V., Hill, C., Achilli, A., Rengo, C., Clarke, D., Meehan, W., Blackburn, J., Semino, O., Scozzari, R., Cruciani, F., Taha, A., Shaari, N.K., Raja, J.M., Ismail, P., Zainuddin, Z., Goodwin, W., Bulbeck, D., Bandelt, H.J., Oppenheimer, S., Torroni, A., and Richards, M. 2005. Single, rapid coastal settlement of Asia revealed by analysis of complete mitochondrial genomes. *Science* **308** (5724):1034–1036.

Makrides, M., Gibson, R.A., Udell, T., and Ried, K. 2005. Supplementation of infant formula with long-chain polyunsaturated fatty acids does not influence the growth of term infants. *Am J Clin Nutr* **81** (5):1094–1101.

Manica, A., Prugnolle, F., and Balloux, F. 2005. Geography is a better determinant of human genetic differentiation than ethnicity. *Hum Genet* **118** (3–4):366–371.

Mann, G.V., Shaffer, R.D., Anderson, R.S., and Sandstead, H.H. 1964. Cardiovascular disease in the Masai. *J Atheroscler Res* **4**:289–312.

Mann, G.V., Shaffer, R.D., and Rich, A. 1965. Physical fitness and immunity to heart-disease in Masai. *Lancet* **2** (7426):1308–1310.

Mann, G.V., Spoerry, A., Gray, M., and Jarashow, D. 1972. Atherosclerosis in the Masai. *Am J Epidemiol* **95** (1):26–37.

Marean, C.W., Bar-Matthews, M., Bernatchez, J., Fisher, E., Goldberg, P., Herries, A.I., Jacobs, Z., Jerardino, A., Karkanas, P., Minichillo, T., Nilssen, P.J., Thompson, E., Watts, I., and Williams, H.M. 2007. Early human use of marine resources and pigment in South Africa during the Middle Pleistocene. *Nature* **449** (7164):905–908.

Mayer, K. and Seeger, W. 2008. Fish oil in critical illness. *Curr Opin Clin Nutr Metab Care* **11** (2):121–127.

McCann, J.C. and Ames, B.N. 2005. Is docosahexaenoic acid, an n-3 long-chain polyunsaturated fatty acid, required for development of normal brain function? An overview of evidence from cognitive and behavioral tests in humans and animals. *Am J Clin Nutr* **82** (2): 281–295.

McCann, J.C. and Ames, B.N. 2008. Is there convincing biological or behavioral evidence linking vitamin D deficiency to brain dysfunction? *FASEB J* **22** (4): 982–1001.

McGrane, M.M. 2007. Vitamin A regulation of gene expression: Molecular mechanism of a prototype gene. *J Nutr Biochem* **18** (8):497–508.

McGrath, J.J., Feron, F.P., Burne, T.H., kay-Sim, A., and Eyles, D.W. 2004. Vitamin D3 – Implications for brain development. *J Steroid Biochem Mol Biol* **89–90** (1–5):557–560.

McIntyre, R.S., Soczynska, J.K., Konarski, J.Z., Woldeyohannes, H.O., Law, C.W., Miranda, A., Fulgosi, D., and Kennedy, S.H. 2007. Should depressive syndromes be reclassified as "metabolic syndrome Type II"? *Ann Clin Psychiatry* **19** (4):257–264.

Mensink, R.P., Zock, P.L., Kester, A.D., and Katan, M.B. 2003. Effects of dietary fatty acids and carbohydrates on the ratio of serum total to HDL cholesterol and on serum lipids and apolipoproteins: A meta-analysis of 60 controlled trials. *Am J Clin Nutr* **77** (5):1146–1155.

Moran, C., Sandoval, T., Duque, X., Gonzalez, S., Moran, S., and Bermudez, J.A. 2006. Increased insulin levels independent of gestational overweight in women with preeclampsia. *Arch Med Res* **37** (6):749–754.

Morgan, D.K. and Whitelaw, E. 2008. The case for transgenerational epigenetic inheritance in humans. *Mamm Genome* **19** (6):394–397.

Mozaffarian, D. 2008. Fish and n-3 fatty acids for the prevention of fatal coronary heart disease and sudden cardiac death. *Am J Clin Nutr* **87** (6):S1991–S1996.

Mozaffarian, D. and Rimm, E.B. 2006. Fish intake, contaminants, and human health: Evaluating the risks and the benefits. *JAMA* **296** (15):1885–1899.

Mozaffarian, D. and Willett, W.C. 2007. Trans fatty acids and cardiovascular risk: A unique cardiometabolic imprint? *Curr Atheroscler Rep* **9** (6):486–493.

Muskiet, F.A., Fokkema, M.R., Schaafsma, A., Boersma, E.R., and Crawford, M.A. 2004. Is docosahexaenoic acid (DHA) essential? Lessons from DHA status regulation, our ancient diet, epidemiology and randomized controlled trials. *J Nutr* **134** (1):183–186.

Muskiet, F.A., Kuipers, R.S., Smit, E.N., and Joordens, J.C. 2007. The basis of recommendations for docosahexaenoic and arachidonic acids in infant formula: Absolute or relative standards? *Am J Clin Nutr* **86** (6): 1802–1803.

Muskiet, F.A., van Goor, S.A., Kuipers, R.S., Velzing-Aarts, F.V., Smit, E.N., Bouwstra, H., Dijck-Brouwer, D.A., Boersma, E.R., and Hadders-Algra, M. 2006. Long-chain polyunsaturated fatty acids in maternal and infant nutrition. *Prostaglandins Leukot Essent Fatty Acids* **75** (3):135–144.

Muskiet, F.A.J. 2006. Adaptation to the conditions of existence. *Ned Tijdschr Klin Chem Labgeneesk* **31**:187–193.

Neel, J.V. 1999. Diabetes mellitus: A "thrifty" genotype rendered detrimental by "progress"? *Bull World Health Organ* **77** (8):694–703.

Nesse, R.M. and Williams, G.C. 1998. Evolution and the origins of disease. *Sci Am* **279** (5):86–93.

Nesse, R.M. and Williams, G.C. 1994. *Why We Get Sick: The New Science of Darwinian Medicine*. New York: Vintage Books.

O'Keefe, J.H. Jr., Cordain, L., Jones, P.G., and Abuissa, H. 2006. Coronary artery disease prognosis and C-reactive protein levels improve in proportion to percent lowering of low-density lipoprotein. *Am J Cardiol* **98** (1): 135–139.

Ocke, M.C., Hulshof, K.F.A.M., and Breedveld, B.C. 2004. *Zo eten jongvolwassenen in Nederland. Resultaten van de Voedselconsumptiepeiling 2003.*

Ozturk, H., Durga, J., van de Rest, O., and Verhoef, P. 2005. The MTHFR 677 CT genotype modifies the relation of folate intake and status with plasma H. cysteine in middle-aged and elderly people. *Ned Tijdschr Klin Chem Labgeneesk* **30**:208–217.

Pasinetti, G.M. and Eberstein, J.A. 2008. Metabolic syndrome and the role of dietary lifestyles in Alzheimer's disease. *J Neurochem* **106** (4):1503–1514.

Perry, G.H., Dominy, N.J., Claw, K.G., Lee, A.S., Fiegler, H., Redon, R., Werner, J., Villanea, F.A., Mountain, J.L., Misra, R., Carter, N.P., Lee, C., and Stone, A.C. 2007. Diet and the evolution of human amylase gene copy number variation. *Nat Genet* **39** (10): 1256–1260.

Prentice, A.M., Rayco-Solon, P., and Moore, S.E. 2005. Insights from the developing world: Thrifty genotypes and thrifty phenotypes. *Proc Nutr Soc* **64** (2):153–161.

Psota, T.L., Gebauer, S.K., and Kris-Etherton, P. 2006. Dietary omega-3 fatty acid intake and cardiovascular risk. *Am J Cardiol* **98** (4A):3i–18i.

Ramachandran, S., Deshpande, O., Roseman, C.C., Rosenberg, N.A., Feldman, M.W., and Cavalli-Sforza, L.L. 2005. Support from the relationship of genetic and geographic distance in human populations for a serial founder effect originating in Africa. *Proc Natl Acad Sci U S A* **102** (44):15942–15947.

Rao, J.S., Ertley, R.N., DeMar, J.C. Jr., Rapoport, S.I., Bazinet, R.P., and Lee, H.J. 2007. Dietary n-3 PUFA deprivation alters expression of enzymes of the arachidonic and docosahexaenoic acid cascades in rat frontal cortex. *Mol Psychiatry* **12** (2):151–157.

Reaven, G.M. 2005. The insulin resistance syndrome: Definition and dietary approaches to treatment. *Annu Rev Nutr* **25**:391–406.

Rigby, J.F. 1995. *A fossil Cocus nucifera L fruit from the latest Pliocene of Queensland, Australia.*

Rosenberg, N.A., Pritchard, J.K., Weber, J.L., Cann, H.M., Kidd, K.K., Zhivotovsky, L.A., and Feldman, M.W. 2002. Genetic structure of human populations. *Science* **298** (5602):2381–2385.

Ross, B.M., Seguin, J., and Sieswerda, L.E. 2007. Omega-3 fatty acids as treatments for mental illness: Which disorder and which f'atty acid? *Lipids Health Dis* **6**:21.

Roubenoff, R. 2007. Physical activity, inflammation, and muscle loss. *Nutr Rev* **65** (12 Pt 2):S208–S212.

Ruan, C., Liu, X., Man, H., Ma, X., Lu, G., Duan, G., DeFrancesco, C.A., and Connor, W.E. 1995. Milk composition in women from five different regions of China: The great diversity of milk fatty acids. *J Nutr* **125** (12):2993–2998.

Sackett, D.L., Strauss, S.E., Scott Richardson, W., Rosenberg, W., and Hayness, R.B. 2000. *Evidence-Based Medicine. How to Practice and Teach EBM.* 2nd ed. Edinburgh: Churchill Livingstone.

Schaeffer, L., Gohlke, H., Muller, M., Heid, I.M., Palmer, L.J., Kompauer, I., Demmelmair, H., Illig, T., Koletzko, B., and Heinrich, J. 2006. Common genetic variants of the FADS1 FADS2 gene cluster and their reconstructed haplotypes are associated with the fatty acid composition in phospholipids. *Hum Mol Genet* **15** (11): 1745–1756.

Schroeder, K.B., Schurr, T.G., Long, J.C., Rosenberg, N.A., Crawford, M.H., Tarskaia, L.A., Osipova, L.P., Zhadanov, S.I., and Smith, D.G. 2007. A private allele ubiquitous in the Americas. *Biol Lett* **3** (2):218–223.

Serhan, C.N. and Chiang, N. 2008. Endogenous pro-resolving and anti-inflammatory lipid mediators: A new pharmacologic genus. *Br J Pharmacol* **153** (Suppl. 1): S200–S215.

Serhan, C.N., Chiang, N., and Van Dyke, T.E. 2008. Resolving inflammation: Dual anti-inflammatory and pro-resolution lipid mediators. *Nat Rev Immunol* **8** (5): 349–361.

Sesso, H.D., Buring, J.E., Christen, W.G., Kurth, T., Belanger, C., MacFadyen, J., Bubes, V., Manson, J.E., Glynn, R.J., and Gaziano, J.M. 2008. Vitamins E and C in the prevention of cardiovascular disease in men: The Physicians' Health Study II randomized controlled trial 2. *JAMA* **300** (18):2123–2133.

Sijbrands, E.J., Westendorp, R.G., Defesche, J.C., de Meier, P.H., Smelt, A.H., and Kastelein, J.J. 2001. Mortality over two centuries in large pedigree with familial hypercholesterolaemia: Family tree mortality study. *BMJ* **322** (7293):1019–1023.

Simmer, K., Patole, S.K., and Rao, S.C. 2008a. Long-chain polyunsaturated fatty acid supplementation in infants born at term. *Cochrane Database Syst Rev* **1**:CD000376.

Simmer, K., Schulzke, S.M., and Patole, S. 2008b. Long-chain polyunsaturated fatty acid supplementation in preterm infants. *Cochrane Database Syst Rev* **1**: CD000375.

Simopoulos, A.P. 2008. The importance of the omega-6/omega-3 fatty acid ratio in cardiovascular disease and other chronic diseases. *Exp Biol Med (Maywood)* **233** (6): 674–688.

Sinclair, A.J., Begg, D., Mathai, M., and Weisinger, R.S. 2007. Omega 3 fatty acids and the brain: Review of studies in depression. *Asia Pac J Clin Nutr* **16** (Suppl. 1): 391–397.

Smit, E.N., Martini, I.A., Kemperman, R.F., Schaafsma, A., Muskiet, F.A., and Boersma, E.R. 2003. Fatty acids in formulae for term infants: Compliance of present recommendations with the actual human milk fatty acid composition of geographically different populations. *Acta Paediatr* **92** (7):790–796.

Smit, E.N., Oelen, E.A., Seerat, E., Muskiet, F.A., and Boersma, E.R. 2000. Breast milk docosahexaenoic acid (DHA) correlates with DHA status of malnourished infants. *Arch Dis Child* **82** (6):493–494.

Smithers, L.G., Gibson, R.A., McPhee, A., and Makrides, M. 2008. Effect of long-chain polyunsaturated fatty acid supplementation of preterm infants on disease risk and neurodevelopment: A systematic review of randomized controlled trials. *Am J Clin Nutr* **87** (4): 912–920.

Stringer, C. 2000. Palaeoanthropology. Coasting out of Africa. *Nature* **405** (6782): pp. 24–25, 27.

Stringer, C. 2003. Human evolution: Out of Ethiopia. *Nature* **423** (6941):692–693, 695.

Sun, C.Q., O'Connor, C.J., and Roberton, A.M. 2002. The antimicrobial properties of milkfat after partial hydrolysis by calf pregastric lipase. *Chem Biol Interact* **140** (no. 2):185–198.

The Wellcome Trust Case Control Consortium. 2007. Genome-wide association study of 14,000 cases of seven common diseases and 3,000 shared controls. *Nature* **447** (7145):661–678.

Van den Berg, S.W., Dollé, M.E.T., and Boer, J.M.A., 2008. *Genetic Contribution to Obesity: A Literature Review. RIVM Report 350020005/2007.*

van Goor, S.A., Dijck-Brouwer, D.A., Hadders-Algra, M., Doornbos, B., Erwich, J.J., Schaafsma, A., and Muskiet, F.A. 2009. Human milk arachidonic acid and docosahexaenoic acid contents increase following supplementation during pregnancy and lactation. *Prostaglandins Leukot Essent Fatty Acids* **80** (1):65–69.

van Goor, S.A., Smit, E.N., Schaafsma, A., Dijck-Brouwer, D.A., and Muskiet, F.A. 2008. Milk of women with lifetime consumption of the recommended daily intake of fish fatty acids should constitute the basis for the DHA contents of infant formula. *J Perinat Med* **36** (6):548–549.

van Vliet, J., Oates, N.A., and Whitelaw, E. 2007. Epigenetic mechanisms in the context of complex diseases. *Cell Mol Life Sci* **64** (12):1531–1538.

Venter, J.C., di Porzio, U., Robinson, D.A., Shreeve, S.M., Lai, J., Kerlavage, A.R., Fracek, S.P. Jr.,

LENTES, K.U., AND FRASER, C.M. 1988. Evolution of neurotransmitter receptor systems. *Prog Neurobiol* **30** (2–3):105–169.

VERTUANI, S., ANGUSTI, A., AND MANFREDINI, S. 2004. The antioxidants and pro-antioxidants network: An overview. *Curr Pharm Des* **10** (14):1677–1694.

VIETH, R. 2006. "What is the optimal vitamin D status for health? *Prog Biophys Mol Biol* **92** (1):26–32.

VIETH, R., BISCHOFF-FERRARI, H., BOUCHER, B.J., WSON-HUGHES, B., GARLAND, C.F., HEANEY, R.P., HOLICK, M.F., HOLLIS, B.W., LAMBERG-ALLARDT, C., MCGRATH, J.J., NORMAN, A.W., SCRAGG, R., WHITING, S.J., WILLETT, W.C., AND ZITTERMANN, A. 2007. The urgent need to recommend an intake of vitamin D that is effective. *Am J Clin Nutr* **85** (3):649–650.

VOLEK, J.S., FERNANDEZ, M.L., FEINMAN, R.D., AND PHINNEY, S.D. 2008a. Dietary carbohydrate restriction induces a unique metabolic state positively affecting atherogenic dyslipidemia, fatty acid partitioning, and metabolic syndrome. *Prog Lipid Res* **47** (5):307–318.

VOLEK, J.S., PHINNEY, S.D., FORSYTHE, C.E., QUANN, E.E., WOOD, R.J., PUGLISI, M.J., KRAEMER, W.J., BIBUS, D.M., FERNANDEZ, M.L., AND FEINMAN, R.D. 2008b. Carbohydrate restriction has a more favorable impact on the metabolic syndrome than a low fat diet. *Lipids* **44** (4):297–309.

WANG, S., LEWIS, C.M., JAKOBSSON, M., RAMACHANDRAN, S., RAY, N., BEDOYA, G., ROJAS, W., PARRA, M.V., MOLINA, J.A., GALLO, C., MAZZOTTI, G., POLETTI, G., HILL, K., HURTADO, A.M., LABUDA, D., KLITZ, W., BARRANTES, R., BORTOLINI, M.C., SALZANO, F.M., PETZL-ERLER, M.L., TSUNETO, L.T., LLOP, E., ROTHHAMMER, F., EXCOFFIER, L., FELDMAN, M.W., ROSENBERG, N.A., AND RUIZ-LINARES, A. 2007. Genetic variation and population structure in native Americans. *PLoS Genet* **3** (11):e185.

WATERLAND, R.A. AND JIRTLE, R.L. 2004. Early nutrition, epigenetic changes at transposons and imprinted genes, and enhanced susceptibility to adult chronic diseases. *Nutrition* **20** (1):63–68.

WATERLAND, R.A. AND MICHELS, K.B. 2007. Epigenetic epidemiology of the developmental origins hypothesis. *Annu Rev Nutr* **27**:363–388.

WCR/AICR. 2007. *World Cancer Research Fund/American Institute for Cancer Research Food, Nutrition, Physical Activity and the Prevention of Cancer, a Global Perspective Washington D.C. AICR, 2007.*

WEBER, W.W. 1999. Populations and genetic polymorphisms. *Mol Diagn* **4** (4):299–307.

WEINBERG, S.L. 2004. The diet-heart hypothesis: A critique. *J Am Coll Cardiol* **43** (5):731–733.

WENDEL, M., PAUL, R., AND HELLER, A.R. 2007. Lipoproteins in inflammation and sepsis. II. Clinical aspects. *Intensive Care Med* **33** (1):25–35.

WESTMAN, E.C., FEINMAN, R.D., MAVROPOULOS, J.C., VERNON, M.C., VOLEK, J.S., WORTMAN, J.A., YANCY, W.S., AND PHINNEY, S.D. 2007. Low-carbohydrate nutrition and metabolism. *Am J Clin Nutr* **86** (2):276–284.

WHITE, T.D., ASFAW, B., DEGUSTA, D., GILBERT, H., RICHARDS, G.D., SUWA, G., AND HOWELL, F.C. 2003. Pleistocene *H. sapiens* from Middle Awash, Ethiopia. *Nature* **423** (6941):742–747.

WILLETT, W.C. 2002. Balancing life-style and genomics research for disease prevention. *Science* **296** (5568): 695–698.

WILLIAMS, G.C. AND NESSE, R.M. 1991. The dawn of Darwinian medicine. *Q Rev Biol* **66** (1):1–22.

YATES, A.A., SCHLICKER, S.A., AND SUITOR, C.W. 1998. Dietary Reference Intakes: The new basis for recommendations for calcium and related nutrients, B vitamins, and choline. *J Am Diet Assoc* **98** (6):699–706.

YOKOYAMA, M., ORIGASA, H., MATSUZAKI, M., MATSUZAWA, Y., SAITO, Y., ISHIKAWA, Y., OIKAWA, S., SASAKI, J., HISHIDA, H., ITAKURA, H., KITA, T., KITABATAKE, A., NAKAYA, N., SAKATA, T., SHIMADA, K., AND SHIRATO, K. 2007. Effects of eicosapentaenoic acid on major coronary events in hypercholesterolaemic patients (JELIS): A randomised open-label, blinded endpoint analysis. *Lancet* **369** (9567):1090–1098.

YOUNGSON, N.A. AND WHITELAW, E. 2008. Transgenerational epigenetic effects. *Annu Rev Genomics Hum Genet* **9**:233–257.

ZHIVOTOVSKY, L.A., ROSENBERG, N.A., AND FELDMAN, M.W. 2003. Features of evolution and expansion of modern humans, inferred from genomewide microsatellite markers. *Am J Hum Genet* **72** (5):1171–1186.

APPENDIX

Abbreviations

AA, arachidonic acid

ALA, alpha-linolenic acid

bya, billion years ago

DHA, docosahexaenoic acid

en%, energy %

EPA, eicosapentaenoic acid

FADS1, fatty acid desaturase 1 (delta-5 desaturase)

FADS2, fatty acid desaturase 2 (delta-6 desaturase)

HDL, high density lipoprotein

LA, linoleic acid

Long-chain PUFA, long-chain polyunsaturated fatty acids

LDL, low density lipoprotein

MTHFR, methylenetetrahydrofolate reductase

mya, million years ago

PPAR, peroxisomal proliferators activated receptor

PUFA, polyunsaturated fatty acid

RXR, retinoic acid receptor

RXR, retinoid X receptor

UVB, ultraviolet-B

VDR, vitamin D receptor

CHAPTER 6

THYROID HORMONE, IODINE AND HUMAN BRAIN EVOLUTION

Sebastiano Venturi and Michel E. Bégin

INTRODUCTION

Evolution of the hominin lineage is marked by progressive brain expansion and complexity concomitant with coordinated changes in other morphological and behavioral traits that characterize speciation events. In addition to gene variation, changes in climate, habitat, and diet are well-recognized environmental stimuli for evolutionary change. Iodine is an environmental stimulus to which living organisms react, a point particularly evident in amphibian metamorphosis and potentially also in hominin evolution. *In toto*, selection pressures effecting evolutionary change involve biological mechanisms permitting adaptation and evolution under changing environmental conditions. A common biological control mechanism could potentially coordinate a suite of physiological, morphological, and behavioral changes as important as brain evolution. We contend here that such a mechanism was hormonal and that thyroid hormone and iodine were pivotal components of such a mechanism.

The principal fossil sites of hominins correlated in space and time with volcanic and fissural local or nearby iodine sources (Borensztejn, 2005). In vertebrates, iodine is incorporated into thyroid hormone in the thyroid gland. Crockford (2003, 2008) provided solid evidence that changing thyroid function, specifically rhythms of thyroid hormone secretion, is crucial for speciation events taking place over decades. The same thyroid hormone mechanism can be applied to the process of humanizing australopithecines. Here, we postulate a link between thyroid function, iodine, and evolutionary changes as they apply to the evolution of hominins and, more specifically, the large brain of *Homo sapiens*. We emphasize changes in habitat and the connection between enhanced dietary availability of iodine, selenium, and polyunsaturated fatty acids as brain-selective nutrients necessary for thyroid function and hominin brain expansion.

THYROID HORMONE METABOLISM AND FUNCTION

Thyroid hormone is formed from iodine atoms attaching to two *tyrosines* and has the same structure in all organisms. In all vertebrates, thyroid hormone is manufactured and stored in the thyroid gland. Thyroid hormone takes two forms that have slightly different actions and effects depending on the number or placement of the attached iodine atoms. Unless otherwise stated, thyroid hormone will refer to its two main forms: *thyroxine* (with four atoms of iodine, also called T4), and *triiodothyronine* (with three atoms of iodine, also

Human Brain Evolution, Edited by Stephen C. Cunnane and Kathlyn M. Stewart

called T3). Of particular relevance to evolutionary processes, if thyroid hormone is consumed in food, it can be absorbed intact through the digestive tract.

During their adaptation to terrestrial life, the primitive marine chordates started to use T4 in order to transport T3 and iodide into the cells. In vertebrates, T3 is the most active thyroid hormone form in metamorphosis and in thermogenesis, permitting better adaptation to terrestrial environments (freshwaters, atmosphere, gravity, temperature, and diet). Iodine and thyroid hormone with its nuclear thyroid hormone receptors are the essential factors in amphibian metamorphosis, transforming the aquatic and vegetarian tadpole into a more complex terrestrial and carnivorous frog. This new hormonal action of T3 was made possible by the evolution of nuclear thyroid hormone receptors.

Thyroid hormone, and therefore iodine, has unique biological attributes that are highly relevant from an evolutionary perspective. Thyroid hormone influences many biological functions in a time- and dose-dependent manner (Hadley, 2000; Hulbert, 2000; Crockford, 2003, 2008). It controls brain and body growth, metamorphosis of body forms, all aspects of reproduction, and all the steps involved in basic metabolism. It regulates many key biochemical reactions, especially protein synthesis, enzymatic activity, and hormonal activity. In addition, thyroid hormone synchronizes the body's response to stress.

Iodine and thyroid hormone have direct effects on brain expansion. For example, Roth (1946) demonstrated brain hypertrophy of the frog tadpole by iodine injection. With a frog tadpole, if one grafts an additional thyroid gland to a frog tadpole, its brain hypertrophies so much so that the cranium can sometimes burst (Borensztejn, 2005). On the other hand, the brain atrophies in tadpoles in which the thyroid gland is removed. Thyroid hormone, specifically T3, is crucial to communication within the brain because it is required for the proper functioning of brain synapses and for the control of brain-specific genes (Jones et al., 2005). Cooperation between thyroid hormone, selenium, and the ω3 polyunsaturated fatty acid, *docosahexaenoic acid (DHA)*, is also necessary for brain development and function (Cunnane, 2005; see also Chapters 3 and 4 in this book). More precisely, DHA is a major brain cell membrane component and is needed for the production of *transthyretin*, which is the protein carrier transporting thyroid hormone to the brain (Episkopou et al., 1993; Horrobin, 1997, 2001 Horrobin and Bennett, 1999; Kitajka et al., 2002). Selenium is required for the conversion of T4 in the brain into T3 by the *deiodinase D2 enzyme* whose activity depends on the presence of selenium. This relationship between thyroid hormone, DHA, and selenium means that sufficient amounts of DHA, selenium and thyroid hormone, specifically T3, must be present for optimal brain function.

Via nuclear thyroid hormone receptors, thyroid hormone also coordinates the expression of the genes that control development and reproductive functions in animals. Thyroid hormone controls cell division and differentiation, thereby orchestrating the timing and duration of development. Thyroid hormone required by a human embryo for body and brain growth comes initially from its mother directly through the placenta. Because a human embryo needs thyroid hormone in order to grow, its provision by the mother controls the growth of one generation to the next in a way that is independent of the genetic makeup of the embryo (Crockford, 2003, 2008). As a consequence, thyroid hormone connects individuals to the environment and to the generations that came before and after in a way that is partially independent of specific gene action. In effect, thyroid function appears to be central to translating environmental factors into physiological and developmental responses that lead to effective genetic variations that, in turn, mediate speciation events.

Thyroid hormone appears responsible for coordinating other hormonal responses necessary for an individual to adapt and to evolve to changing environmental conditions. Through hormonal interaction and interdependence, thyroid hormone is the only factor known to link the morphological, reproductive, and behavioral traits that change in coordinated fashion over evolutionary time. Many effects of thyroid hormone involve the action of genes, which means that changing hormone levels can have the same effect as a mutation to the genes themselves (Crockford, 2003, 2008).

Cyclicity and pulsatile secretion are well-established characteristics of thyroid hormone production, with rhythms and patterns that are species-specific during early development (Eales, 1997; Manzon and Youson, 1997; Crockford, 2008). Fluctuations of thyroid hormone concentration in blood and tissues that result from rhythmic secretion appear to be critical to brain growth and development. These thyroid rhythms and the modular effect of thyroid hormone on other hormones produced by the pineal gland, hypothalamus, pituitary, adrenal, and gonads, give the thyroid a crucial pacemaker role exerted by no other organ. Given the time- and dose-dependent relationship that exists between thyroid hormone and other hormones, as a pacemaker hormone, thyroid hormone coordinates the adaptive response of the body to short- and long-term environmental changes, a process that may explain evolutionary change.

Thyroid hormone secretion is controlled by complex interactions between the cells of the *suprachiasmatic nucleus* of the anterior hypothalamus and the neurohormone *melatonin*, produced by the pineal gland. The suprachiasmatic nucleus is responsible for controlling endogenous circadian rhythms. Its neurohormonal activities regulate many different body functions over a 24-h period. Light, temperature, and emotional signals are transmitted to the pineal gland, which responds with release of melatonin. This pathway can be increased or overridden by electrical and/or hormonal output from the suprachiasmatic nucleus (Hadley, 2000; Wright, 2002). Thus, hormonal stimulation from the suprachiasmatic nucleus or the pineal gland stimulates pulsatile secretion of *thyrotropin-releasing hormone* from the hypothalamus, which stimulates release of thyroid stimulating hormone from the pituitary gland. This stimulates pulsatile release of thyroid hormone from the thyroid gland. In addition, two direct neural mechanisms have been demonstrated: (1) the direct connection between the suprachiasmatic nucleus and the retina, such that the suprachiasmatic nucleus and its neurohormones can be governed by melatonin rhythms released by the pineal gland or stimulated directly by retinal nerves (Scheer et al., 2001; Reppert and Weaver, 2002); and (2) the direct connection between the suprachiasmatic nucleus and the thyroid gland, which can stimulate or dampen thyroid function independently of changing thyroid stimulating hormone concentrations (Kalsbeek et al., 2000; Young et al., 2005). Since electrical signals from the retina and suprachiasmatic nucleus are produced intermittently, thyroid hormone is also secreted in a rhythmic manner as a result of direct nerve stimulation.

Rhythms of thyroid hormone secretion have a genetic basis, but the precise mechanism is not well understood. Many genes governing the rhythmic pattern of thyroid hormone secretion are located in brain cells and include the clock and rhythm genes, those controlling the *deiodinase* enzymes (necessary for the conversion of T4 into its derivatives), and the gene-modulating factors (corepressors or coactivators, *retinoid X receptor*, thyroid hormone receptor) required for normal gene expression and regulation. The genetic constitution and physical relationship of these cells to one another are jointly responsible for the precise profile of individually unique and species-specific thyroid hormone rhythms. A shift in thyroid hormone rhythms can modify gene expression, that is, the amount of a specific protein produced by a specific gene, as well as proteins of many other genes at the same time, including those involved in fetal development, postnatal growth, and behavior.

FETAL DEVELOPMENT

Thyroid hormone is crucial to fetal development. Thyroid hormone supplied by the mother is required for embryonic growth at all stages, particularly for brain tissue (Yen, 2001; Jones et al., 2005). For example, the migration and maturation of emerging brain cells (including *epidermal* and *choroidal pigment cells*, neurons and glia of the peripheral nervous system, neuroendocrine and inner ear sensory cells) are controlled by thyroid hormone in a time- and dose-dependent manner (Barres et al., 1994; Cowling et al., 1994; Crockford, 2008). T3 influences the transcription of a wide variety of genes, including nerve and epidermal growth factors and a large number of crucial brain function proteins (Oppenheimer and Schwartz, 1997; Köhrle, 2000; Anderson et al., 2003; Jones et al., 2005). In the rat, both T4 and T3 are essential for *oligodendrocyte* differentiation, axonal myelination, dendritic and axonal growth, neurotransmitter regulation, and synaptogenesis in the central nervous system (Anderson et al., 2000; Chan and Kilby, 2000; Dubois-Dalcq and Murray, 2000; Park et al., 2001; Smallridge and Ladenson, 2001; Anderson et al., 2003; Lavado-Autric et al., 2003; Garcia-Segura and McCarthy, 2004; Jones et al., 2005).

Virtually all target genes, cells, and tissues that require thyroid hormone during the embryonic and postnatal growth periods respond to it in a dose- and time-dependent manner (Lavado-Autric et al., 2003; Zoeller, 2003; Garcia-Segura and McCarthy, 2004). This suggests that precision in timing and absolute thyroid hormone circulating levels must be critical to species-specific growth and body functions, and therefore, that the rhythms of thyroid hormone secretion are probably species-specific (Crockford, 2008). Because thyroid hormone is crucial to fetal development, when it passes from the mother through the placenta in a species-specific rhythm, it results in the fetus growing at a precise species-specific rate. The precise endocrine physiology of the mother controls the early development of offspring, and continues to influence growth until they are born (Burrow, 1997; Piosik et al., 1997; Wilson and McNabb, 1997; Chan and Kilby, 2000).

ANTIOXIDANT ACTIVITY OF IODINE

Aside from its role in thyroid hormone production, iodine, as *iodide* (I-), is essential for all aerobic organisms because it has an antioxidant activity which protects against damage caused by *reactive oxygen species*. Küpper et al. (2008) showed that iodide scavenges reactive oxygen species in algae and that iodide was the first inorganic antioxidant to be described in a living system, a hypothesis first proposed by Venturi (1985) and by Venturi and Venturi (1999). A biochemical mechanism of iodides as antioxidants is proposed in Table 6.1. The primary event in the use of iodides by cells was the development of their capacity to collect iodide ion and bind it to the amino acids – *tyrosine* and *histidine* – forming *iodoproteins* and to polyunsaturated fatty acids forming *iodolipids*. In vertebrate cells, iodide acts as an electron donor in the presence of *hydrogen peroxide* and the *peroxidase* enzyme. The remaining iodine atom readily iodinates tyrosine, histidine, and certain specific lipids forming iodocompounds such as iodotyrosines, iodohistidines, iodo-

TABLE 6.1 Proposed antioxidant biochemical mechanism of iodides

$2\ I^- \rightarrow I_2 + 2\ e-$ (electrons) = –0.54 Volt;
$2\ I^- +$ Peroxidase $+ H_2O_2 + 2$ Tyrosine $\rightarrow 2$ Iodo-Tyrosine $+ H_2O + 2\ e-$ (antioxidants);
$2\ e- + H_2O_2 + 2\ H+$ (of intracellular water solution) $\rightarrow 2\ H_2O$

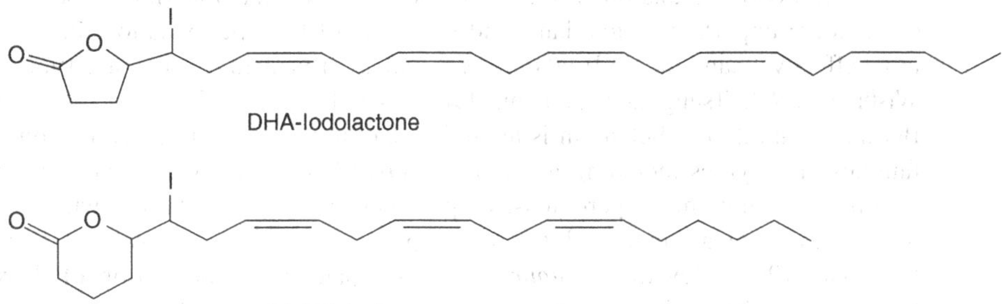

Figure 6.1 Structure formulae of DHA-iodolactone (5-iodo-4-hydroxy-7,10,13,16,19-docosapentaenoic acid, gamma-lactone) and AA-iodolactone (6-iodo-5-hydroxy-8,11,14-eicosatrienoic acid, delta-iodolactone).

thyronines (including thyroid hormone), iodoproteins, iodolipids, and iodocarbons (Gelb et al., 1962; Banerjee et al., 1985; Aceves et al., 2005). The effects of iodine may be better understood when its hormonal and nonhormonal actions are considered separately.

Iodide, as a primary antioxidant, may play a key role in the protection of vulnerable brain polyunsaturated fatty acids from lipid peroxidation. The existence of an evolutionary antioxidant biochemical cooperation between polyunsaturated fatty acids, selenium, and iodides was first suggested by Venturi and colleagues (Venturi and Venturi, 1999; Cocchi and Venturi, 2000). *Arachidonic acid* (AA) and DHA are two polyunsaturated fatty acids essential for cell membrane structure and function of the brain, and can be iodinated by two mechanisms. One mechanism consists of the addition of iodine to double bonds of DHA and arachidonic acid, making them less reactive to free oxygen radicals. In the second mechanism, in the presence of iodides and hydrogen peroxide, peroxidase catalyzes the *iodolactonization* of DHA and arachidonic acid. Arachidonic acid can then be converted to *6-iodo-5-hydroxy-eicosatrienoic acid, delta-iodolactone*, as well as various *omega-lactones*. Transformation of DHA into *5-iodo-4-hydroxy-docosapentaenoic acid, gamma-lactone*, has also been reported (Fig. 6.1). The requirements for DHA and arachidonic acid iodolactone formation are (1) the enzyme – *iodo-peroxidase*, (2) an elevated concentration of iodide, and (3) hydrogen peroxide. These conditions can be met in various tissues, including the thyroid gland and extrathyroidal tissues including *choriod plexus* in the brain.

Multiple iodolipid classes comprising iodolactones and *iodoaldehydes* (*α-iodohexadecanal*) have been identified that have structural and metabolic functions in plant, animal, and human cells (Cocchi and Venturi, 2000; Dembitsky and Tolstikov, 2003). For example, arachidonic acid iodolactones specifically inhibit signal transduction pathways induced by local growth factors such as *epidermal growth factor* and *basic fibroblast growth factor* (Chazenbalk et al., 1988; Tramontano et al., 1989). Delta-iodolactones, at physiological concentrations, have antiproliferative effects (Banerjee et al., 1985; Pisarev et al., 1988; Dugrillon, 1996; Cann et al., 2000; Cocchi and Venturi, 2000; Venturi et al., 2000a,b; Aceves et al., 2005) and could be intermediates in iodine-induced autoregulation of cell proliferation, especially of the thyroid gland (Pisarev et al., 1988, 1992). This inhibition could suppress goiter formation and induce its shrinking (Pisarev et al., 1988, 1992). Interestingly, *2-iodohexadecanal* biosynthesis involves the addition of iodine to thyroid and brain *plasmalogens* (Pereira et al., 1990). It was also suggested that iodolipids could play a role in the transport of iodide and in T4 formation and secretion (Pereira et al., 1990).

T4, *reverse-T3* and other iodothyronines (T2, T1), may function as iodine transporters, and are important antioxidants and inhibitors of lipid peroxidation. In fact, they are more effective than vitamin E, *glutathione*, and ascorbic acid (Cash et al., 1966; Ware and Wishner, 1968; Tseng and Latham, 1984; Winkler et al., 2000; Oziol et al., 2001; Berking et al., 2005). Selenium is an antioxidant mineral and an essential component of families of enzymes including *glutathione peroxidases*, deioidinases, and *glutathione-S-transferases*. Glutathione peroxidases repair damaged cell membranes while glutathione S-transferases repair damaged DNA and prevents mutations (Low and Berry, 1996; Stadtman, 1996). Several *selenoproteins* participate in the protection of thyroid cells from damage by hydrogen peroxide produced during thyroid hormone biosynthesis. Extrathyroidal or peripheral T4 metabolism is mediated by three deiodinases (*type I deiodinase [D1], type II deiodinase [D2]* and *type III deiodinase [D3]*). The distribution of deiodinase enzymes varies between tissues and each has a distinct developmental profile (Hulbert, 2000; Beckett and Arthur, 2005; Bianco and Kim, 2006). D1 and D2 catalyze the conversion of T4 into T3. D2 is very active in the brain. D3 catalyzes the inactivation of T4 into reverse -T3 and of T3 into 3,3-T2. In this way, D3 delivers peripheral cells with one or two atoms of iodide per molecule of T4, thereby allowing iodides to exert their antioxidant activity. D3 protein is also expressed by granulocytes and monocarboxylate transporter 8, a very active and specific T4 transporter, is also present at the site of inflammation (Friesema et al., 2003; Stone, 1988). Iodide efficiently scavenges reactive oxygen species in human blood cells (Küpper et al., 2008) and contributes to regulating the inflammatory response (Stone, 1988).

Benefical effects of the antioxidant activity of iodides have been reported for many chronic diseases including cancer, arteriosclerosis, cataract, cardiovascular, and articular diseases (Eskin, 1970, 1977; Elstner et al., 1985; Venturi et al., 1987; Buchberger et al., 1991; Rieger et al., 1995; Funahashi et al., 1996, 2001; Cann et al., 2000; Winkler et al., 2000; Venturi et al., 2000b; Venturi, 2001; Smyth, 2003a,b; Kessler, 2004; Muranov et al., 2004; Szybinski et al., 2004; Aceves et al., 2005; Abnet et al., 2006; Golkowski et al., 2007). Katamine et al. (1985) demonstrated that dietary iodides protect rat brain cells from lipid peroxidation. Liu (2000a,b) showed that reactive oxygen species and lipid peroxidation increase in iodine-deficient rats and children.

DIETARY SOURCES OF IODINE

Oceans and seawater are the principal reservoirs of iodine. Underground iodine is distributed into the atmosphere by volcanic activity and fissural fault emissions. Such activities were characteristic of the Pliocene and affected not only most of Europe (Western and Eastern Mediterranean, Greece, Italy, the Massif Central, Kaiserstuhl, Rhenish Schistous Massif, Czechoslovakia, etc.) but also America (Western United States, Patagonia) and Africa (East Africa Rift Valley, South Africa Great Dyke) (Borensztejn, 2005). Once in the atmosphere, iodine incorporates into rain and is thereafter carried into streams, rivers, ponds, lakes, and then to the ocean. Thus, iodine distribution is closely associated with the water cycle. Because iodine is volatile, it is constantly emitted back into the atmosphere.

The sea is rich in iodine (containing about 60 μg/L in coastal seawaters) since this is where most of the iodine accumulates after being washed away by rain or removed from the soil by glaciation. In the open ocean, the total iodine concentration is around 0.06 ppm (Elderfield and Truesdale,1980). The major iodine species in coastal seawaters are *iodate* (IO3-) and iodide, along with smaller concentrations of molecular iodine, *hypoiodous acid*, and iodinated organic compounds (Truesdale et al., 1995).

Iodine required for the production of thyroid hormone is derived from dietary sources. Algal *phytoplankton*, the basis of marine food chain, are biological accumulators of iodides, selenium, and polyunsaturated fatty acids (Venturi and Venturi, 1999; Cocchi and Venturi, 2000; Küpper et al., 2008). Brown algae (seaweed) accumulate iodine to more than 30,000 times the concentration of this element in seawater, up to levels as high as 1–3% of dry weight (Colin et al., 2003; Teas et al., 2004). Iodine concentration decreases stepwise from seawater to estuary (about 5 μg/L) and river sources (less than 0.2 μg/L in some Triassic mountain regions of northern Italy).

The distribution of iodine varies geographically: maritime regions are most likely to be iodine-rich while inland and mountainous regions are most likely to be iodine-deficient. Marine fish contain 40–100 times more iodine than foods of terrestrial origin (Bernard, 1939; Dahl and Meltzer, 2009). Saltwater fish such as herring contain iodine at about 500–800 μg/kg compared to freshwater (i.e., trout) about 20 μg/kg (Venturi and Venturi, 1999; Venturi et al., 2000a,b, 2003). On land, iodine is taken up by plants. Herbivorous animals get their iodine from plants or by drinking iodine-containing water. Carnivores acquire iodine mostly from thyroid hormone that is present in the blood and tissues (thyroid gland, liver, kidneys, brain, bone marrow) of the prey animals (including their eggs) they eat and also from iodine in drinking water. Food resources on the littoral provide the richest dietary source of iodine (fish, shellfish, etc.; see also Chapter 3) and preformed thyroid hormone (mammals, birds, reptiles, amphibians, and both primitive and advanced forms of fish).

IODINE DEFICIENCY DISORDERS

Thyroid hormone deficiency can be caused by iodine deficiency or by a genetic defect in thyroid hormone production or in thyroid function. Prolonged dietary iodine deficiency results in chronic reduction of thyroid hormone and clinical symptoms of hypothyroidism. Iodine deficiency leads to inadequate production of thyroid hormone that is indispensable for brain growth and development (Delange, 2000, 2001), but the precise mechanism by which iodine deficiency impairs brain development is not yet known. Hence, hypothyroidism produces a wide spectrum of pathological effects ranging, in increasing order of severity, from lethargy, dwarfism, deaf-mutism, to mental retardation and *cretinism*.

Deficiency of iodine, as an antioxidant, causes damage in developing embryos and in their brain tissues and results in a global loss of 10–15 IQ points at a population level. Iodine deficiency is the world's single most significant cause of preventable brain damage and mental retardation (Dunn and Delange, 2001). Maternal *hypothyroxinaemia* during early pregnancy is a key factor in the development of the brain damage in the cretin. Calvo et al. (1990) demonstrated that maternal T4 (but not T3) plays a crucial role in protecting the fetal brain from damage caused by hypothyroidism. Hence, in this case, we would emphasize the effective difference between T4 and T3 in brain cell metabolism, and in particular, the effectiveness of the fourth iodine atom of T4. In pregnant women, iodine deficiency causes abortions and stillbirths. This damage seems not to be caused by thyroid hormone deficiency, but rather by iodine deficiency per se (Wolff, 1964; Goethe et al., 1999).

Dobson (1998) proposed that Neanderthals suffered iodine-deficiency disorders (see section Phase 3: From *H. erectus* to early *H. sapiens*). Recently, Obendorf et al. (2008) hypothesized that *H. floresiensis*, a pygmy-sized, microcephalic hominin who lived from 95,000 to 13,000 years ago on the Indonesian island of Flores, were *myxoedematous endemic cretins*, who are born without a functioning thyroid. Their congenital hypothyroidism leads to severe dwarfism, mental retardation, and reduced brain size.

Nowadays, iodine-deficient humans, like endemic cretins, suffer physical, neurological, mental, immune, and reproductive diseases. In collaboration with U.S. and Pisa University researchers, 15 cases of endemic cretinism were studied in the territory of Montefeltro, in the central Apennines of Italy (Venturi, 1985; Donati et al., 1989, 1992). In the Montefeltro region, endemic cretinism was still present in the 1980s, with goiter prevalence of 55% and mean urinary iodine level of 39 µg/g of creatinine. Human cretins represented about 0.4 % of the overall population. Clinical and biochemical features of patients with myxedematous and *neurologic cretinism* were studied. All the cases of myxedematous cretinism had some neurologic disorders (hyperreflexia, increased muscle tone, disorder of gait, Babinski sign, hypoacusia, and mental impairment). Some cases of *microcephaly* were also observed (Venturi, 1985). These findings suggested that brain damage reflected a diffuse insult to the developing fetal nervous system. An important degree of immune deficiency was also observed in iodine-deficient schoolchildren in the Montefeltro region, but they had normal values of T4, T3, and thyroid-stimulating hormone (Venturi, 1985; Marani and Venturi, 1985; Marani et al., 1986).

According to current global WHO statistics, more than 3 billion people live in iodine-deficient countries. Based on the National Health and Nutrition Examination Surveys data in the United States, moderate to severe iodine deficiency is currently present in 11.7% of the U.S. population, with a clear increasing trend over the past 20 years,

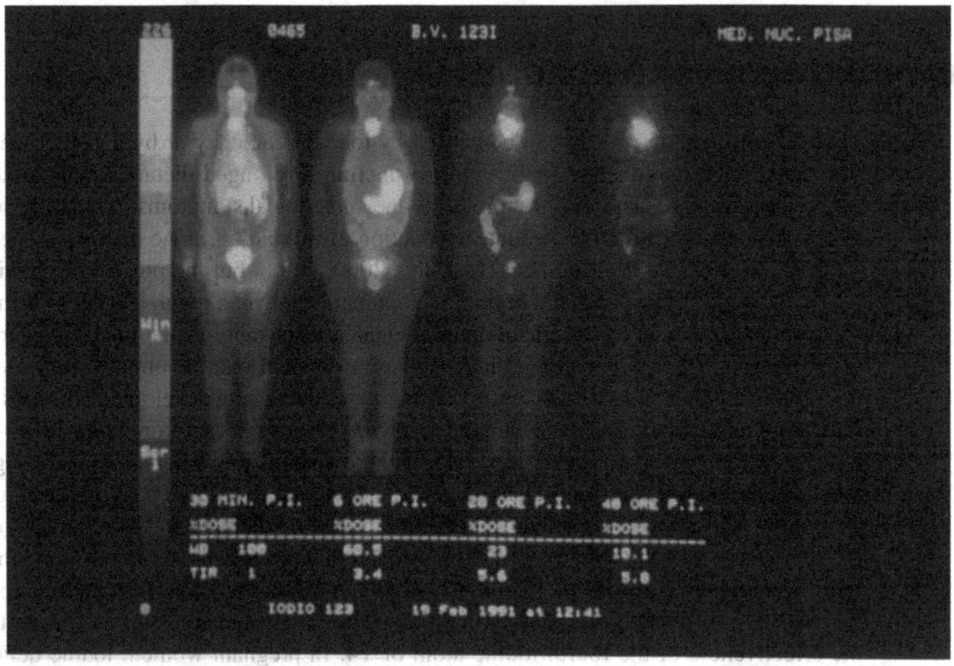

Figure 6.2 Sequence of 123-iodide total-body scintiscans of a woman after intravenous injection of 123-iodide (half-life: 13 h); (from left) respectively at 30 min, and at 6, 20, and 48 h. The highest and rapid concentration of radioiodide (in white) is evident in gastric mucosa of the stomach, salivary glands, and oral mucosa. In gastric mucosa of the stomach, 131-iodide (half-life: 8 days) persists in scintiscans for more than 72 h. In the thyroid, iodide concentration is more progressive, as in a reservoir (from 1% [after 30 min] to 5.8% [after 48 h] of the total injected dose). Here, iodide concentration by the mammary gland is not evident because this woman was not pregnant or lactating. A high excretion of radioiodide is observed in the urine. (See color insert.)

caused by reduced iodized table salt usage (Hollowell et al., 1998; Dasgupta et al., 2008; Grzesiuk, 2009). The U.S. Food and Nutrition Board (2001) recommended daily allowance of iodine ranges from 150 µg/d for adult humans to 290 µg/d for lactating mothers. However, the thyroid gland needs no more than 70 µg/d to synthesize the requisite daily amounts of T4 and T3. These higher recommended daily allowance levels of iodine seem necessary for optimal function of a number of body systems, including choroid plexus and cerebrospinal fluid, eye, gastric mucosa, thymus, and salivary glands (Miller, 2006; Venturi and Venturi, 2007).

In humans, the total body content is about 20–50 mg of which about 50–70% iodine is nonhormonal and is concentrated in extrathyroidal tissues. For all its importance in thyroid function, more iodine is actually located *outside* the thyroid gland than within it (Venturi et al., 1993; Venturi and Venturi, 1999, 2007). Active iodide transport is facilitated by three transporters: *sodium-iodide symporter*, *pendrin*, and the recently described *apical iodide transporter*. All three transporters are expressed in the thyroid gland and in extrathyroidal tissues (Rodriguez et al., 2002; Burbridge et al., 2005). Iodine is present, in different concentrations, in every organ and tissue of the human body, not just the thyroid gland (Fig. 6.2). Organs transiently concentrating iodide includes white blood cells, salivary and lacrimal glands, choroid plexus, eye, renal cortex, pancreas, liver, gastric, small and large intestinal mucosa, nasopharynx, skin, adrenal cortex, mammary gland, placenta, uterus, and ovary (Brown-Grant, 1961). Recently, Thrall et al. (2009) reported that fetal stomach showed the highest accumulation of radioiodides compared to thyroid tissues.

HUMAN BRAIN EVOLUTION

Human brain evolution can be viewed as a sequence of speciation events divided into four major broad phases (Cunnane, 2005). Four broad phases of human brain evolution are proposed in Table 6.2, with the major species involved and the corresponding changes in diet, habitat and brain size. The link between them is briefly described below with emphasis on the potentially important role of thyroid function.

Phase 1: Australopithecines

Hominins evolved from the australopithecines, which were bipedal ancestors appearring in the fossil record at ca. 4.4 million years ago in East Africa, Tchad and South Africa (Ehrlich, 2000). According to available specimens, the australopithecine brain size evolved on average from 450 g to 700 g in *H. habilis*. This increase in brain size corresponds to 40% increase in the *relative* brain size ("encephalization quotient," [EQ]) because brain growth exceeded body growth. EQ expresses the proportion of brain weight relative to body weight, which gives a more accurate measure of brain expansion than brain size alone (Cunnane, 2005). Australopithecine habitat is thought to have changed from closed forests or forests mixed with bush (Reed, 1997; deMenocal, 2004) to woodland habitats. As evidenced by the wear patterns of their teeth, their predominant diet was probably mixed with fruits and nuts.

New foods associated with a changing habitat (i.e., insects and grubs, bird eggs and fledgling birds, small mammals, reptiles, and amphibians) as determined by carbon isotope analysis (Crockford, 2008), necessarily involved the consumption of significant amounts of thyroid hormone and higher iodine intakes. Continued consumption of large quantities of foods containing thyroid hormone could have had a major impact on their populations, especially on females of reproductive age. High enough dietary doses of thyroid hormone

TABLE 6.2 The four broad phases of human brain evolution with corresponding habitats and diets

Approximate duration[a]	Hominin species involved[a]	Habitat	Diet	Brain size[a] Weight[b] (grams)	Relative EQ[b]
Phase 1: 1.5 million years (from 4.4 to 2.5 million years ago)	Australopithecines	From closed woodland forests or forests mixed with bush to woodland with small animals; shores	Fruits, nuts, small animals, and probably crustaceans	450	42
Phase 2: 1 million years (from 2.5 to 1.7 million years ago)	*H. habilis*	More open habitats with abundant bovid species; shores	Vegetation, small animals, scavenged carcasses of large species; shellfish and fish	700	58
Phase 3: 400,000 years (from 1.5 to 0.4 million years ago)	*H. erectus* to early *H. sapiens*	Savanna-type; artic steppe; tundra; mountains; shores	Raw meat from medium- and large-sized terrestrial mammals; shellfish and fish	1100	76
Phase 4: 30,000 years (from 100,000 years ago to present)	Early to present *H. sapiens*	Shores	Raw and/or cooked marine foods; meat from terrestrial animals	1360	100[c]

[a] From Cunnane (2005).

[b] Averaged values from Cunnane (2005) in which values for living *H. sapiens* are set at 100 and comparisons become percentages of the human value.

[c] For early *H. sapiens* (100,000–10,000 years ago): 102.

could even have resulted in birth defects (Weetman, 1997; Arem, 1999; Porterfield, 2000; Crockford, 2002, 2003; Cudd et al., 2002; Rovet, 2004). Thus, as a consequence of dietary changes associated with colonization of new habitats, it is possible that disruptively high levels of thyroid hormone intake by certain australopithecine clades may have altered thyroid hormone rhythms in the offspring. The thyroid hormone pattern produced by the australopithecine mother could have induced slight differences in brain development during embryonic growth of the fetus. As a consequence, slight individual variations in thyroid hormone rhythms were bound to occur due to mutations in the genes of the biological clock and rhythm cells, or to gradual changes in the relationship between these cells. Plausibly, higher exposure to thyroid hormone and to iodine, and changes in thyroid hormone rhythms could have induced changes in embryonic brain cellular architecture sufficient to stimulate brain expansion concomitant with a prolongation of fetal development, thereby extending the duration of neurogenesis (Smith, 1992; Finlay and Darlington, 1995; McKinney, 1998).

Australopithecines with a physiological tolerance to excessive dietary thyroid hormone and with offspring with skeletal changes toward primitive bipedalism seem to have formed a new colonizing population. Indeed, colonizing a new habitat may offer both many advantages and new stresses that have to be dealt with by each colonizing individual.

The new habitat would have preferentially attracted physiologically stress-tolerant individuals over less stress-tolerant ones. As a consequence, the colonizing group consisted of stress-tolerant individuals with similar particular thyroid hormone rhythms, which become established within their descendants. Over many generations and with continuous exposure to higher dietary iodine and thyroid hormone from small animals and amphibians, the descendants would have undergone changes in morphology and behavioral traits leading the way for the emergence of *H. habilis*.

Phase 2: *H. habilis*

The second phase of human brain evolution is represented by *H. habilis*, currently still recognized as the first hominin and the first tool user. Diverging from a late australopithecine, possibly *Australopithecus afarensis* or one of a closely related species, *H. habilis* appeared about 2.3 million years ago and lived for roughly 600,000 years in somewhat drier and more exposed habitats where bovid species were plentiful (Reed, 1997). Not unlike australopithecines, their diet probably consisted principally of vegetation, some small animals, and scavenged carcasses of large species, the latter providing a substantial amount of bone marrow and brain (Fleagle, 1999; Crockford, 2003, 2008). Such a diet would have been proportionately richer in *brain-selective nutrients* including the polyunsaturated fatty acids – arachidonic acid and DHA, both of which were necessary for brain development and function (see Chapters 3 and 4 in this book; Horrobin, 2001). These conditions would have favored higher thyroid hormone intake and therefore the emergence of more stress-tolerant individuals with a somewhat larger brain who colonized new open habitats.

H. habilis and its close relatives were the first human ancestors to show traces of asymmetry in the brain's left hemisphere that is associated with development of areas for speech and language (Ehrlich, 2000). Compared with the australopithecines, change in *H. habilis*' relative brain size was minimal because body and brain growth occurred in parallel (Cunnane, 2005). With their slightly larger brain, the effects of their diet on brain development in *H. habilis* may have been less dramatic than those on early australopithecine, but still advantageous in such a way as to support more manual dexterity for tool manufacture and use, as well as better visuospatial skills, foresight capacities, and decision-making abilities. Indeed, DHA is a major component of the neocortex (O'Brian and Sampson, 1965; Svennerholm, 1968) and is required for the production of *transthyretin* (Kitajka et al., 2002), which transports thyroid hormone to the brain. It is also possible that its combination with thyroid hormone resulted in better functioning of brain cellular membranes (Hulbert, 2000; Horrocks and Farooqui, 2004) and in greater cognitive development that would have conferred survival advantages.

Phase 3: From *H. erectus* to early *H. sapiens*

H. erectus was a hunting hominin with a larger body and brain than *H. habilis* and represents the start of the third major speciation event in the hominin lineage. During this phase, the main change in brain size corresponded to a roughly 60% increase in EQ because brain growth exceeded body growth. *H. erectus* hunted a wide range of medium- and large-sized terrestrial mammals (Ehrlich, 2000; Crockford, 2003, 2008). As primary predators of such prey, *H. erectus* would have had ready access to iodine and thyroid hormone-rich organs such as thyroid gland, liver, kidney, and brain. As a consequence, increased consumption of dietary iodine and thyroid hormone would have triggered further shifts in thyroid hormone rhythms and stimulated the continuing enlargement of the brain and changes in body proportions. Individuals with thyroid hormone rhythms that were adapted to increased consumption of dietary thyroid hormone, larger body size, more complex brain, increased manual dexterity, and better reproduction became the first hominin to move beyond Africa. *H. erectus* fossils have been found in southern (Indonesia) and western Asia (Georgia). Some paleoanthropologists have referred to the initial African form of this hominin as *Homo ergaster* and to the Asian form as *H. erectus*, but recent fossil records confirm that they probably formed a single species (Asfaw et al., 2002). In any event, this hominin was the first to live successfully in both Asian and African habitats and appears to be the direct ancestor of *H. sapiens*.

Within the same phase, the next major step in the evolution of the hominin lineage was the evolution of *H. erectus* into contemporaneous but distinct forms of *H. sapiens*, represented by *H. heidelbergensis, H. neanderthalensis* and, eventually, early *H. sapiens*. Many paleoanthropologists consider *H. heidelbergensis* to be the common ancestor of *H. neanderthalensis* in Eurasia, and of early *H. sapiens* in Africa. The emergence of the large-bodied and large-brained *H. heidelbergensis* coincided with a major global climate change that brought colder temperatures and drier Pleistocene environments leading to important changes in many habitats and animal populations around the world. For hominins, such changes probably meant less scavenging and more hunting inland as well as more fishing on the shorelines. Because of the intensified hunting and fishing activities, theses hominins could have increased their relative proportions of both exogenous thyroid hormone, iodine, and DHA compared to their predecessors. Again, individuals with the particular thyroid hormone rhythms adapted to higher levels of thyroid hormone without disruption of reproductive function would have been favored. Conversely, individuals with cold-sensitive thyroid hormone rhythms would have been eliminated from the populations (Crockford, 2003, 2008). *H. neanderthalensis* colonized Arctic steppes and tundra and, since thyroid hormone controls the body temperature, must have been cold-tolerant variants with particular thyroid hormone rhythms originating from the *H. heidelbergensis* populations. Neanderthals had brain sizes larger but also differently shaped than humans. Their diet appears to have been composed mainly of raw red meat because far fewer edible plants would have been available (Mellars, 1996; Balter et al., 2001). Plausibly, they also practiced foraging and fishing along shorelines. Thus, Neanderthals were exposed to high levels of dietary thyroid hormone and iodine.

A different perspective suggests that some Neanderthals actually suffered iodine-deficiency disorders probably caused by inland environment or by a genetic difference of their thyroid compared to the thyroid of modern *H. sapiens* (Dobson, 1998). These two scenarios, based on high intake of thyroid hormone for one, or on iodine deficiency for the other, appear incompatible. One plausible explanation for their apparent divergence may be that, as the climate got colder, Neanderthal populations probably migrated inland and/or higher up in the mountains distant from coastal habitats rich in thyroid hormone and an iodine-rich diet, such as shorelines near hunting grounds. Some of them, especially those living in iodine-deficient inland and mountainous regions, could have suffered from endemic cretinism due to iodine deficiency. On the other hand, hyperthyroidism could be another possibility since severe chronic hyperthyroidism has been shown to produce cretinoid offspring in rats (Waterman, 1958).

Four arguments suggest that *H. neanderthalensis* and *H. sapiens* were distinct species. First, Neanderthals had faster postnatal growth rates of tooth enamel than do modern humans (Ramirez Rozzi, 2002; Crockford, 2008). Second, Neanderthals had a faster craniofacial growth rate during early childhood growth periods than that of humans and chimpanzees (Minugh-Purvis, 2002; Williams et al., 2002). Third, analysis of mitochondrial DNA sequences suggests that Neanderthals were genetically distinct and reproductively isolated from humans for a considerable length of time (Lindahl, 1997; Krings et al., 1997; Fabre et al., 2009). Finally, Neanderthals may have had a turnover rate for thyroid hormone similar to that of carnivorous modern dogs and cats (about 12–13 h) compared to about 7 days for modern humans (Kaptein et al., 1994). If true, a faster turnover rate for thyroid hormone implies that Neanderthals possessed a distinctly different thyroid hormone rhythm, a criterion for distinguishing discrete species (Crockford, 2008).

Phase 4: From early to present day *H. sapiens*

The fourth and last phase toward humans concerns the great leap in creativity that took place from early *H. sapiens* to present *H. sapiens*, starting around 100,000 years ago. Brain size, while it provides a foundation from which creativity might develop, differs from creativity. Creativity probably depends more on increasing complexity in neuronal connections. During this phase, the brain and skull changed shape together along with the addition of neurological pathways were added that had not been present before (Parker and McKinney, 1999; McKinney, 2002), circuits which led to art, religion, music, warfare, and cultural diversity.

Modern *H. sapiens* (Cro-Magnon type) succeeded *H. heidelbergensis* as a successful, widely distributed species. Modern *H. sapiens* expanded their ecological niche to include less hospitable habitats, and migrated from East Africa to the Middle East, to South Africa, to Europe, to Central Asia then toward the New World (Ehrlich, 2000; Borensztejn, 2005). This geographic expansion of human distribution may have been made possible by enhanced trade, hunting skills and organization, regular use of fire for cooking, and by following coasts and river valleys. Migrating groups may have begun to take iodine or its source materials with them to the interior. They probably had the technology to store fish and other marine foods for future use, perhaps by drying, or salting. At that time, marine shells are found several hundred kilometers from their sources (Ehrlich, 2000). Initially, transport of shellfish could have occurred due to cultural preference for certain foods from the sea. Later, regular trade may have developed. If so, then a continual supply became ubiquitously available after 30,000 years ago (Ehrlich, 2000; Borensztejn, 2005). Other possible sources of iodine on land include the return of saltwater salmon to streams that had been blocked by glaciers. Dillehay et al. (2008) found from the analysis of archeological sites that the earliest Americans used marine algae (richest in iodine) and other marine resources for food and medicine. Evidence that the first Americans came by sea from Asia and spread throughout the New World by at least 14,000 years ago, supports the hypothesis that they took the coastal route rather than traveling inland. Significant quantities of iodides, thyroid hormone, and polyunsaturated fatty acids are found in marine fish, in eggs of freshwater and teleost fishes, and in egg yolk from land animals.

THYROID HORMONE, IODINE, AND HUMAN BRAIN EVOLUTION

Human dependence on dietary sources of iodine and thyroid hormone suggests that the evolutionary changes responsible for physiological differences between early and present-day *H. sapiens* may have occurred in an iodine-rich and therefore probably coastal or estuarine environment (see also Chapter 3). Anatomically modern human remains dating from 120,000 to 100,000 years ago have been found in South Africa, Israel, and Northwestern Africa (Dobson, 1998). All these early modern human sites are close to coastal iodine-rich food resources.

Shore-based habitats would have created a cognitive niche for early humans (Crawford and Marsh, 1989; Cunnane, 2005). For the most part, life on freshwater shore-based habitats could have been one of relatively low stress around food resources, with hunting as a play (Cunnane, 2005). They would have had more time to devote to planning, communication, and coordination with others in their group, and, by doing so, an enriched opportunity to form a cognitive niche. As populations grew and group sizes increased, they would have gained superior capacity for information storage and greater manipulation of the environment within and outside of groups. This increased capacity in knowledge and manipulation would have been accompanied by changes in brain organization that allowed them to become culturally more advanced and much more innovative. Both cultural and genetic changes would have been advantageous, especially the genes controlling brain development that are thyroid hormone dependent, because the selection pressures would have favored the thyroid hormone rhythm that worked best with the new living conditions.

Ongoing proximity to riverine, lacustrine, and maritime diets is crucial for optimal brain development and function (Crawford and Marsh, 1989; Cunnane et al., 1993; Ellis, 1993; Stewart, 1994, 1996; Walter et al., 2000; Broadhurst et al., 2002; Cunnane, 2005). As early as 4 million years ago, marine organisms such as crustaceans appear to have been an important food for some australopithecines (Walter et al., 2000). Based on fish fossil sites, Stewart (1994, 1996) demonstrated that catfish and perch fishing were practiced by *H. habilis* 2 million years ago. Marine food resources (shellfish, crustaceans, fish, mollusks, frogs, reptiles, aquatic birds, eggs, marine plants and seaweed, etc.) provide a rich source of brain-selective nutrients, including preformed DHA, iodine, selenium, and other trace elements (copper, zinc) and minerals, especially iron, which are in short supply on land.

Cordain et al. (2005) recently reported that the profound changes in diet that began with the introduction of agriculture and animal husbandry approximately 10,000 years ago, occurred too recently on an evolutionary time scale for the human genome to adjust (see also Chapter 5). In conjunction with this discordance between hunter-gatherer and fisher-gatherer societies on the one hand and the nutritional patterns of contemporary Western populations on the other, many of the so-called degenerative diseases of civilization have emerged. Cordain et al. (2005) suggests that micronutrient density (including iodine) was probably one change, among other dietary variations including fiber and polyunsaturated fatty acids contents, introduced during the neolithic and industrial periods which have altered simultaneously crucial nutritional characteristics of ancestral human diets. The evolutionary collision of our ancient genome with the poorer nutritional qualities of recently introduced foods may underlie many of the chronic diseases of Western civilization.

CONCLUSION

Iodine and thyroid hormone are crucial for brain development and functioning throughout human evolution. Hormonal and nonhormonal actions of iodine may provide the basis for a biological mechanism explaining evolutionary change in response to changing environmental conditions. Although much is known about how it might work, the precise biochemical mechanisms, including quantitative measurements of thyroid hormone rhythms, remain yet to be established.

Adequate iodine uptake and thyroid function are essential for normal human development and brain functioning throughout life (Loosen, 1992; Bégin et al., 2008). Indeed, about one-fifth of the total human population consumes diets that are inadequate in supporting brain development. Human communities in inland or mountainous regions of Europe, India, Southeast Asia, North and South Americas, and Africa, are at grave risk of iodine deficiency. Compared to other common food groups (fish, eggs, pulses, cereals, meat, nuts, vegetables, fruits, and milk), shellfish is best able to meet the adult daily requirement for the brain-selective minerals. Thus, regular inclusion of any quantity of shellfish in the diet would help to improve the supply of these minerals, thereby contributing to optimal human brain function. In view of the high risk of iodine deficiency and of the insufficient intake of foods that are good sources of iodine (shellfish, fish, eggs, meat, milk), governments legislate supplementation of iodine, as iodized table salt, in the human food supply. However, salt intake has been decreasing in the last decades due to the perceived risk of excess dietary salt for hypertension. Ironically, food groups consisting of cereals, vegetables, and fruits, presently promoted as the basis of a healthy diet, are very likely to be deficient in iodine.

Although vegetables and fruits have been widely promoted as being protective against many chronic and degenerative diseases, such as cancer and cardiovascular, the efficacy of vegetable antioxidants (i.e., *carotenoids*, vitamins A, C, and E) in these diseases has not been recently supported by statistical data. Furthermore, their utility in prevention of these diseases has not been recently confirmed by epidemiological data (Morris and Carson, 2003; Bjelakovic et al., 2004, 2007; Hung et al., 2004; Lin et al., 2005; Sato et al., 2005; Tsubono et al., 2005). Moreover, vegetables and fruits contain goiterogens that deplete the body of iodine. This problem can be avoided by reducing intake of iodine-depleting foods while increasing consumption of iodine-containing foods such as shellfish, fish, and/or eggs.

Presently, humans are living beyond the optimal nutrient limits for intake of brain-selective nutrients, particularly iodine and iron. Changes in dietary habits are necessary by making marine foods and/or iodinated supplements more widely available, or we put ourselves at risk of reducing human cognitive capacity over a short period of time.

REFERENCES

Abnet, C.C., Fan, J.H., Kamangar, F. et al. 2006. Self-reported goiter is associated with a significantly increased risk of gastric non cardia adenocarcinoma in a large population-based Chinese cohort. *International Journal of Cancer* **119**:1508–1510.

Aceves, C., Anguiano, B., and Delgado, G. 2005. Is Iodine a gatekeeper of the integrity of the mammary gland? *Journal of Mammary Gland Biology and Neoplasia* **10**:189–196.

Anderson, G.W., Mariash, C.N., and Oppenheimer, J.H. 2000. Molecular actions of thyroid hormone. In *Werner and Ingbar's The Thyroid, Eighth Edition*, ed. L.D. Braverman and R.D. Utiger, pp. 174–195. Philadelphia: Lippincott Williams & Wilkins.

Anderson, G.W., Schoonover, C.M., and Jones, S.A. 2003. Control of thyroid hormone action in the developing rat brain. *Thyroid* **13**:1039–1056.

Arem, R. 1999. *The Thyroid Solution*. New York: Ballantine Books.

Asfaw, B., Gilbert, W.H., Beyene, Y. et al. 2002. Remains of *Homo erectus* from Bouri, Middle Awash, Ethiopia. *Nature* **416**:317–320.

Balter, V., Person, A.S., Labourdette, N. et al. 2001. Les Néandertaliens étaient-ils essentiellement carnivores? Résultats préliminaires sur les teneurs en Sr et en Ba de la palébiocénose mammalienne de Saint-Césaire. *Comptes Rendus de l'Académie des Sciences, Series IIA. Earth and Planetary Science* **332**:59–65.

Banerjee, R.K., Bose, A.K., Chakraborty, T.K. et al. 1985. Peroxidase-catalysed iodotyrosine formation in dispersed cells of mouse extrathyroidal tissues. *Journal of Endocrinology* **106**:159–165.

Barres, B.A., Lazar, M.A., and Raff, M.C. 1994. A novel role for thyroid hormone, glucocorticoids and retinoic acid in timing oligodendrocyte development. *Development* **120**:1097–1108.

Beckett, G.J. and Arthur J.R. 2005. Selenium and endocrine systems. *Journal of Endocrinology* **184**:455–465.

Bégin, M.E., Langlois, M.F., Lorrain, D. et al. 2008. Thyroid function and cognition during aging. *Current Gerontology and Geriatrics Research* doi:10.1155/2008/474868.

Berking, S., Czech, N., Gerharz, M. et al. 2005. A newly discovered oxidant defence system and its involvement in the development of *Aurelia aurita* (Scyphozoa Cnidaria): Reactive oxygen species and elemental iodine control medusa formation. *International Journal of Development Biology* **49**:969–976.

BERNARD, L.M. 1939. *Le mirage de l'iode*. Société d'éditions géographiques, maritimes et coloniales, Paris.

BIANCO, A.C. AND KIM, B.W. 2006. Deiodinases: Implications of the local control of thyroid hormone action. *Journal of Clinical Investigation* **116**:2571–2579.

BJELAKOVIC, G., NIKOLOVA, D., GLUUD, L.L. ET AL. 2007. Mortality in randomized trials of antioxidant supplements for primary and secondary prevention: Systematic review and meta-analysis. *Journal of the American Medical Association* **297**:842–857.

BJELAKOVIC, G., NIKOLOVA, D., SIMONETTI, R.G. ET AL. 2004. Antioxidant supplements for prevention of gastrointestinal cancers: a systematic review and meta-analysis. *Lancet* **364**:1219–1228.

BORENSZTEJN, S. 2005. A probabilistic interaction model between the environmental evolution and the biological evolution. http://dinosaurs.ifrance.com/eightiodineand-homosapienssapienscurrentsettlement.htm

BROADHURST, C.L., WANG, Y., CRAWFORD, M.A. ET AL. 2002. Brain-specific lipids from marine, lacustrine, or terrestrial food resources: potential impact on early African *Homo sapiens*. *Comparative Biochemistry and Physiology (Part B)* **131**:653–673.

BROWN-GRANT, K. 1961. Extrathyroidal iodide concentrating mechanisms. *Physiological Review* **41**:189–213.

BUCHBERGER, W., WINKLER, R., MOSER, M. ET AL. 1991. Influence of iodide on cataractogenesis in emory mice. *Ophthalmic Research* **23**:303–308.

BURBRIDGE, E., NAWOOR, Z., SMITH, D.F. ET AL. 2005. Expression of iodide transporters in human placental tissue. *Endocrine Abstracts* **9**:49.

BURROW, G.N. 1997. Editorial: Mothers are important! *Endocrinology* **138**:3–4.

CALVO, R., OBREGON, M.J., RUIZ DE ONA, C. ET AL. 1990. Congenital hypothyroidism, as studied in rats: Crucial role of maternal thyroxine but not of 3,5,3′-triiodothyronine in the protection of the fetal brain. *Journal of Clinical Investigation* **86**:889–899.

CANN, S.A., VAN NETTEN, J.P., AND VAN NETTEN, C. 2000. Hypothesis: iodine, selenium and the development of breast cancer. *Cancer Causes Control* **11**:121–127.

CASH, W.D., GARDY, M., CARLSON, H.E. ET AL. 1966. Mitochondrial swelling and lipid peroxidation studies with mixture of thyroxine and micromolar concentration of certain metal ions. *Journal of Biological Chemistry* **241**:1745–1750.

CHAN, S. AND KILBY, M.D. 2000. Review: Thyroid hormone and central nervous system development. *Journal of Endocrinology* **165**:1–8.

CHAZENBALK, G.D., VALSECCHI, R.M., KRAWIEC, L. ET AL. 1988. Thyroid autoregulation: Inhibitory effects of iodinated derivatives of arachidonic acid on iodine metabolism. *Prostaglandins* **36**:163–172.

COCCHI, M. AND VENTURI, S. 2000. Iodide, antioxidant function and omega-6 and omega-3 fatty acids: A new hypothesis of a biochemical cooperation? *Progress in Nutrition* **2**:15–19.

COLIN, C., LEBLANC, C., WAGNER, E. ET AL. 2003. The brown algal kelp *Laminaria digitata* features distinct bromoperoxidase and iodoperoxidase activities. *Journal of Biological Chemistry* **278**:23545–23552.

CORDAIN, L., EATON, S.B., SEBASTIAN, A. ET AL. 2005. Origins and evolution of the Western diet: Health implications for the 21st century. *American Journal of Clinical Nutrition* **81**:341–354.

COWLING, K., ROBBINS, R.J., HAIGH, G.R. ET AL. (1994) Coat color genetics of *Peromyscus*: IV. *Variable white, a new dominant mutation in the deer mouse. Journal of Heredity* **85**:48–52.

CRAWFORD, M.A. AND MARSH, D. 1989. *The Driving Force: Food in Evolution and the Future*. London: William Heinemann.

CROCKFORD, S.J. 2002. Thyroid hormone in Neandertal evolution: A natural or a pathological role? *The Geographical Review* **92**:73–88.

CROCKFORD, S.J. 2003. Thyroid rhythm phenotypes and hominid evolution: A new paradigm implicates pulsatile hormone secretion in speciation and adaptation changes. *Comparative Biochemistry and Physiology (Part A)* **135**:105–129.

CROCKFORD, S.J. 2008. *Rhythms of life: Thyroid Hormone & the Origin of Species*. Victoria: Trafford Publishing.

CUDD, T.A., CHEN, W.-J.A., AND WEST, R.W. 2002. Fetal and maternal thyroid hormone responses to ethanol exposure during the third trimester equivalent of gestation in sheep. *Alcoholism: Clinical and Experimental Research* **26**:53–58.

CUNNANE, S.C. 2005. *Survival of the Fattest: The Key to Human Brain Evolution*. Singapore: World Scientific.

CUNNANE, S.C., HARBIGE, L.S., AND CRAWFORD, M.A. 1993. The importance of energy and nutrient supply in human brain evolution. *Nutrition and Health* **9**:219–235.

DAHL, L. AND MELTZER, H.M. 2009. Iodine content of foods and diets. In *Comprehensive Handbook of Iodine: Nutritional, Biochemical, Pathological and Therapeutic Aspects*, ed. V.R. Preedy, G.N. Burrow, AND R. Watson, pp. 345–352. Amsterdam: Academic Press.

DASGUPTA, P.K., LIU, Y., JASON, V., AND DYKE, J.V. 2008. Iodine nutrition: Iodine content of iodized salt in the United States. *Environmental Science & Technology*, **42**:1315–1323.

DELANGE, F. 2000. The role of iodine in brain development. *Proceedings of the Nutritional Society* **59**:75–79.

DELANGE, F. 2001. Iodine deficiency as a cause of brain damage. *Postgraduate Medical Journal* **77**:217–220.

DEMBITSKY, V.M. AND TOLSTIKOV, G.A. 2003. *Natural Halogenated Organic Compounds*. Novosibirsk: Nauka.

DEMENOCAL, P.B. 2004. African climate change and faunal evolution during the Pliocene-Pleistocene. *Earth and Platnetary science Letters* **6976**:1–22.

DILLEHAY, T.D., RAMÍREZ, C., PINO, M. ET AL. 2008. Monte verde: Seaweed, food, medicine, AND the peopling of South America. *Science* **320**:784–786.

DOBSON, J.E. 1998. The iodine factor in health and evolution. *The Geographical Review* **88**:1–28.

DONATI, L., ANTONELLI, A., VENTURI, S. ET AL. 1992. Clinical picture of endemic cretinism in central Apennines (Montefeltro). *Thyroid* **2**:283–290.

DONATI, L., VENTURI, S., ANDREANI, M. ET AL. 1989. Audiological and neuropsychological development study in a sample of school children from a low-iodine area of the Central Apennines that is endemic for cretinism. *Minerva Endocrinology* **14**:99–103.

DUBOIS-DALCQ, M. AND MURRAY, K. 2000. Why are growth factors important in oligodendrocyte physiology? *Pathological Biology* **48**:80–86.

DUGRILLON, A. 1996. Iodolactones and iodoaldehydes mediators of iodine in thyroid autoregulation. *Experimental and Clinical Endocrinology & Diabetes* **104**:41–145.

DUNN, J.T. AND DELANGE, F. 2001. Damaged reproduction: The most important consequence of iodine deficiency. *Journal of Clinical Endocrinology & Metabolism* **86**:2360–2363.

EALES, J.G. 1997. Iodine metabolism and thyroid-related functions in organisms lacking thyroid follicles: Are thyroid hormones also vitamins? *Proceedings of the Society for Experimental Biology and Medicine* **214**: 302–317.

EHRLICH, P.R. 2000. *Human Natures: Genes, Cultures and the Human Prospect.* New York: Penguin Books.

ELDERFIELD, H. AND TRUESDALE, V.W. 1980. On the biophilic nature of iodine in seawater. *Earth and Planetary Science Letters* **50**:105–114.

ELLIS, D.V. 1993. Wetlands or aquatic ape? Availability of food resources. *Nutrition and Health* **9**:205–217.

ELSTNER, E.F., ADAMCZYK, R., KROME, R. ET AL. 1985. The uptake of potassium iodide and its effects as an antioxidant in isolated rabbit eyes. *Ophthalmologica* **191**: 122–126.

EPISKOPOU, F., MAEDA, S., NISHIGUCHI, S. ET AL. 1993. Disruption of the transthyretin gene results in mice with depressed levels of plasma retinol and thyroid hormone. *Proceedings of the National Academy of Sciences of the United States of America* **90**:2375–2379.

ESKIN, B.A. 1970. Iodine metabolism and breast cancer. *Transactions of the New York Academy of Sciences* **32**:911–947.

ESKIN, B.A. 1977. Iodine and mammary cancer. *Advances in Experimental Medicine and Biology* **91**:293–304.

FABRE, V., CONDEMI, S., DEGIOANNI, A. 2009. Genetic evidence of geographical groups among Neanderthals. *PLoS ONE* **4** (4):e5151, doi:10.1371/journal.pone.0005151.

FINLAY, B. L. AND DARLINGTRON, R.B. 1995. Linked regularities in the development and evolution of mammalian brains. *Science* **268**:1578–1584.

FLEAGLE, J.G. 1999. *Primate Adaptation and Evolution*. 2nd ed. San Diego: Academic.

FRIESEMA, E.C.H., GANGULY, S., ABDALLA, A. ET AL. 2003. Identification of monocarboxylate transporter 8 as a specific thyroid hormone transporter. *Journal of Biological Chemistry* **278**:40128–40135.

FUNAHASHI, H., IMAI, T., MASE, T. ET AL. 2001. Seaweed prevents breast cancer? *Japanese Journal of Cancer Research* **92**:483–487.

FUNAHASHI, H., IMAI, T., AND TANAKA, Y. 1996. Suppressive effect of iodine on DMBA-induced breast tumor growth in the rat. *Journal of Surgical Oncology* **61**:209–213.

GARCIA-SEGURA, L.M. AND MCCARTHY, M.M. 2004. Role of glia in neuroendocrine function. *Endocrinology* **145**: 1082–1086.

GELB, A., HOROWITZ, M.I., AND HOLLANDER, F. 1962. Iodine binding by proteins in canine gastric mucus. *Nature* **195**:575–577.

GOETHE, S., WANG, Z., NG, L. ET AL. 1999. Mice devoid of all known thyroid hormone receptors are viable but exhibit disorders of the pituitary-thyroid axis, growth, and bone maturation. *Genes & Development* **13**:1329–1341.

GOŁKOWSKI, F., SZYBIŃSKI, Z., RACHTAN, J. ET AL. 2007. Iodine prophylaxis-the protective factor against stomach cancer in iodine deficient areas. *European Journal of Nutrition* **46**:251–256.

GRZESIUK, W.M. 2009. Iodization of salt in Poland: Past and the future. In *Comprehensive Handbook of Iodine: Nutritional, Biochemical, Pathological and Therapeutic Aspects*, ed. V.R. Preedy, G.N. Burrow, AND R. Watson, pp. 811–823. Amsterdam: Academic Press.

HADLEY, M.F. 2000. *Endocrinology*. 5th ed. Englewood Cliffs: Prentice-Hall.

HOLLOWELL, J.G., STAEHLING, N.W., HANNON, W.H. ET AL. 1998. Iodine nutrition in the United States. Trends and public health implications: Iodine excretion data from National Health and Nutrition Examination Surveys I and III (1971–1974 and 1988–1994). *Journal of Clinical Endocrinology & Metabolism* **83**:3401–3408.

HORROBIN, D. 1997. Fatty acids, phospholipids, and schizophrenia. In *Handbook of Fatty Acid Biology* (ed. S. Yehuda and D.I. Mostofsky), pp. 245–256. Toronto: Humana Press.

HORROBIN, D. 2001. *The Madness of Adam and Eve: How Schizophrenia Shaped Humanity*. London: Bantam.

HORROBIN, D. AND BENNETT, C.N. 1999. Depression and bipolar disorder: Relationship to impaired fatty acid and phospholipids metabolism and to diabetes, cardiovascular disease, immunological abnormalities, cancer, aging and osteopososis: possible candidate genes. *Prostaglandins Leukotrienes & Essential Fatty Acids* **60**:111–167.

HORROCKS, L.A. AND FAROOQUI, A.A. 2004. Docosahexaenoic acid in the diet: Its importance in maintenance and restoration of neural membrane function. *Prostaglandin Leukotriene & Essential Fatty Acids* **70**:361–372.

HULBERT, A.J. 2000. Thyroid hormones and their effects: A new perspective. *Biological Review* **75**:519–631.

HUNG, H.C., JOSHIPURA, K.J., JIANG, R., ET AL. 2004. Fruit and vegetable intake and risk of major chronic disease. *Journal of the National Cancer Institute* **96**:1577–1584.

JONES, S.A., THOEMKE, L.R., AND ANDERSON, G.W. 2005. The role of thyroid hormone in fetal and neonatal brain development. *Current Opinion in Endocrinology and Diabetes* **12**:10–16.

KALSBEEK, A., FLIERS, E., FRANKE, A.N. ET AL. 2000. Functional connections between the suprachisamatic nuleus and the thyroid gland as revealed by lesioning and viral tracing techniques in the rat. *Endocrinology* **141**:3832–3841.

KAPTEIN, E.M., HAYS, M.T., AND FERGUSON, D.C. 1994. Thyroid hormone metabolism: A comparative evaluation. In *Veterinary clinics of North America, Small Animal Practice 24*, ed. D.C. Ferguson, pp. 431–463. Philadelphia: W.B. Saunders.

KATAMINE, S., HOSHINO, N., TOTSUKA, K. ET AL. 1985. Effects of the long-term feeding of high-iodine eggs on lipid metabolism and thyroid function in rats. *Journal of Nutritional Science and Vitaminology* **31**:339–345.

KESSLER, J.H. 2004. The effect of supraphysiologic levels of iodine on patients with cyclic mastalgia. *Breast Journal* **10**:328–336.

KITAJKA, K., PUSKAS, L.G., ZVARA, A. ET AL. 2002. The role of n-3 polyunsaturated fatty acids in brain: Modulation of rat brain gene expression by dietary n-3 fatty acids. *Proceedings of the National Academy of Sciences of the United States of America* **99**:2619–2624.

KÖHRLE, J. 2000. Thyroid hormone metabolism and action in the brain and pituitairy. *Acta Medica Austriaca* **17**:1–7.

KRINGS, M., STONE, A., SCHMITZ, R.W. ET AL. 1997. Neandertal DNA sequences and the origin of modern humans. *Cell* **90**:19–30.

KÜPPER, F.C., CARPENTER, L.J., MCFIGGANS, G.B. ET AL. 2008. Iodide accumulation provides kelp with an inorganic antioxidant impacting atmospheric chemistry. *Proceedings of the National Academy of Sciences of the United States of America* **13** (105):6954–6958.

LAVADO-AUTRIC, R., AUSO, E., GARCIA-VELASCO, J.V. ET AL. 2003. Early maternal hypothyrozinemais alters histogenesis and cerebral cortex cytoarchitecture of the progeny. *Journal of Clinical Investigation* **111**:1073–1082.

LIN, J., ZHANG, S.M., COOK, N.R. ET AL. 2005. Dietary intakes of fruit, vegetables, and fiber, and risk of colorectal cancer in a prospective cohort of women (UnitedStates). *Cancer Causes & Control* **16**:225–233.

LINDAHL, T. 1997. Facts and artifacts of ancient DNA. *Cell* **90**:1–3.

LIU, Y. 2000a. Study on antioxidative unbalance and structural damage in iodine deficient rats. 12th International Thyroid Congress, Kyoto, October 22–27, p. 559.

LIU, Y. 2000b. The damage role of free radical in iodine deficient children.12th International Thyroid Congress, Kyoto, October 22–27, p. 560.

LOOSEN, P.T. 1992. Effects of thyroid hormones on central nervous system in aging. *Psychoneuroendocrinology* **17**:355–374.

LOW, S. AND BERRY, M.J. 1996. Knowing when not to stop: selenocysteine incorporation in eukaryotes. *Trends in Biochemical Sciences* **21**:203–208.

MANZON, R.G. AND YOUSON, J.H. 1997. The effects of exogenous thyroxine (T4) or triiodothyronine (T3), in the presence and absence of potassium perchlorate, on the incidence of metamorphosis and on serum T4 and T3 concentrations in larval sea lampreys (*Petromyzon marinus* L.). *General Comparative Endocrinology* **106**: 211–220.

MARANI, L. AND VENTURI, S. 1985. Iodine and delayed immunity. *Minerva Medicine* **77**:805–809.

MARANI, L., VENTURI, S., AND MASALA, R. 1986. Role of iodine in delayed immune response. *Israel Journal of Medical Sciences* **2**:864.

MCKINNEY, M.L. 1998. The juvenilized ape myth – Our "overdeveloped" brain. *Bioscience* **48**:109–116.

MCKINNEY, M.L. 2002. Brain evolution by stretching the global mitotic clock of development. In *Human Evolution Through Developmental Change*, ed. M. Minugh-Purvis and K. McNamara, pp. 173–188. Baltimore: Johns Hopkins University Press.

MELLARS, P. 1996. *The Neandertal Legacy: An Archaeological Perspective from Western Europe*. Princeton: Princeton University Press.

MILLER, D.W. 2006. Extrathyroidal benefits of iodine. *Journal of the American Physicians and Surgeons* **11**: 106–110.

MINUGH-PURVIS, N. 2002. Heterochronic change in the neurocranium and the emergence of modern humans. In *Human Evolution through Developmental Change*, ed. N. Minugh-Purvis and K. McNamara, pp. 479–498. Baltimore: Johns Hopkins University Press.

MORRIS, C.D. AND CARSON, S. 2003. Routine vitamin supplementation to prevent cardiovascular disease: A summary of the evidence for the U.S. Preventive Services Task Force. *Annals of Internal Medicine* **139**: 56–70.

MURANOV, K., POLIANSKY, N., WINKLER, R. ET AL. 2004. Protection by iodide of lens from selenite-induced cataract. *Graefe's Archive for Clinical and Experimental Ophthalmology* **242**:146–151.

OBENDORF, P.J., OXNARD, C.E., AND KEFFORD, B.J. 2008. Are the small human-like fossils found on Flores human endemic cretins? *Proceedings of Biological Sciences* **7** (75):1287–1296.

O'BRIEN, J.S. AND SAMPSON, E.L. 1965. Fatty acid and fatty aldehyde composition of the major brain lipids in normal human gray matter, white matter, and myelin. *Journal of Lipid Research* **6**:545–551.

OPPENHEIMER, J.H. AND SCHWARTZ, H.L. 1997. Molecular basis of thyroid hormone dependent brain development. *Endocrine Review* **18**:462–475.

OZIOL, L., FAURE, P., VERGELY, C. ET AL. (2001) In vitro free radical scavenging capacity of thyroid hormones and structural analogues. *Journal of Endocrinology*, **170**, 197–206.

PARK, S.K., SOLOMON, D., AND VARTANIAN, T. 2001. Growth factor control of CNS myelination. *Developmental Neuroscience* **23**:327–337.

PARKER, S.T. AND MCKINNEY, M. 1999. *Origins of Intelligence: The Evolution of Cognitive Development in Monkeys, Apes, and Humans*. Baltimore: Johns Hopkins University Press.

PEREIRA, A., BRAEKMAN, J.-C., DUMONT, J.E. ET AL. 1990. Identification of a major iodolipid from the horse thyroid gland as 2-iodohexadecanal. *The Journal of Biological Chemistry* **265**:17018–17025.

PIOSIK, P.S., VAN GROENIGEN, M., VAN DOORN, J. ET AL. 1997. Effects of maternal thyroid status on thyroid hormones and growth in congenitally hypothyroid goat fetuses during the second half of gestation. *Endocrinology* **138**:5–11.

Pisarev, M.A., Bocanera, L.V., Chester, H.A. et al. 1992. Effect of iodoarachidonates on thyroid FRTL-5 cells growth. *Hormone and Metabolic Research* **24**: 558–561.

Pisarev, M.A., Chazenbalk, G.D., Valsecchi, R.M. et al. 1988. Thyroid autoregulation. Inhibition of goiter growth and of cyclic AMP formation in rat thyroid by iodinated derivatives of arachidonic acid. *Journal of Endocrinological Investigation* **11**:669–674.

Porterfield, S.P. 2000. Thyroidal dysfunction and environmental chemicals – Potential impact on brain development. *Environmental Health Perspectives* **108**: 433–438.

Ramirez Rozzi, F. 2002. Enamel microstructure in hominids. In *Human Evolution through Developmental Change*, ed. N. Minugh-Purvis and K. McNamara, pp. 319–348. Baltimore: Johns Hopkins University Press.

Reed, L.E. 1997. Early hominid evolution and ecological change through the African Plio-Pleistocene. *Journal of Human Evolution* **32**:289–322.

Reppert, S.M. and Weaver, D.R. 2002. Coordination of circadian timing in mammals. *Nature* **418**:935–941.

Rieger G., Winkler R., Buchberger, W. et al. 1995. Iodine distribution in a porcine eye model following iontophoresis. *Ophthalmologica* **209**:84–87.

Rodriguez, A.M., Perron, B., Lacroix, L. et al. 2002. Identification and characterization of a putative human iodide transporter located at the apical membrane of thyrocytes. *Journal of Clinical Endocrinology & Metabolism* **87**:3500–3503.

Roth, P. 1946. Contribution à l'étude de l'action de la thyroxine et des substances antagonistes dans la métamorphose des Batraciens Anoures. *Mémoires du Muséum (Nouvelle Série)* **21**:175–273.

Rovet, J.F. 2004. Neurodevelopmental consequences of maternal hypothyroidism during pregnancy. 76th Annual meeting of the American Thyroid Association, Vancouver. Abstract 88. *Thyroid* **14**:710.

Sato, Y., Tsubono, Y., Nakaya, N. et al. 2005. Fruit and vegetable consumption and risk of colorectal cancer in Japan: The Miyagi Cohort Study. *Public Health Nutrition* **8**:309–314.

Scheer, F.A.J.L., Ter Horst, G.J., van der Vliet, J. et al. 2001. Phsiological and anatomic evidence for regulation of the heart by suparchiasmatic nucleus in rats. *American Journal of Physiology Heart and Circulatory Physiology* **280**:H1391–H1399.

Smallridge, R.C. and Ladenson, P.W. 2001. Hypothyroidism in pregnancy: Consequences to neonatal health. *Journal of Clinical Endocrinology and Metabolism* **86**:2349–s2353.

Smith, H. 1992. Life history and the evolution of human maturation. *Evolutionary Anthropology* **1**:134–142.

Smyth, P.P. 2003a. The thyroid, iodine and breast cancer. *Breast Cancer Research* **5**:235–238.

Smyth, P.P. 2003b. Role of iodine in antioxidant defence in thyroid and breast disease. *Biofactors* **19**:121–130.

Stadtman, T.C. 1996. Selenocysteine. *Annual Review of Biochemistry* **65**:83–100.

Stewart, K.M. 1994. Early hominid utilisation of fish resources and implications for seasonality and behaviour. *Journal of Human Evolution* **27**:229–245.

Stewart, K.M. 1996. A report on the fish remains from Beds I and II sites, Olduvai Gorge, Tanzania, *Darmstadter Beitrage zur Naturgeschichte* **6**:263–269.

Stone, O.J. 1988. The role of the primitive sea in the natural selection of iodides as a regulating factor in inflammation. *Medical Hypotheses* **25**:125–129.

Svennerholm, L. 1968. Distribution and fatty acid composition of phosphoglycerides in normal human brain. *Journal of Lipid Research* **9**:570–579.

Szybinski, Z., Rachtan, J., Huszno, B. et al. 2004. Is iodine-prophylaxis a protective factor against gastric cancer in iodine-deficient areas? 30th Annual Meeting of European Thyroid Association. Istanbul, Turkey, September 18–22, p. 111 (Abstract No. 257).

Teas, J., Pino, S., Critchley, A. et al. 2004. Variability of iodine content in common commercially available edible seaweeds. *Thyroid* **14**:836–841.

Thrall, K.D., Sasser, L.B., Creim, J.A. et al. 2009. Studies supporting the development of a physiologically based pharmacokinetic (PBPK) model for methyl iodide: Pharmacokinetics of sodium iodide (NaI) in pregnant rabbits. *Inhalation Toxicology* **31**:1–5.

Tramontano, D., Veneziani, B.M., Lombardi, A. et al. 1989. Iodine inhibits the proliferation of rat thyroid cells in culture. *Endocrinology* **125**:984–992.

Truesdale, V.W., Luther, G.W. III, and Canosa-Mas, C. 1995. Molecular iodine reduction in seawater: An improved rate equation considering organic compounds. *Marine Chemistry* **48**:143–150.

Tseng, Y.L. and Latham, K.R. 1984. Iodothyronines: Oxidative deiodination by hemoglobin and inhibition of lipid peroxidation. *Lipids* **19**:96–102.

Tsubono, Y., Otani, T., Kobayashi, M., et al. 2005. No association between fruit or vegetable consumption and the risk of colorectal cancer in Japan. *British Journal of Cancer* **92**:1782–1784.

U.S. Food and Nutrition Board; Institute of Medicine. 2001. *Dietary Reference Intakes for Vitamin A, Vitamin K, Arsenic, Boron, Chromium, Copper, Iodine, Iron, Manganese, Molybdenum, Nickel, Silicon, Vanadium and Zinc.* Washington, DC: National Academy Press.

Venturi, S. 1985. Preliminari ad uno studio sui rapporti tra cancro gastrico e carenza alimentare iodica: prospettive specifiche di prevenzione. Editor USL n.1, Regione Marche, Novafeltria.

Venturi, S. 2001. Is there a role for iodine in breast diseases? *The Breast* **10**:379–382.

Venturi, S., Donati, F.M., Venturi, A. et al. 2000a. Environmental iodine deficiency: A challenge to the evolution of terrestrial life? *Thyroid* **10**:727–729.

Venturi, S., Donati, F.M., Venturi, A. et al. 2000b. Role of iodine in evolution and carcinogenesis of thyroid, breast and stomach. *Advances in Clinical Pathology* **4**:11–17.

Venturi, S., Grossi, L., Marra, G.A. et al. 2003. Iodine, helicobacter pylori, stomach cancer and evolution. *European EPI-Marker* **7**:1–7.

VENTURI, S., MARANI, L., AND MAGNI, M.A. 1987. Syndrome thyro-gastric: Study on the relation between alimentary iodine deficiency and gastric. In *Epidemiology, Prevention and Early Detection of Gastric Cancer*, ed. G. Ghironzi, C.E. Jackson, and B.M. Schuman, pp. 272–281. Rome: CIC Edit.

VENTURI, S., VENTURI, A., CIMINI, D. ET AL. 1993. A new hypothesis: Iodine and gastric cancer. *European Journal of Cancer Prevention* **2**:17–23.

VENTURI, S. AND VENTURI, M. 1999. Iodide, thyroid and stomach carcinogenesis: Evolutionary story of a primitive antioxidant? *European Journal of Endocrinology* **140**:371–372.

VENTURI, S. AND VENTURI, M. 2007. Evolution of dietary antioxidant defenses. *European Epi-Marker* **1**:1–12.

WALTER, R.C., BUFFLER, R.T., BRUGGEMANN, J.H., ET AL. 2000. Early human occupation of the Red Sea coast of Eritrea during the last interglacial. *Nature* **405**:65–69.

WARE, C.M. AND WISHNER, L.A. 1968. The lipid antioxidant properties of iodine compounds. *Lipids* **3**:182–183.

WATERMAN, A.J. 1958. Development of the thyroid-pituitary system in warm-blooded amniotes. In *Comparative Endocrinology*, ed. A. Gordman, pp. 351–367. New York: John Wiley & Sons.

WEETMAN, A.P. 1997. Hypothyroidism: Screening and subclinical disease. *British Medical Journal* **314**:1175–1178.

WILLIAMS, F.L., GODFREY, L.R., AND SUTHERLAND, M.R. 2002. Heterochrony and the evolution of Neanderthal and modern craniofacial form. In *Human Evolution through Developmental Change*, ed. N. Minugh-Purvis and K. McNamara, pp. 405–441. Baltimore, MD: Johns Hopkins University Press.

WILSON, C.M. AND MCNABB, F.M.A. 1997. Maternal thyroid hormones in Japanese quail eggs and their influence on embryonic development. *General and Comparative endocrinology* **107**:153–165.

WINKLER, R., GRIEBENOW, S., AND WONISCH, W. 2000. Effect of iodide on total antioxidant status of human serum. *Cell Biochemistry and Function* **18**:143–146.

WOLFF, J. 1964. Transport of iodide and other anions in the thyroid gland. *Physiological Review* **44**:45–90.

WRIGHT, M.L. 2002. Melatonin, diel rhythms, and metamorphosis in anuran amphibians. *General and Comparative Endocrinology* **126**:251–254.

YEN, P.M. 2001. Physiological and molecular basis of thyroid hormone actions. *Physiological Reviews* **81**:1097–1142.

YOUNG, J.B., BÜRGI-SAVILLE, M.D., BÜRGI, U. ET AL. 2005. Sympathetic nervous system activity in rat thyroid: Potential role in goitrogenesis. *American Journal of Physiology: Endocrinology & Metabolism* **288**:E861–E867.

ZOELLER, R.T. 2003. Transplacental thyroxine and fetal brain development. *Journal of Clinical Investigation* **111**:954–957.

CHAPTER 7

FOOD FOR THOUGHT: THE ROLE OF COASTLINES AND AQUATIC RESOURCES IN HUMAN EVOLUTION

Jon M. Erlandson

INTRODUCTION

As the twentieth century unfurled, archaeology gradually emerged as an increasingly sophisticated and interdisciplinary science for studying the human past. Growing from antiquarian roots, archaeologists worked to construct culture histories for regions around the world. As knowledge accumulated, new dating techniques provided a deeper understanding of our long biological and cultural evolution, and archaeologists identified the broad patterns of human (pre)history. One of these patterns was a recent and dramatic explosion in the evidence for marine and aquatic resource use, coincident with the stabilization of global sea level and the spread of agriculture (e.g., Washburn and Lancaster, 1968; Yesner, 1987). To explain this supposedly late development of fishing and maritime economies, archaeologists and anthropologists constructed models and theories that proposed that shellfish, fish, and other marine or aquatic resources were low-ranked or even "starvation" foods turned to only after more productive terrestrial resources were depleted (see Hogg et al., 1971; Cohen, 1977; Osborn, 1977). Such models were widely accepted, despite repeated warnings that the archaeology of coastlines was severely biased by postglacial sea level rise and dramatic changes in coastal geography throughout the Pleistocene (e.g., Shepard, 1964; Emery and Edwards, 1966; Kraft et al., 1983; Van Andel, 1989).

In 2001, hoping to help reverse decades of anthropological theory marginalizing the significance of coastlines and aquatic habitats in human evolution, I reviewed the archaeological evidence for aquatic resource use by our Pleistocene ancestors (Erlandson, 2001). I argued that the record of Pleistocene marine and freshwater fishing was fundamentally biased by global sea level rise of ~120 m between 20,000 and 6000 years ago, which flooded the shorelines and nearshore lowlands where Pleistocene coastal populations would have lived. I proposed that marine and other aquatic resources played a considerably more significant role in human evolution (see also Bailey, 2004) than previously recognized, with certain "low-tech" aquatic foraging strategies (shellfish collecting, simple shallow-water fishing, shoreline scavenging, etc.) being relatively ancient, but "high-tech" strategies (seafaring and pelagic hunting, harpoons, fishing with hook-and-line, nets, or weirs, etc.) expanding greatly after the appearance and rapid geographic expansion of

Human Brain Evolution, Edited by Stephen C. Cunnane and Kathlyn M. Stewart anatomically modern humans (*Homo sapiens sapiens*). I saw this as an intellectual middle ground between relatively polarized perspectives of coastlines I had earlier characterized as "Gardens of Eden" (Sauer, 1962) versus "Gates of Hell" (Washburn and Lancaster, 1968; Osborn, 1977) models for our human evolution (Erlandson, 1994).

As I was synthesizing the global archaeological literature for my "Aquatic Adaptations" paper, I was largely unaware that scientists applying evolutionary perspectives to the study of human nutrition, physiology, and molecular genetics were working on independent but parallel paths to gain valuable insights into the role of aquatic ecosystems and resources in human evolution. Other scholars studying the ecology and paleoecology of coastlines were also making discoveries that contributed to a better understanding of the contexts under which early fishing economies, coastal migrations, and maritime voyaging may have developed. These other scientific developments provide important support for fragmentary archaeological records, as well as independent lines of evidence for a significant role of coastlines and aquatic ecosystems in human evolution. In the present chapter, I look briefly at those parallel developments, before updating my 2001 paper with additional archaeological evidence for the Pleistocene use of marine, estuarine, and freshwater fishing.

FOOD FOR THOUGHT

Humans evolved in an environmental crucible marked by dramatic fluctuations between glacial and interglacial conditions. The Pleistocene was marked by numerous cycles of climate and sea level change, with massive glaciers forming and melting, lake levels rising and falling, land bridges emerging from the sea and flooding again, and major changes in the distribution of ecosystems and their associated floral and faunal communities. After sea levels rose ~125 m in the last 20,000 years (Shackleton, 1987), humans now live in an interglacial era of unusually high sea level, with roughly 90% of human history in coastal settings now beneath the sea (Bailey et al., 2007). In some areas, coastlines have shifted laterally hundreds of kilometers since the Last Glacial Maximum (LGM), with modern shorelines now situated adjacent to archaeological sites that were once far removed from Pleistocene coastlines. These and other preservation problems (see Erlandson, 2001) limit what we know about the archaeology of human settlement and fishing in coastal, riverine, and lacustrine habitats around the world. To understand the history of hominin fishing and aquatic foraging, archaeologists rely primarily on the study of those rare sites occupied during full interglacial periods when sea levels were higher than today, on coastlines with steep bathymetry and minimal shoreline changes, or on proxy evidence for coastal economies. Some of these proxies are archaeological in nature, including evidence for island colonization and seafaring, the use of shell beads, and isotopic signatures of marine or aquatic diet in hominin fossils. Others come from entirely different sources but provide strong circumstantial evidence for the importance of fish and fishing in human evolution.

In discussing the history of early hominin migrations, Gamble (1994) listed 10 key habitats our ancestors encountered as they spread out of Africa – none of them aquatic. Clearly, however, early *Homo erectus* populations crossed rivers and other water "barriers" in migrating from Africa to Southeast Asia well over a million years ago. To this circumstantial evidence, we can now add genetic evidence for human migration patterns gathered from the study of DNA samples from a diverse array of modern humans living in regions around the world. Beginning about 70,000 years ago, the demographic expansion of anatomically modern humans (AMH) out of Africa initiated one of the most momentous migrations in human history, the initial steps in the spread of humans to the far reaches of the globe (see Wells, 2002 and Oppenheimer, 2003 for popular accounts). One of the earliest successful migration corridors involved in the migration of AMH out of Africa is sometimes referred to as the Southern Dispersal Route, starting from the Red Sea area in northeast Africa and following the south coast of Asia eastward (see Stringer, 2000; Field and Lahr, 2005). There is currently little archaeological evidence from South Asian coastlines to support the theory, but studies of DNA among living peoples of South Asia suggest the existence of such a route (e.g., Endicott et al., 2003; Macaulay et al., 2004).

The Pleistocene peopling of Island Southeast Asia, Australia, New Guinea, and western Melanesia also provides important supporting evidence that such a migration took place, facilitated (at least at its eastern end) by systematic seafaring and maritime voyaging that began roughly 50,000 ± 5000 years ago. Recent genetic studies have also proposed that maritime peoples from East Asia followed North Pacific coastlines into the Americas (see Kemp et al., 2007). Such maritime migrations imply a substantial commitment to marine resources and ecosystems, although the specific nature of such early coastal or fishing economies can only be understood with archaeological data (see below).

Ecological evidence has also been proffered in support of the importance of coastlines in human evolution and migrations. For instance, coastlines are important ecotones, with relatively high biodiversity where a combination of aquatic and terrestrial resources would have been available to ancient humans. It seems unlikely that our ancestors would have ignored coastlines and coastal resources as some archaeologists (e.g., Osborn, 1977) once suggested (see also Chapters 8 and 10). In arid zones such as the Arabian Peninsula, Faure et al. (2002) argued that the hydrology of shorelines associated with lower sea levels would have provided springs attractive to early humans moving out of Africa. Field and Lahr (2005) modeled potential migration corridors associated with the Southern Dispersal Route, arguing that coastlines generally would have offered the most optimal routes. Bulbeck (2007) argued that estuaries along the south coast of Asia would have provided particularly "sweet spots" of biological productivity and biodiversity attractive to AMH migrating out of Africa. Mannino and Thomas (2002) proposed that easily collected shellfish would have attracted and supported coastal peoples, but that the progressive depletion of shellfish beds may have encouraged AMH populations to continue migrating in search of pristine coastal ecosystems. Finally, Erlandson et al. (2007) argued that the high productivity and ecological similarity of kelp forests may have facilitated the migration of maritime peoples from East Asia into the Americas after about 16,000 years ago, following linear and unobstructed travel corridors entirely at sea level. Kelp forests around the Pacific Rim provided rich three-dimensional habitats for a diverse array of shellfish, fish, sea mammals, seabirds, and seaweeds, and also could have reduced wave energy and provided holdfasts for nearshore seafaring and fishing peoples (Erlandson et al., 2007).

HUMAN NUTRITION AND PHYSIOLOGY

The emergence of evolutionary perspectives in human nutrition and physiology is also providing fundamental insights into the importance of aquatic ecosystems in the evolution of the genus *Homo*. For instance, hominins evolved a body cooling system heavily reliant on sweating, which puts a premium on proximity to freshwater sources and a need for regular replenishment of sodium (Kempf, 2009). One of the most fundamental questions related to human evolution is how and why our large and energy-sapping brain developed over the past 2.5 million years. According to Cunnane (2005: 37), the fossil record shows that this encephalization process was punctuated, with marked expansions of brain-to-body-size ratios in the transitions from australopithecines to *Homo habilis* (~40%) and from *H. erectus* to *Homo sapiens* (~62%). The development and maintenance of larger brains resulted in relatively unique physiological and nutritional requirements that have led some researchers to propose that aquatic habitats may have been more important in human evolution than once thought. In a series of provocative publications, for instance, Broadhurst, Crawford, Cunnane, and their colleagues (see Cunnane et al., 1993; Broadhurst et al., 1998, 2002; Crawford et al., 1999) argued that the normal development and operation of the large human brain depends on rich and relatively stable sources of energy (especially fat) and key nutrients such as iodine, selenium, zinc, iron, copper, and specialized long-chain fatty acids (docosahexaenoic acid, arachidonic acid, etc.). They argue that savanna ecosystems – often proposed as the primary cradle of early human evolution – are unlikely ecological contexts for such encephalization because they are deficient in key nutrients crucial to normal brain development and function.

Noting that shellfish and fish from lake, river, estuarine, and marine ecosystems provide an optimal mix of the nutrients required, they propose a "shore-based scenario" of human evolution (e.g., Broadhurst et al., 1998; Cunnane, 2005; see also Chapters 2 and 3 in this book). In their model, shellfish, fish, and other aquatic foods are literally foods for thought, which contributed to the long-term success of our ancestors. Carlson and Kingston (2007a,b) have contested the need for aquatic resources to provide long-chain fatty acids, and Cordain et al. (2002) noted that the brains and other internal organs of savanna herbivores could have provided a source of key nutrients needed for brain development. Still, aquatic foods provide an optimal source of the broader range of "brain-specific nutrients" required in the development and full function of large hominin brains.

One way or the other, our hominin ancestors were closely tethered to aquatic habitats for a variety of reasons including the need for freshwater, ecological diversity, dietary stability, salt, and brain-specific nutrients. Living in proximity to freshwater and/or marine habitats – where a variety of predators and carrion-feeders were also feeding on fish, shellfish, seabirds or waterfowl, aquatic mammals, and seaweeds – they would have had ample opportunity to experiment with aquatic foods. In those parts of their geographic range where marine, estuarine, or freshwater resources were abundant and accessible, it seems highly improbable that our increasingly brainy ancestors would have ignored them. If encephalization and the efficient operation of the growing hominin brain were linked to such foods, then aquatic resources were a significant component of human diets for 2 million years or more (see also Chapter 8).

The long history of human physical and cultural evolution has been marked by geographic expansion, increased population densities, intelligence, adaptive flexibility, and technological complexity. If these genetic, ecological, and physiological perspectives correctly predict an important role for lakeshores, rivers, and coastlines in human evolution, what does the archaeological record tell us about the deep history of human fishing and coastal settlement?

ARCHAEOLOGICAL EVIDENCE FOR THE ANTIQUITY OF FISHING

For the purposes of this chapter, I define fishing very broadly, including a variety of activities related to foraging, hunting, or fishing in aquatic habitats: from shellfish collecting, scavenging for beached carcasses, and the collection of seaweeds or other aquatic plants, to the use of sophisticated technologies to capture fish or sea mammals in deeper waters. In this sense, fishing includes a diverse array of subsistence-related activities, some of which require virtually no specialized technology, while others require costly and sophisticated equipment ranging from boats to nets, hook-and-line, and other specialized knowledge, materials, and composite technologies. On one end of this spectrum are fishing activities engaged in by numerous nonhuman predators or carrion-feeders; on the other end are the highly complex and specialized whaling activities of indigenous peoples in some parts of the world.

Reducing the enormous diversity of past human fishing activities into dichotomies of simple versus complex, of course, obscures a continuous range of behaviors (see Erlandson, 2001: 289–300). Nonetheless, gathering intertidal plants or sessile shellfish (mussels, limpets, abalones, etc.), capturing fish trapped by the drying of shallow ponds, or scavenging stranded carcasses off the shore are "low-tech" (or often "no-tech") activities available to virtually all hominins, including the old and the very young. The Olduwan and Acheulean technologies used by early hominids in terrestrial settings could also have been adapted for capturing and processing aquatic fauna, from spearing fish in shallow water to butchering stranded carcasses. After about 100,000 years ago, however, AMH added a series of more sophisticated fishing technologies (e.g., seaworthy boats, harpoons, nets, and other fishing gear) to the human technological repertoire, many of which required other advances such as woven technologies, bone tools, and complex composite tools. These technological developments allowed a more efficient and intensive reliance on fishing that, when combined with terrestrial resources, may have helped fuel the demographic and geographic expansion of AMH out of Africa and around the world (Erlandson, 2001).

As noted earlier, the archaeological record of the deep history of human fishing and coastal settlement is fundamentally flawed due to postglacial sea level rise, changes in coastal geography, coastal erosion, and other problems (see also Chapter 10, fig. 10.2). As a result, any evidence for human fishing more than about 8,000 to 10,000 years old is likely to represent the tip of a proverbial iceberg. In an earlier paper (Erlandson, 2001), I compiled a long list of Pleistocene archaeological sites around the world that contained possible evidence[1] for marine fishing in the form of aquatic animal remains, fishing technologies, or evidence for seafaring. A comprehensive review of the archaeological literature would be repetitive here and is beyond the scope of this chapter. Instead, I provide an overview, focusing on evidence for Pleistocene fishing and seafaring around the world that has come to light since 2000.

Unequivocal archaeological evidence for the exploitation of aquatic resources by *H. habilis* and *H. erectus* is still relatively limited. Stewart (1994: 235–242; Chapter 8) has analyzed several Olduvai Gorge fish bone assemblages dating between about 1,900,000 and 800,000 years old. Stewart argues that early hominins ate freshwater fish such as catfish (*Clarias*) that could be captured with little or no technology in the shallow waters of East African lakes or streams. Southeast Asia has also provided evidence for shellfish gathering by *H. erectus*, in the form of a pile of freshwater oyster shells found in Kao Pah Nam Cave in Thailand dated to about 700,000 years ago (Pope, 1989). Evidence for remarkably early maritime activity has also come from the Southeast Asian island of Flores, where Morwood et al. (1998, 1999) reported evidence that *H. erectus* had reached Flores by about 800,000 ± 50,000 years ago. Colonizing Flores appears to have required several short sea crossings, which led Bednarik (2003) to propose that *H. erectus* had relatively sophisticated seafaring capabilities. Flores and nearby islands were also colonized by several land animal species that are good swimmers, however, and the lack of physical evidence that *H. erectus* ever made the longer voyages required to settle Australia or New Guinea suggests that their maritime capabilities were limited to relatively simple rafts or other flotation devices.

Similar to the evidence for hominin fishing in East African lakes, Stuart et al. (1993) reported the remains of freshwater fish at the Lower Paleolithic site of Hoxne in England, tentatively dated between about 350,000 and 300,000 years ago. Limited evidence for marine fishing has also been reported for the Paleolithic sites of Terra Amata and Mas des Caves on the south coast of France and several other Mediterranean caves (see Cleyet-Merle and Madelaine, 1995; Erlandson, 2001). The Neanderthals, a regional population of archaic *H. sapiens* (*Homo neandertalensis* to some scholars) that lived in western Eurasia from approximately 300,000 to 30,000 years ago, offer further evidence for the use of marine resources. Neanderthals are a particularly interesting case since they appear to have had brains slightly larger on average than modern humans, although their brain-to-body-size ratio may have been slightly lower than AMH (Cunnane, 2005). Archaeological evidence from Gibraltar, Italy, and other sites around the Mediterranean suggest that some Neanderthal populations collected shellfish between at least 125,000 and 30,000 years ago (see Garrod et al., 1928; Waechter, 1964; Stiner, 1994). Stringer et al., (2008) recently reported that Neanderthals at Gibraltar also scavenged for the beached carcasses of seals and other marine animals, a pattern not unlike that reported for AMH at Middle Stone Age sites in South Africa (see below). The evidence for Neanderthal fishing along European shores is interesting, in part because isotopic studies suggest that Neanderthal diets were dominated by terrestrial game while Upper Paleolithic peoples relied more heavily on marine foods (Richards et al., 2001; Pettitt et al., 2003). A larger sample of Neanderthal and Upper Paleolithic skeletons needs to be analyzed before the nature of aquatic resource use among Neanderthals and early AMH is fully understood (Pearson, 2007). So far, however, no evidence for specialized fishing technologies has been identified in any sites associated with archaic *H. sapiens*, *H. erectus*, or *H. habilis*.

With the appearance of AMH, evidence for the use of aquatic resources seems to increase significantly (McBrearty and Brooks, 2000; Erlandson, 2001). The earliest evidence comes from the Pinnacle Point Caves complex in South Africa, where a Middle Stone Age layer dated to about 160,000 years ago contains large numbers of rocky intertidal shellfish (Marean et al., 2007). Archaeological research in South Africa has focused on numerous shell middens that have produced the remains of marine shellfish, seals, penguins, and other seabirds – as well as considerable evidence for the use of red ochre, shell beads, bone tools, and other evidence of "modern" intellectual behavior (see also Chapter 10). These include the well known cave sites at Klasies River Mouth, Blombos, Die Kelders, and Pinnacle Point, as well as several open sites in the West Cape region such as Ysterfontein, Saldanha Bay, and Hoodjies Punt (see Singer and Wymer, 1982; Parkington, 2003; Klein et al., 2004).

Until recently, the only coastal Middle Stone Age site in South Africa that had produced sizable number of finfish remains was Blombos Cave, where bone and stone points dated to about 75,000 years ago were found associated with the bones of numerous nearshore marine fish (Henshilwood and Sealy, 1997). Initial excavations at Klasies River Mouth Caves produced only very small numbers of fish bones (Singer and Wymer, 1982), but more recent work utilizing fine screens recovered thousands of marine fish bones in relatively small samples from the Middle Stone Age strata (Driesch, 2004). On the Semliki River in Zaire, two Middle Stone Age sites at Katanda have produced several beautifully made barbed bone harpoons dated to about 80,000 years ago and associated with numerous bones of large freshwater fish (Yellen et al., 1995). Along with the bone points from Blombos, the Katanda harpoons represent the earliest known composite fishing technologies in the world, sophisticated aquatic hunting tools unmatched by anything produced by our archaic ancestors.

Beads made from marine shells have been found in even earlier Middle Stone Age sites in North and South Africa and the Levant dated between about 125,000 and 75,000 years ago (see Henshilwood et al., 2004; Vanhaeren et al., 2006; Bouzouggar et al., 2007). Such ornaments, often linked to the emergence of a higher level of symbolic behavior among AMH, may also reflect the growing importance of the sea to our ancestors. In fact, almost everywhere that substantial archaeological evidence for AMH occupations has been found in general proximity to ancient coastlines, shell beads or other ornaments have also been found, serving as markers of the presence of AMH and their connections to sometimes distant coasts (e.g., Erlandson et al., 2005; Fitzgerald et al., 2005; Balme and Morse, 2006). At the Western Australian site of Riwi, for instance, tusk shell beads found in strata dated to ~32,000 years ago appear to have been transported between 300 and 500 km from contemporary shorelines.

Similar evidence comes from many sites in Upper Paleolithic Europe, including Ũcagizli Cave along the Turkish Coast and the Ksar Akil site in Lebanon where shell beads and marine shellfish remains were recovered from early Upper Paleolithic levels dated to ~40,000 years ago (Stiner et al., 2002; Kuhn et al., 2009). Marine shellfish and fish remains have also been found in many Upper Paleolithic sites along the southern and southwestern shores of Europe (e.g., Kuhn and Stiner, 1998), and thousands of bones from freshwater fish were found at the Ohalo II site on the Sea of Gallilee, dated to about 23,000 years ago (Nadel and Zaidner, 2002; Nadel et al., 2004).

Other than the short voyages that took *H. erectus* to Flores and some other islands of Wallacea, the earliest evidence for Pleistocene seafaring and fishing comes from Australia, Island Southeast Asia, and western Melanesia. Migrating from Southeast Asia to Australia and New Guinea required several maritime voyages up to 80–90 km long (Allen and O'Connell, 2008; Erlandson, in press) and reaching the Solomon Islands and Bismarck Archipelago required several additional voyages of even greater length (Irwin, 1992). A growing number of sites in Island Southeast Asia, Australia, and Western Melanesia, dating between roughly 45,000 and 15,000 years ago, provide evidence of seafaring or early marine and freshwater fishing. Jerimalai Cave in East Timor has produced marine fish, shellfish, and turtle remains from strata dated between about 45,000 and 40,000 years ago, including numerous bones from pelagic tuna (Ono et al., in press). Other evidence for marine shellfish use comes from the Aru Islands in the Torres Strait region (O'Connor et al., 2005), the Talaud Islands between 35,000–32,000 years ago and 22,000–17,000 years ago (Ono et al., in press), as well as other sites that remained within reach of coastal habitats between about 35,000 and 15,000 years ago (e.g., Bellwood et al., 1998; O'Connor et al., 2002). Several other sites that were further removed from coastal habitats during this period of time have produced freshwater shellfish and fish remains, including the Leang Burung site (e.g., Glover, 1981; Ono et al., in press).

Archaeological evidence from the Ryukyu Islands, which form an arc between Taiwan and Japan, suggests that early maritime peoples continued to make substantial sea voyages as they migrated northward along the coast of East Asia, beginning at least 35,000 years ago (Erlandson, in press). Although the oldest shell middens in Japan date to roughly 10,000 years ago, there is evidence for marine voyages over 30 km long at least 25,000 years ago to obtain obsidian from Kozushima Island (Erlandson and Fitzpatrick, 2006). By the LGM, maritime peoples in the Japanese Archipelago appear to have adapted to the relatively cool waters of the North Pacific, and it seems increasingly likely that they followed the Kurile Islands and the coastlines and kelp forests of Northeast Asia and Beringia into the New World (Erlandson, 2002; Erlandson et al., 2007). Rising sea levels would have flooded most of the archaeological evidence for such a coastal migration, but some recent studies of ancient and modern human DNA provide some support for such a scenario (see Kemp et al., 2007).

Further evidence comes from within the Americas, where steep bathymetry has moderated the effects of postglacial sea level rise on coastal landscapes, and the antiquity of fishing and seafaring has been pushed back into the Pleistocene along the Pacific Coast. In North America, where the earliest humans were long thought to be terrestrial big-game hunters, evidence for fishing has been found at a variety of interior and coastal sites dating to the terminal Pleistocene (Erlandson et al., 2008a). For instance, California's Northern Channel Islands were colonized by maritime Paleoindians at least 13,000 years ago, and several shell middens have recently been identified that date between about 12,000 and 10,000 years ago. On Cedros Island off the west coast of Baja California, Des Lauriers (2006) has also identified Terminal Pleistocene middens with shellfish, fish, sea turtle, and sea mammal remains. Several other Baja California and Pacific Northwest sites have produced evidence for early fishing in freshwater or marine settings (see Erlandson, 2001; Erlandson et al., 2008b).

In South America, even earlier evidence for the use of marine and estuarine resources comes from the Monte Verde II site in Chile, where kelp and several other seaweeds have been identified in a peri-coastal encampment dated to roughly 14,000 years ago (Dillehay et al., 2008; Erlandson et al., 2008a). This site, older than the terrestrial big-game hunters of North America's Clovis Complex, has bolstered the coastal migration theory and the kelp highway hypothesis. At Pedra Pintada Cave in Amazonian Brazil, Roosevelt et al. (1996) recovered the remains of freshwater fish and shellfish in Paleoindian strata dated to ~13,000 years ago. Other evidence for early fishing comes from the Peruvian Coast, where several Terminal Pleistocene middens have produced the remains of shellfish, finfish, or seabirds (Sandweiss, 2008).

CONCLUSIONS

Throughout the twentieth century, a significant role for fishing or aquatic foraging in human evolution was generally dismissed, with most anthropologists considering intensive fishing to be a post-Pleistocene development. Because the discipline of archaeology emerged during a full interglacial period when sea levels were among the highest of the Pleistocene, however, the current evidence for Pleistocene marine resource use may represent the tip of the proverbial iceberg. Today, a growing body of evidence from archaeology, paleoecology, molecular genetics, and evolutionary perspectives in human physiology and nutrition, suggests that fishing played an important role in the evolution of the genus *Homo*, the emergence of *H. sapiens sapiens*, and the relatively rapid spread of AMH around the globe.

In the East African rift zone, hominins appear to have consumed fish for more than 2 million years (see Chapter 8). An early spread of *H. erectus* through Africa and Eurasia required the ability to cross rivers and other water barriers, including short sea crossings in Island Southeast Asia (Erlandson, 2001, in press; Bednarik, 2003). Archaeological evidence suggests that Neanderthals also collected shellfish and sea mammal carcasses from Mediterranean coastlines. With the appearance of AMH, archaeological evidence for the development of more widespread, intensive, and diversified fishing increases, often facilitated by the use of new aquatic technologies. It now seems likely that fishing – including shellfish gathering, fin-fishing, aquatic mammal and bird hunting – provided a broad economic foundation for the human demographic and geographic expansion of the last 100,000 years. The development of seaworthy watercraft capable of long-range voyaging (>50 km), beginning roughly 50,000 years ago in Island Southeast Asia and greater Australia, facilitated the spread of coastal and maritime peoples around the world, allowing the colonization of islands and continents not previously inhabited by humans.

These new data summarized in this volume provide food for thought for all those interested in human evolution and the deep history of fishing and human migrations, including archaeologists and other scientists working in coastal, island, river, and lake systems around the world. I hope they will also inspire those who synthesize the long history of human evolution to more completely consider the role of aquatic resources and ecosystems in our past. Is it possible, for instance, that an intensification of fishing and aquatic resource use during the Middle Stone Age and Upper Paleolithic facilitated the "creative explosion" that accompanied the appearance and spread of AMH? Only further research can answer that question.

ACKNOWLEDGMENTS

I am grateful to Kathlyn Stewart and Stephen Cunnane for inviting me to participate in this volume and for their editorial work in reviewing my chapter and preparing a final version for publication. I also thank Erica Kempf and Rintaro Ono for permission to cite manuscripts that are currently under review or in press.

NOTES

1. Even where the remains of marine shells or bones are present, we must carefully consider whether they were deposited by hominin behavior or by natural biological or geological processes.

REFERENCES

Allen, J. and O'Connell, J.F. 2008. Getting from Sunda to Sahul. In *Islands of Inquiry: Colonisation, Seafaring, and the Archaeology of Maritime Landscapes*, ed. G. Clark, F. Leach, and S. O'Connor, pp. 31–46. Canberra: Australian National University Press.

Bailey, G.N. 2004. World prehistory from the margins: The role of coastlines in human evolution. *Journal of Interdisciplinary Studies in History and Archaeology* **1**:39–50.

Bailey, G.N., Flemming, N.C., King, G. et al. 2007. Coastlines, submerged landscapes and human evolution: The Red Sea Basin and the Farasan Islands. *Journal of Island and Coastal Archaeology* **2**:127–160.

Balme, J. and Morse, K. 2006. Shell beads and social behavior in Pleistocene Australia. *Antiquity* **80**:799–811.

Bednarik, R.G. 2003. Seafaring in the Pleistocene. *Cambridge Archaeological Journal* **13**:41–66.

Bellwood, P., Nitihaminoto, G., Irwin, G. et al. 1998. 35,000 years of prehistory in the northern Moluccas. *Modern Quaternary Studies in Southeast Asia* **15**:233–275.

Bouzouggar, A., Barton, N., Vanhaeren, M. et al. 2007. 82,000-year-old shell beads from North Africa and implications for the origins of modern human behavior. *Proceedings of the National Academy of Sciences of the United States of America* **104**:9965–9969.

Broadhurst, C.L., Cunnane, S.C., and Crawford, M.A. 1998. Rift Valley lake fish and shellfish provided brain-specific nutrition for early *Homo*. *British Journal of Nutrition* **79**:3–21.

Broadhurst, C.L., Wang, Y., Crawford, M.A. et al. (2002) Brain-specific lipids from marine, lacustrine, or terrestrial food resources: Potential impact on early African *Homo sapiens*. *Comparative Biochemistry and Physiology Part B* **131**:653–673.

Bulbeck, D. 2007. Where river meets sea: A parsimonious model for *Homo sapiens'* colonization of the Indian Ocean rim and Sahul. *Current Anthropology* **48**:315–321.

Carlson, B.A. and Kingston, J.D. 2007a. Docosahexaenoic acid, the aquatic diet, and hominin encephalization: Difficulties in establishing evolutionary links. *American Journal of Human Biology* **19**:132–141.

Carlson, B.A. and Kingston, J.D. 2007b. Docosahexaenoic acid biosynthesis and dietary contingency: Encephalization without aquatic constraint. *American Journal of Human Biology* **19**:585–588.

Cleyet-Merle, J. and Madelaine, S. 1995. Inland evidence of human sea coast exploitation in Palaeolithic

France. In *Man and Sea in the Mesolithic*, ed. A. Fischer, pp. 303–308. Oxford: Oxbow Monographs.

COHEN, M.N. 1977. *The Food Crisis in Prehistory: Overpopulation and the Origin of Agriculture*. New York: Yale University Press.

CORDAIN, L., WATKINS, B.A., AND MANN, N.J. 2002. Fatty acid composition and energy density of foods available to African hominids: Evolutionary implications for human brain development. *World Review of Nutrition and Dietetics* **90**:144–161.

CRAWFORD, M.A., BLOOM, M., BROADHURST, C.L. ET AL. 1999. Evidence for the unique function of docosahexaenoic acid during the evolution of the modern hominid brain. *Lipids* **34**:S39–S47.

CUNNANE, S.C. 2005. *Survival of the Fattest: The Key to Human Brain Evolution*. London: World Scientific.

CUNNANE, S.C., HARBIGE, L.S., AND CRAWFORD, M.A. 1993. The importance of energy and nutrient supply in human brain evolution. *Nutrition and Health* **9**:219–235.

DES LAURIERS, M.R. 2006. Terminal Pleistocene and Early Holocene occupations of Isla de Cedros, Baja California, Mexico. *Journal of Island and Coastal Archaeology* **1**:255–270.

DILLEHAY, T.D., RAMIREZ, C., PINO, M. ET AL. 2008. Monte Verde: Seaweed, food, medicine, and the peopling of South America. *Science* **320**:784–786.

DRIESCH, A. VON DEN. 2004. The Middle Stone Age fish fauna from the Klasies River main site, South Africa. *Anthropozoologica* **39** (2):33–59.

EMERY, K.O. AND EDWARDS, R.L. 1966. Archaeological potential of the Atlantic continental shelf. *American Antiquity* **31**:733–737.

ENDICOTT, P., GILBERT, M.T.P., STRINGER, C., ET AL. 2003. The genetic origins of the Andaman Islanders. *American Journal of Human Genetics* **72**:178–184.

ERLANDSON, J.M. 1994. *Early Hunter-Gatherers of the California Coast*. New York: Plenum.

ERLANDSON, J.M. 2001. The archaeology of aquatic adaptations: Paradigms for a new millennium. *Journal of Archaeological Research* **9**:287–350.

ERLANDSON, J.M. 2002. Anatomically modern humans, maritime adaptations, and the peopling of the New World. In *The First Americans: The Pleistocene Colonization of the New World*, ed. N. Jablonski, pp. 59–92. San Francisco: Memoirs of the California Academy of Sciences.

ERLANDSON, J.M. In press. Neptune's children: The origins and evolution of seafaring. In *Global Origins and Development of Seafaring*, ed. A. Anderson, J. Barrett, and K. Boyle, Cambridge: McDonald Institute for Archaeological Research, Cambridge University.

ERLANDSON, J.M., BRAJE, T.J., AND GRAHAM, M.H. 2008a. How old is MVII?: Seaweeds, shorelines, and chronology at Monte Verde, Chile. *Journal of Island and Coastal Archaeology* **3**:277–281.

ERLANDSON, J.M., MOSS, M.L., AND DES LAURIERS, M. 2008b. Living on the edge: Early maritime cultures of the Pacific Coast of North America. *Quaternary Science Reviews* **27**:2232–2245.

ERLANDSON, J.M. AND FITZPATRICK, S. 2006. Oceans, islands, and coasts: Current perspectives on the role of the sea in human prehistory. *Journal of Island and Coastal Archaeology* **1**:5–33.

ERLANDSON, J.M., GRAHAM, M.H., BOURQUE, B.J. ET AL. 2007. The kelp highway hypothesis: Marine ecology, the coastal migration theory, and the peopling of the Americas. *Journal of Island and Coastal Archaeology* **2**:161–174.

ERLANDSON, J.M., MACKO, M.E., KOERPER, H. ET AL. 2005. The antiquity of *Olivella* shell beads at CA-ORA-64: AMS radiocarbon dated between 9420 and 7780 cal BP. *Journal of Archaeological Science* **32**:393–398.

FAURE, H., WALTER, R.C. AND GRANT, D.R. 2002. The coastal oasis: Ice age springs on emerged continental shelves. *Global and Planetary Change* **33**:47–56.

FIELD, J.S. AND LAHR, M.M. 2005. Assessment of the Southern Dispersal Route: GIS-based analyses of potential routes at Oxygen Isotopic Stage 4. *Journal of World Archaeology* **19**:1–45.

FITZGERALD, R.T., JONES, T.L., AND SCHROTH, A. 2005. Ancient long distance trade in western North America: New AMS radiocarbon dates from southern California. *Journal of Archaeological Science* **32**:423–434.

GAMBLE, C. 1994. *Timewalkers: The Prehistory of Global Colonization*. Cambridge: Harvard University Press.

GARROD, D.A.E, BUXTON, L.H.D., SMITH, G.E. ET AL. 1928. Excavations of a Mousterian rock-shelter at Devil's Tower, Gibraltar. *Journal of the Royal Anthropological Institute of Great Britain and Ireland* **58**:33–113.

GLOVER, I.C. 1981. Leang Burung 2: An Upper Paleolithic rock shelter in south Sulawesi, Indonesia. *Modern Quaternary Research in Southeast Asia* **6**:1–38.

HENSHILWOOD, C. AND SEALY, J. 1997. Bone artefacts from the Middle Stone Age at Blombos Cave, Southern Cape, South Africa. *Current Anthropology* **38**:890–895.

HENSHILWOOD, C., D'ERRICO, F., VANHAEREN, M. ET AL. 2004. Middle Stone Age shell beads from South Africa. *Science* **304**:404.

HOGG, T.C., SMITH, C.L., AND DAVIS, W.A. 1971. *Man in the Marine Environment*. Corvallis, OR: Marine Anthropological Research Unit Report 1, Oregon State University.

IRWIN, G. 1992. *The Prehistoric Exploration and Colonisation of the Pacific*. Cambridge: Cambridge University Press.

KEMP, B.M., MALHI, R.S., MCDONOUGH, J. ET AL. 2007. Genetic analysis of Early Holocene skeletal remains from Alaska and its implications for the settlement of the Americas. *American Journal of Physical Anthropology* **132**:605–621.

KEMPF, E. 2009. Patterns of water use in primates. *Folia Primatologica* **80**:275–294.

KLEIN, R.G., AVERY, G., CRUZ-URIBE, K. ET AL. 2004. The Ysterfontein 1 Middle Stone Age site, South Africa, and early human exploitation of coastal resources. *Proceedings of the National Academy of Sciences of the United States of America* **101**:5708–5715.

KRAFT, J.C., BELLKNAP, D.F., AND KAYAN, I. 1983. Potentials of discovery of human occupation sites on the continental shelves and nearshore coastal zone. In *Quaternary Coastlines and Marine Archaeology*, ed. P.M. Masters and N.C. Flemming, pp. 87–120. New York: Academic Press.

Kuhn, S.L. and Stiner, M.C. 1998. The earliest Aurignacian of Riparo Mochi (Liguria, Italy). *Current Anthropology* **39**:S175–S189.

Kuhn, S.L., Stiner, M.C., Güleç, E. et al. 2009. The early Upper Paleolithic occupations of Üçağizli Cave (Hatay, Turkey). *Journal of Human Evolution* **56**: 87–113.

Macaulay, V., Hill, C., Achilli, A. et al. 2004. Single, rapid coastal settlement of Asia revealed by analysis of complete mitochondrial genomes. *Science* **308**: 1034–1036.

Mannino, M.A. and Thomas, K.D. 2002. Depletion of a resource? The impact of prehistoric human foraging on intertidal mollusc communities and its significance for human settlement, mobility and dispersal. *World Archaeology* **33**:452–474.

Marean, C.W, Bar-Matthews, M., Bernatchez, J. et al. 2007. Early human use of marine resources and pigment in South Africa during the Middle Pleistocene. *Nature* **449**:905–908.

McBrearty, S. and Brooks, A.S. 2000. The revolution that was not: A new interpretation of the origin of modern human behaviour. *Journal of Human Evolution* **39**: 453–563.

Morwood, M.J., O'Sullivan, P.B., Aziz, F. et al. 1998. Fission-track ages of stone tools and fossils on the East Indonesian island of Flores. *Nature* **392**:173–176.

Morwood, M.J., Aziz, F., O'Sullivan, P. et al. 1999. Archaeological and palaeontological research in Central Flores, East Indonesia: Results of fieldwork 1997–98. *Antiquity* **73**:273–286.

Nadel, D. and Zaidner, Y. 2002. Upper Pleistocene and mid-Holocene net sinkers from the Sea of Galilee, Israel. *Journal of the Israel Prehistoric Society* **32**:49–71.

Nadel, D., Weiss, E., Simchoni, O. et al. 2004. Stone Age hut in Israel yields world's oldest evidence of bedding. *Proceedings of the National Academy of Sciences of the United States of America* **101**:6821–6826.

O'Connor, S., Spriggs, M., and Veth, P. 2002. Excavation at Lene Hara establishes occupation in East Timor at least 30,000 to 35,000 years on: Results of recent fieldwork. *Antiquity* **76**:45–50.

O'Connor, S., Spriggs, M., and Veth, P. 2005. On the cultural history of the Aru Islands: Some conclusions. In *The Archaeology of the Aru Islands, Eastern Indonesia*, ed. S. O'Connor, M. Spriggs, and P.M. Veth, pp. 307–314. Terra Australis, Australian National University #22. Canberra: Research School of Pacific and Asian Studies.

Ono, R., Soegondoh, S., and Yoneda, M. In press. Changing marine exploitation during Late Pleistocene in northern Wallacea: Shellfish remains from Leang Sarru rockshelter in Talaud Islands. *Asian Perspectives.*

Oppenheimer, S. 2003. *The Real Eve: Modern Man's Journey Out of Africa.* New York: Carroll & Graf.

Osborn, A. 1977. Strandloopers, mermaids, and other fairy tales: Ecological determinants of marine resource utilization – The Peruvian case. In *For Theory Building in Archaeology*, ed. L.R. Binford, pp. 157–205. New York: Academic Press.

Parkington, J. 2003. Middens and moderns: Shellfishing and the Middle Stone Age of the Western Cape, South Africa. *South African Journal of Science* **99**:243–247.

Pearson, J.A. 2007. Hunters, fishers and scavengers: A review of the isotope evidence for Neanderthal diet. *Before Farming* **2**:1–16.

Pettitt, P.B., Richards, M., Maggi, R. et al. 2003. The Gravettian burial known as the Prince ("Il Principe"): New evidence for his age and diet. *Antiquity* **77**:15–19.

Pope, G. 1989. Bamboo and human evolution. *Natural History*, October, pp. 49–57.

Richards, M.P., Pettitt, P.B. Stiner, M.C. et al. 2001. Stable isotope evidence for increasing dietary breadth in the European mid-Upper Paleolithic. *Proceedings of the National Academy of Sciences of the United States of America* **98** (11):6528–6532.

Roosevelt, A.C., Lima da Costa, M., Lopes Machado, C. et al. 1996. Paleoindian cave dwellers in the Amazon: The peopling of the Americas. *Science* **272**:373–384.

Sandweiss, D.H. 2008. Early fishing societies in western South America. *The Handbook of South American Archaeology*, ed. H. Silverman and W.H. Isbell, pp. 145–156. , New York: Springer.

Sauer, C.O. 1962. Seashore – Primitive home of man? *Proceedings of the American Philosophical Society* **106**: 41–47.

Shackleton, N.J. 1987. Oxygen isotopes, ice volume, and sea level. *Quaternary Science Reviews* **6**:183–190.

Shepard, F.P. 1964. Sea level changes in the past 6,000 years: Possible archaeological significance. *Science* **143**:574–576.

Singer, R. and Wymer, J. 1982. *The Middle Stone Age at Klasies River Mouth in South Africa.* Chicago: University of Chicago Press.

Stewart, K.M. 1994. Early hominid utilization of fish resources and implications for seasonality and behavior. *Journal of Human Evolution* **27**:229–245.

Stiner, M.C. 1994. *Honor Among Thieves: A Zooarchaeological Study of Neandertal Ecology.* Princeton, NJ: Princeton University Press.

Stiner, M.C., Pehleven, C., Sagir, M. et al. 2002. Zooarchaeological studies at Üçağizli Cave: Preliminary results on Paleolithic subsistence and shell ornaments. In *Arkeometri, Sonuclari, Toplantisi*, pp. 29–35. Ankara: T.C. Kültür Bakanliüi.

Stringer, C. 2000. Coasting out of Africa. *Nature* **405**:24–27.

Stringer, C.B., Finlayson, J.C., Barton, R.N.E. et al. 2008. Neanderthal exploitation of marine mammals in Gibraltar. *Proceedings of the National Academy of Sciences of the United States of America* **105** (38): 14319–14324.

Stuart, A.J., Wolff, R.G., and Lister, R.G. 1993. Fossil vertebrates. In *The Lower Paleolithic Site at Hoxne, England*, ed. R. Singer, B.G. Gladfelter, and J.J. Wymer, pp. 163–206. Chicago: University of Chicago Press.

Van Andel, T. 1989. Late Quaternary sea-level changes and archaeology. *Antiquity* **63**:733–745.

Vanhaeren, M., d'Errico, F., Stringer, S. et al. 2006. Middle Paleolithic shell beads in Israel and Algeria. *Science* **312**:1785–1788.

WAECHTER, J.D. 1964. The excavation of Gorham's Cave, Gibraltar, 1951–54. *Bulletin of the Institute of Archaeology* 4:189–221.

WASHBURN, S.L. AND LANCASTER, C.S. 1968. The evolution of hunting. In *Man the Hunter*, ed. R.B. Lee and I. DeVore, pp. 293–303. Chicago: Aldine.

WELLS, S. 2002. *The Journey of Man: A Genetic Odyssey*. Princeton, NJ: Princeton University Press.

YELLEN, J.E., BROOKS, A.S., CORNELISSEN, E. ET AL. 1995. A Middle Stone Age worked bone industry from Katanda, Upper Semliki Valley, Zaire. *Science* **268**: 553–556.

YESNER, D.R. 1987. Life in the "Garden of Eden": Constraints of marine diets for human societies. In *Food and Evolution*, ed. M. Harris and E. Ross, pp. 285–310. Philadelphia: Temple University Press.

CHAPTER 8

THE CASE FOR EXPLOITATION OF WETLANDS ENVIRONMENTS AND FOODS BY PRE-SAPIENS HOMININS

Kathlyn M. Stewart

INTRODUCTION

A considerable literature exists on different selective pressures that may have driven the trend to encephalization in hominins, including climate change, social interaction, complex feeding patterns, locomotion, and mental mapping. Diet, or more particularly the enhanced quality of diet, has also been seen both as a major selective force and/or a releaser in brain growth (e.g., Aiello and Wheeler, 1995; Aiello, 1997; Cordain et al., 2001; Cunnane and Crawford, 2003; Chapter 3 in this book). Many studies have pointed out that modern human babies have a high requirement for two long-chained polyunsaturated fatty acids (PUFA) – docosahexaenoic acid and arachidonic acid (DHA and AA) – to fuel normal growth of the fetal and infant human brain (e.g., Crawford et al., 1992, 1997; Farquharson et al., 1992; Cunnane and Crawford, 2003). These fatty acids are also needed for the ongoing maintenance of the adult brain. PUFAs are dietarily "essential," that is, some of the lipids which comprise the mammalian central nervous system cannot be synthesized and must come from external food sources. Researchers agree that consistent consumption of sources of essential fatty acids are needed to maintain the AA:DHA balance in the developing brain of infants (e.g., Crawford et al., 1992; Cunnane et al., 1993; Broadhurst et al., 1998, 2002; Cunnane, 2005).

Given these specific requirements, early hominins must have evolved in an environment which could provide consistent amounts of food resources with abundant dietary long-chained PUFA to fuel the developing brain (Aiello and Wheeler, 1995; Aiello 1997; Broadhurst et al., 1998; Cordain et al., 2001; Cunnane, 2005; Crawford, 2006). Considerable discussion exists about the exact sources of these high-quality foods, with candidates ranging from freshwater or marine fish and shellfish, mammal meat, and plant foods. Plant foods are generally rejected as the sole source of essential fatty acids, because those with sufficiently high calories contain almost no DHA or AA (e.g., Salem in Cordain et al.,

Human Brain Evolution, Edited by Stephen C. Cunnane and Kathlyn M. Stewart

2001; Cunnane, 2005). Meat from mammals is another source of these fatty acids; however, the fats available in the soft tissues of mammals and other land-based animals are less accessible, contain lower amounts of long-chain PUFA (e.g., Broadhurst et al., 1998; Cunnane, 2005; Chapters 2 and 3 in this book), and evidence presented below suggests that *routine* procurement and consumption of mammal meat was a later hominin adaptation. Tropical and freshwater fish have long-chained PUFA ratios more similar to that of the human brain than other food sources, making them an ideal source of AA and especially DHA (e.g., Broadhurst et al., 1998; Chapters 2 and 3 in this book). It is more likely therefore that hominins evolved in freshwater and/or marine biomes where fish and/or shellfish could provide abundant and consistent essential fatty acids for the developing brain.

The emergence of *Homo* at about 2.4 Ma with their already encephalized brains, smaller teeth, full bipedalism, and rough toolmaking abilities, indicates these hominins were already consistently accessing high-quality foods containing essential fatty acids. There is however no direct evidence that *Homo* at this time was *consistently* accessing fish or shellfish, or indeed any other source of DHA. In fact, early *Homo* is suggested to be a dietary generalist based on evidence from several sources: stable carbon isotope analysis of teeth enamel, craniodental studies, and evidence from archaeological sites (e.g., Sponheimer and Lee-Thorp, 2003; Peters and Vogel, 2005; Ungar et al., 2006)

Closer examination of sites associated with early *Homo* remains indicates most frequent association with lake margin sites (e.g., Behrensmeyer, 1975; Sikes 1994; Blumenschine et al., 2003), with evidence (when available) that at least some of these are not just death sites, but sites *Homo* frequented in life (e.g., Ashley et al., 2009). A similar pattern of association is also found at some earlier hominin sites, where data are available (e.g., *Sahelanthropus* [Vignaud et al., 2002]). Further, while many researchers have focused on hominin exploitation of terrestrial environments, several lines of evidence strongly point to hominin consumption of aquatic vegetation, including C4 sedges (e.g., van der Merwe et al., 2008). Frequenting of wetlands environments, consistent consumption of aquatic vegetation, and both inadvertent and deliberate ingestion of small fish and aquatic invertebrates, is documented for extant ape taxa. This chapter discusses evidence for these adaptations in early hominins.

Heightened Plio-Pleistocene climatic instability and, in particular, intensified seasonality (e.g., Trauth et al., 2007) must have forced hominins to rely more on wetlands environments for potable water and food. Fish and aquatic invertebrates were available and often abundant in these environments, often in good condition, in sharp contrast to dessicated terrestrial vegetation and fat-depleted mammals. This chapter presents evidence for hominin consumption of these foods, which were sources of essential fatty acids and other "brain-selective" nutrients, which include certain minerals – particularly iodine and iron, vitamins, and omega-3 fatty acids like DHA (see Chapter 3). This pattern of consumption exaptively fulfilled several "preconditions" for encephalization, including the reduction of gut size and the "release" of energy which could be diverted to brain growth (Aiello and Wheeler, 1995; Aiello, 1997).

Plio-Pleistocene precessionally forced wet/dry cycles resulted in the shift from high lake to low lake levels in eastern Africa (Deino et al., 2006; Trauth et al., 2007), causing dramatic lake level fluctuations, and associated decimation of fish and aquatic invertebrate communities. This chapter discusses the implications of these events for encephalization and other physiological and morphological changes in hominins, and ultimately implications for evolutionary change in the hominin lineage.

HOMINID EXPLOITATION OF WETLANDS ENVIRONMENTS AND RESOURCES

Extant Apes

Early (and later) hominins were "colonizers" (Wells and Stock, 2007). Given the environmental aridity of the late Miocene, and degradation and fragmentation of preferred habitats (e.g., Cerling et al., 1993), late Miocene large apes faced competition for habitats and resources. This competition forced some apes into colonization of new habitats; it was therefore a major driving force in the hominin divergence from apes. Successful colonization of these habitats required hominins to possess generalized physiological traits and develop new behavioral and ecological strategies of adaptation (Foley, 1987; Wells and Stock, 2007).

Insight into these strategies comes from field studies of extant large apes. These apes, including chimpanzees (*Pan troglodytes, P. paniscus*), gorillas (*Gorilla gorilla*), and orangutans (*Pongo pygmaeus*), are phylogenetically and anatomically the closest relatives to late Miocene apes (e.g., McBrearty and Jablonski, 2005) and hominins. Hominins probably shared a common ancestor with chimpanzees until a final split later at about 6.3 Ma at the earliest (Patterson et al., 2006), with gorillas and orangutans diverging earlier. As such, extant apes are the closest proxy we have for the behaviors of late Miocene apes and earliest hominins. Behavioral, ecological, and specific morphological (craniodental) data assimilated from field and dental studies on extant apes therefore can provide insights into how the earliest hominins adapted to fluctuating environments and scarce food resources.

All extant large apes inhabit much reduced areas resulting in part from late Tertiary and Quaternary climate change, and in part from ongoing encroachment of human activities, particularly hunting and logging (e.g., Poulsen and Clark, 2004; Rizkalla et al., 2007; Stokes et al., 2008). The apes' habitats include discontinuous evergreen rainforests, woodlands, and occasionally wetlands (chimpanzees), lowland rainforest and swamp forest (bonobo chimpanzees, lowland gorillas, orangutans), and montane forests (mountain gorillas, orangutans) (e.g., Goodall, 1986; Uehara, 1990; Kaplan and Rogers, 1994; McGrew et al., 1996; Wrangham et al., 1996b; Poulsen and Clark, 2004; Knott, 2005).

Most early field studies of apes focused on groups of common chimpanzees in terrestrial rainforest and woodland environments (e.g., Wrangham, 1977; Goodall, 1986). Few early studies were conducted in wetlands environment, partly because of difficult access, leading to the view that chimpanzees and gorillas avoided water (e.g., McGrew, 1977; but see Nishida, 1980; discussion in Fay et al., 1989). "Wetlands" here refers to "areas that are periodically or continuously inundated by shallow water or have saturated soils, and where plant growth and other biological activity are adapted to the wet conditions" (Tooth and McCarthy, 2007:4). This perspective changed when field studies in the late 1980s reported large numbers of lowland gorillas in swamp forests in northern Congo (Fay et al., 1989; Fay and Agnagna, 1992; Blake et al., 1995). While field studies in wetlands environments are still uncommon relative to other studies (e.g., Poulsen and Clark, 2004), there is increasing documentation of both short- and long-term utilization of wetlands (generally swamps) and wetlands resources by gorillas and chimpanzees, particularly bonobos (e.g., Uehara, 1990; Blake et al., 1995; Williamson and Usongo, 1996; Magliocca and Gautier-Hion, 2002; Poulsen and Clark, 2004; Breuer et al., 2005; Stokes et al., 2008).

Most apes have developed strategies for coping with scarcity and abundance of foods, usually on a seasonal basis. Data from field studies clearly show that the preferred foods (items which are "overselected" often relative to their natural abundance [Marshall and Wrangham, 2007]) of modern large apes are primarily fruits and/or leaves/stems from terrestrial forests, a diet which is generally low in fat and high in fiber and carbohydrates; dietary protein content is also low for chimpanzees (e.g., Nishihara, 1995; Remis et al., 2001; Conklin-Brittain et al., 2002; Rogers et al., 2004; Tutin and Fernandez, 2009).

Fallback foods are less preferred resources consumed often in the dry season, when preferred foods are unavailable (Marshall and Wrangham, 2007); they are usually more fibrous, and are lower in carbohydrates and fats. Fallback plant foods are similar for both bonobo chimpanzees and western lowland gorillas, in being more fibrous than those of common chimpanzees, and chiefly include parts of herbaceous plants and higher fiber fruits from both aquatic and terrestrial environments; rare animal foods are also consumed (Badrian et al., 1981; Malenky et al., 1996; Doran et al., 2002).

Swamp food plants are important fallback foods for gorillas and bonobos because of their "abundance, nutritional value and mineral content" (Rogers et al., 2004; see also Uehara, 1990; Nishihara, 1995; Kuroda et al., 1996; Magliocca and Gautier-Hion, 2002). Both gorillas and bonobos seek preferred and fallback swamp vegetation which is high in protein and mineral content, particularly sodium, calcium, and potassium. Sodium content is low in herbivore diets and is deficient in most herbivores (eg Rode et al., 2002); regular sodium replacement is important in primates (Kempf, 2009). "Sodium hunger" in gorillas is met through aquatic plant consumption (see also Kuroda et al., 1996; Magliocca and Gautier-Hion, 2002; Doran et al., 2004). One of the most "heavily consumed" aquatic plants by gorillas and bonobos is *Hydrocharis*, whose roots are high in sodium and other minerals, and protein. *Hydrocharis* roots and other parts were eaten by bonobos, and by gorillas regularly throughout the year in most study sites (Uehara, 1990; Nishihara, 1995; Kuroda et al., 1996; Malenky et al., 1996; Remis et al., 2001; Doran et al., 2002, 2004; Rogers et al., 2004). *Hydrocharis* roots apparently can only be fed on while sitting in water (Kuroda et al., 1996), maybe reflecting a long-term feeding adaptation (Kingdon, 2003), perhaps in response to feeding on wetlands foods.

Selection of these foods in swamps appears to reflect the low sodium and calcium content found in some favored terrestrial foods, including the popular herb *Haumania* (Nishihara, 1995; Malenky et al., 1996; Remis et al., 2001; Magliocca and Gautier-Hion, 2002; Doran et al., 2002, 2004). Many aquatic (and terrestrial) plants are also deliberately selected by large apes for their medicinal properties – several are consumed for gastrointestinal ailments – often the same as those used by modern humans for the same ailments (e.g., Ohigashi et al., 1991; Huffman, 1994; Cousins and Huffman, 2002).

The importance of swamp resources to these large apes is demonstrated by deliberate and regular visits to swamp forests, often some distance away (Nishihara, 1995; Remis, 1997; Magliocca and Hion, 2002). At one study site (Ndoki), gorillas foraged 45% of time in swamp forests and 41% in mixed forests (Malenky et al., 1996). Bonobos at the Wamba study site frequented the swamp forest (and disturbed forest) almost daily, primarily to consume numerous herbaceous plants which were "constantly available" (Hashimoto et al., 1998). A transect through the swamp forest at Wamba found bonobos were consuming parts of 34 different species of swamp-living trees (Idani et al., 1994).

Bonobo chimpanzees occasionally eat fish (e.g., Nishida, 1972; Kano, 1979; de Waal, 1990: 185) and aquatic mollusks and snails (Badrian et al., 1981; Badrian and Malenky, 1984; Susman et al., 1985), presumably to bolster protein, fat, and macro/micronutrient components; fish and invertebrates are high in fats, protein, and minerals including potassium. "Tiny transparent crustaceans" (de Waal and Lanting 1997: 80) are scooped up by hand while in streams, as are small fish (Badrian and Malenky, 1984); the crustaceans are also a popular food with local people. Anecdotal reports state that bonobo chimpanzees spend time digging in mud and pools for "mudfish" (Nishida, 1972), a term which usually refers to *Clarias*, the catfish. *Clarias* burrow in mud when water levels are low (see below). *Clarias* are common in African archaeological sites (Van Neer, 1986; Stewart, 1994), and are hunted by fishermen today. Gorillas reportedly do not eat fish, although anecdotally, they have been seen scooping up tiny aquatic invertebrates (Kuroda et al., 1996). Gorillas inadvertently eat insects and ants when they eat leaves and fruit of terrestrial vegetation (Tutin and Fernandez, 1992), and similarly must inadvertently ingest tiny aquatic shrimp and fish while eating underwater roots of *Hydrocharis* and sedges, or while scooping up water to drink.

The fallback foods of most common chimpanzees include less fibrous vegetation, pith, and occasional small terrestrial animals and invertebrates (e.g., Goodall, 1986; Nishihara, 1995; Reynolds et al., 1998; Conklin-Brittain et al., 2002). Common chimpanzees in terrestrial rainforests consume insufficient protein seasonally (Milton, 1999) and also lack some minerals in their preferred diet (Milton, 1999). While most common chimpanzees do not reportedly frequent wetlands, several groups inhabit "marginal" habitats which are near wetlands. These groups rely on aquatic vegetation as fallback foods, similar to bonobos and lowland gorillas (e.g., Nishida, 1980; Sakamaki, 1998). Algae and aquatic plants are frequently consumed: at Mahale Mountains 16 of 192 wild plant species consumed by chimpanzees were classified as lakeside beach or riverside vegetation (Nishida and Uehara, 1983; Sakamaki, 1998). Papyrus, an abundant wetlands sedge, is eaten by Kanyawara chimpanzees "during periods of extreme scarcity" although it is not a preferred food (Wrangham et al., 1996a). It is also eaten by chimpanzees at Mahale along with other sedges, and other aquatic shore plants (Nishida and Uehara, 1983).

The differences in fallback foods of extant apes are to some extent expressed in their craniodental features (Lambert, 2007; Ungar et al., 2007; Marshall and Wrangham, 2007; Vogel et al., 2008). Therefore, similar features in fossil apes and hominins should reflect a similar fallback diet. However, microwear patterns sometimes point to a different mastication pattern than that for fallback foods (Grine et al., 2006; Ungar et al., 2006). The teeth of common chimpanzeess and bonobos have thin enamel and low cusp relief, which is linked to the consumption of their leafy fallback diet (Vogel et al., 2008). However, unique cusp cresting features of bonobo chimpanzee teeth indicate greater adaptations for shearing, probably due to their greater reliance on tough, fibrous vegetation than common chimpanzees (Kinzey, 1984; Kono and Suwa, 2008). Gorillas also have thin enamel, but molars with steeper cusps and longer crests for shearing (Ungar, 2004), also probably an adaptation to cope with tough, fibrous fallback aquatic and/or terrestrial vegetation. Orangutan teeth have thicker enamel and crenulated occlusal surfaces, linked to more routine consumption of hard-object fallback foods – tough leaves, bark, and nuts (Rodman, 1988; Vogel et al., 2008).

Large populations of western lowland gorillas and smaller populations of common chimpanzees are being increasingly and unexpectedly found in swamp forest environments, particularly in the isolated northern Congo (e.g., Blake et al., 1995; Poulsen and Clark, 2004; Stokes et al., 2008). These populations are not new (e.g., Fay and Agnagna, 1992), but have been increasing in numbers, in part reflecting the need for isolation from encroaching human activities: hunting, logging, and mining (e.g., Blake et al., 1995; Walsh et al., 2003; Stokes et al., 2008). Surveys of nesting behavior of gorillas in northern Congo indicate that these apes are "capable of living in flooded swamps for extended periods without the need to retreat to *terra firma* forest" (Blake et al., 1995:288). Some gorillas and chimpanzees, however, move to *terra firma* during high waters; others stay in flooded areas (Poulsen and Clark, 2004). In times of low waters most apes move to the swamp forest. Both preferred foods and fallback foods are available on both *terra firma* and in the swamp forest (Poulsen and Clark, 2004). The increasing use of swamp forests may reflect the strategy of lowland gorillas to move to familiar fallback foods – and a "fallback" environment – for long periods, when their preferred resources and habitat are inaccessible (i.e., access too dangerous because of humans). Common chimpanzees, which break into smaller foraging groups when preferred resources are scarce (fission-fusion; e.g., Chapman et al., 1993), have clearly adapted this strategy to a swamp environment.

In short, all extant ape taxa consistently visit wetlands to consume vegetation which contain micro- and macro-nutrients – particularly sodium, potassium, and protein – not apparently provided in their terrestrial foods. Many wetlands foods are also selected for medicinal properties. Fish and mollusks are consumed by bonobo chimpanzees, and anecdotally by gorillas, on an occasional basis, perhaps to bolster protein and fats. Apes are also undoubtedly ingesting small aquatic invertebrates and fish as an inadvertent by-product of eating submerged swamp vegetation. Those apes living in isolated swamp forests have apparently coped with long-term inaccessibility (scarcity) of preferred foods and environments by occupying a "fallback" environment, and by relying much more heavily and for longer periods on fallback foods.

Late Miocene Apes

The short-term and long-term exploitation by extant apes of wetlands and their foods, as a fallback to more preferred habitats and resources, may have parallels to the strategies of late Miocene apes and the earliest hominins. Throughout much of the Miocene, large apes were taxonomically diverse and widespread in the tropical and subtropical forests in Europe and Asia, and Africa. Increasing climatic diversity in the mid-late Miocene reduced and fragmented the tropical/subtropical forests, resulting in more open, seasonal environments and different food resources (e.g., Andrews et al., 1997; Bernor, 2007). While early Miocene apes were thought to be largely frugivorous, middle and late Miocene apes showed more craniodental and therefore dietary diversity, in association with changing climate and resources. They probably either incorporated more leaves and stems in their diet or more hard object foods (e.g., Ungar and Kay, 1995; Ungar, 1996). Increasingly large apes were forced into different habitats and niches, and while some adapted, others disappeared. Disappearances followed a climatic gradient from north to south (e.g., Potts, 2004), with later apes living in southern Eurasia and in Africa (e.g., Kordos and Begun, 2002; Liu and Zheng, 2005; Suwa et al., 2007).

Several of the latest known Eurasian apes inhabited swamp forest environments (Potts, 2004), and show adaptations to this environment. Best known is *Oreopithecus*, an ape found in Italy and dating to about 8.5 Ma (Cameron, 2004), which inhabited a swamp environment on a large island for 2 Ma, isolated from outside predators (Cameron, 2004). Among other specializations, its dental morphology indicates a focused possibly leaf-based diet (Carnieri and Mallegni, 2003), although Harrison and Rook (1997: 341) have suggested it may have eaten wetland plants, such as "water lilies, reeds, sedges, cattail, pondweeds, horsetails and stoneworts, all of which are abundantly represented in the pollen spectrum …" Recent microwear analysis of *Oreopithecus* teeth suggests a more abrasive diet than previously thought (Galbany et al., 2005), thus supporting Harrison's suggestion for consumption of sedges and reeds. For such a diet, more specializations for terrestrial locomotion may be expected (Harrison and Rook, 1997), and it is suggested *Oreopithecus* could have been habitually bipedal (e.g., Rook et al., 1999).

Oreopithecus has been viewed as an "endemic insular form" (Köhler et al., 2001). However, while its isolation allowed it to develop specialized adaptations in the absence of predators or much competition, what is important is that *Oreopithecus* – classified as a "stem hominid" (now hominin) (Harrison and Rook, 1997) – was able to successfully adapt both physically and behaviorally to a swamp forest environment for 2 million years. Similarly, other late Miocene ape remains are associated with swamp woodland and forest contexts. These include different species of *Lufengpithecus*, which variously ate hard and/or tough objects, and/or soft foods (Liu and Zheng, 2005). *Dryopithecus* specimens have been recovered from Rudabanya Hungary at about 10 Ma and also from other swampy environments at about this time (Kordos and Begun, 2002). These apes were also apparently adapted to long-term swamp foods and environment, as recent analysis of *Dryopithecus* teeth suggests that its buccal microwear patterns most resembled that of gorillas, which eat "abrasive, tough" plants (Galbany et al., 2005).

As the climatic drying progressed, large apes moved into Africa probably from the Mediterranean (Kordos and Begun, 2002). Some of the apes adapted to wetlands (swamps) would presumably "follow" these environments south as more northern environments fragmented and disappeared. Of the three late Miocene apes known from the eastern rift of Africa, all were associated with or near wetlands, and in more seasonal but still relatively closed terrestrial environments. *Samburupithecus*, at about 9.6 Ma, was associated with swamps with nearby woodlands; near-water and/or aquatic animals made up very roughly 20% of the total fauna (Tsujikawa, 2005). The associated faunas include hippopotami, and two species of thryonomyids, which are cane rats; modern cane rats are associated with low-lying swampy areas (López-Antoñanzas et al., 2004); hominin remains are also frequently associated with cane rat remains (which are today a popular food). *Nakalipithecus* (about 9.8 Ma) may be close to the last common ancestor of humans and large apes (Kunimatsu et al., 2007), and was found in an environment similar to *Samburupithecus* but more forested (Kunimatsu et al., 2007). *Chororapithecus* (10.5–10 Ma) is suggested to be a member of the gorilla clade (Suwa et al., 2007), and is associated with a forested lake margin environment, and a primarily primate fauna (Suwa et al., 2007). These three apes were large-bodied, similar to the size of small to medium gorillas (Bernor, 2007), and like gorillas, probably spent considerable time on the ground.

All three late Miocene African apes had thicker tooth enamel more similar to *Pongo* than *Gorilla* or *Pan*, suggesting a fallback diet of tougher and/or harder foods. The diet of *Samburupithecus* was probably tough vegetation which required considerable processing (Cameron, 2004), while *Chororapithecus* is suggested to have consumed tough fibrous foods, as well as hard or abrasive foods (Suwa et al., 2007). *Nakalipithecus* is interpreted to have a "hard object" fallback diet (Kunimatsu et al., 2007); however, its tooth morphology is similar to the Eurasian late Miocene ape *Ouranopithecus*, whose morphology has recently been reported as adapted to tough vegetation (Merceron et al., 2004).

Therefore *Samburupithecus*, *Nakalipithecus*, and *Chororapithecus* had craniodental adaptations for masticating tough, fibrous vegetation, and the former two were associated with a swampy setting, the latter a lake margin setting. By analogy with extant gorillas and bonobo chimpanzees, and in keeping with craniodental and behavioral adaptations of other late Miocene apes, it seems likely that they were exploiting tough, fibrous vegetation at least occasionally in aquatic settings, possibly both as a secondary food resource, and to fill nutritional deficiencies. It is not a great leap to suggest that the ancestral hominins also occasionally consumed wetlands vegetation for these reasons. Because of disappearing habitats and resources in the late Miocene, they were increasingly forced by competition into wetlands environments.

Following these late Miocene apes, very few ape remains have been recovered in Africa with two exceptions – a 7 Ma molar from Lukeino (e.g., Senut and Pickford,

2001) and Pleistocene *Pan* fossils from the Kapthurin Formation (McBrearty and Jablonski, 2005). Plio-Pleistocene apes apparently lived and died in terrestrial forest environments (as they do today), where fossil preservation and recovery is very poor. This contrasts with the comparatively rich Plio-Pleistocene record for hominins, providing support for the frequenting of wetlands by hominins, where their remains are better preserved.

EARLY HOMININS: COLONIZATION OF NEW ENVIRONMENTS

The Earliest Hominins

An approximate 2.5 million year hiatus occurs between the last Miocene African ape and the earliest hominin, *Sahelanthropus* (Chad, between 6.8 and 7.2 Ma; Brunet et al., 2002; Lebatard et al., 2008), followed by *Orrorin* (Kenya, ca 6 Ma; Senut and Pickford, 2001), *Ardipithecus kadabba* (e.g., Ethiopia [Haile-Selassie et al., 2004a]), *A. ramidus* (Ethiopia ca 4.39 Ma [White et al., 1994, 1995, 2009]) and a fourth hominid from Niger, of which little is yet known. These earliest hominins came from widespread locations, in east, central, and west Africa. This surprisingly widespread dispersal, and particularly the location of two hominins far west of the eastern Rift valley (plus the recovery of the Pliocene hominin *Australopithecus bahrelghazali* from Chad), makes more sense when viewed against the trajectory of a large river system flowed east to west across what is now the sahel zone from the Red Sea mountains probably as far west as the present-day Niger river system, and possibly to the Atlantic. This system existed during at least much of the Miocene (McCauley et al., 1982, 1998). Present-day distribution and taxonomy of some freshwater fish taxa and hippopotami indicate that this river system facilitated movements of fish and other freshwater animals across what are now deserts in northern Africa, until the late Miocene (Boisserie et al., 2005; Otero et al., 2009; Stewart, 2009). Evidence below indicates that several of these earliest hominins were associated with wetland environments. Their widespread remains may result from late Miocene dispersion along this river system, but this hypothesis needs further testing, particularly through recovery of further fossils.

Sahelanthropus frequented wetlands environments. The reconstructed environment of this hominin is "gallery forest at the edge of a lake area," with nearby savanna and grassland (Vignaud et al., 2002:155), a more open environment than with late Miocene apes. The diverse and abundant fish and aquatic reptile fauna associated with the hominin remains suggests a somewhat deep, reasonably well-oxygenated lake with a swampy margin and stands of aquatic plants (Vignaud et al., 2002). "Amphibious" mammals – mainly hippopotami and otters – represent about 28% of the mammal fauna. Detailed studies of *Sahelanthropus*' teeth are not published, but it had thick enamel (Brunet et al., 2002), similar to teeth of orangutans, suggesting that it could eat harder and/or tougher foods as fallback foods (Vogel et al., 2008). Analogy with extant apes suggests that *Sahelanthropus* was able to and probably did occasionally consume lake shore sedges and other available aquatic vegetation.

There is no indication of either carnivore modification to, or fluvial transport of, the hominin bones (Vignaud et al., 2002), so *Sahelanthropus* may well have lived (and died) in this lake margin environment. Of course, the recovery of hominin (or other vertebrate) remains in wetlands environments does not necessarily mean that these hominins lived in or even frequented these environments (Sikes, 1994). Bones are preferentially preserved in lake, river, or fluvial sediments, so almost by default they will be associated with an aquatic setting (Behrensmeyer, 1975; Sikes, 1994). However, while the burial location of the bones may represent their living environment – and often does (Behrensmeyer, 1993) – the bones may also have been transported there from elsewhere by fluvial or channel flow (e.g., Behrensmeyer, 1975, 1982; Lyman, 1994). Or, as has been pointed out by Ward et al. (1999), many of the hominin remains at Kanapoi and perhaps Allia Bay have carnivore damage, so carnivores may have dragged them from one setting to another. However, the remains of many apes and hominins are found associated with swampy or marshy grassland environments, which are unlikely spots to be dragged by predators (except crocodiles), or for lag deposits from fluvial transport. They may actually represent where these hominids frequented in life.

The other earliest hominins are also associated with wetlands, but with the exception of *Ardipithecus kadabba*, there is little evidence they actually inhabited the wetlands. The *A. kadabba* remains are associated with swamps and wet grasslands, and with terrestrial environments reconstructed as forest and/or woodland, possibly at a much higher elevation than today (WoldeGabriel et al., 2001; Haile-Selassie et al., 2004b). Ecomorphological analyses conducted on the associated bovids indicate that 48% were associated with bush, woodland, swamp, and near-water settings (Haile-Selassie et al., 2004b), suggesting the hominin lived in a swamp or near-water community. *A. ramidus* lived in or near "groundwater-supported grassy woodland to forest" (Woldegabriel et al., 2009:65e4), while *Orrorin* is associated with woodlands and/or forest and wet grasslands near a lake setting.

Freshwater Environments and the Spread of C4 Wetlands Grasses

The emergence of the australopithecines is coincident with dramatic environmental change. Extensive ongoing rifting and uplifting in eastern Africa dramatically altered rain shadow patterns (e.g., deMenocal, 1995, 2004; Trauth et al., 2007), resulting in long-term drying throughout the late Miocene and Plio-Pleistocene, with accompanying fragmentation of forests, and in particular, the spread of both aquatic and terrestrial C4 grasses and sedges. C4 grasses are first known in Africa in the late Miocene near the equator (Cerling et al., 1993), with their real impact occurring with their spread into eastern and southern Africa during the Pliocene (Ségalen et al., 2007).

Several hypotheses have been postulated to explain the association between climate change and hominin evolution (see, e.g., Behrensmeyer et al., 1997; Bobe et al., 2002; Bobe and Behrensmeyer, 2004). Dart (1925) was one of the earliest proponents of the savanna hypothesis, which linked the spread of grasslands with the onset of bipedalism, and hominins moving from forest to savanna environments. Fossil evidence of speciation and extinction events associated with periods of climate change are the basis of the faunal turnover hypothesis (e.g., Vrba, 1985, 1988). The variability selection hypothesis focuses instead on the importance of climatic variability in shaping hominin evolution (Potts, 1998). Discussion of these hypotheses and their data is not the focus of this chapter. However, several papers (e.g., Sikes, 1994; Plummer et al., 1999; Lam, 2008) highlight the uncertainties surrounding the most basic assumptions of Plio-Pleistocene hominin habitat, tool use, and diet, despite the plethora of studies and investigations devoted to resolving these uncertainties. The data presented in the chapter in part complements many earlier studies by focusing on the spread of C4 *wetlands* sedges and grasses (Greb et al., 2006), in association with the formation of new lakes and lake environments. This chapter proposes that early hominins colonized and later radiated in these environments.

The ongoing rifting and uplifting in eastern and southern Africa directly or indirectly resulted in the formation of new lake basins. This event was one of the most important catalysts to late Tertiary/Quaternary evolutionary change in aquatic and semiaquatic vertebrates and invertebrates (see also Beadle, 1981: 24). Newly formed basins included the large Rift lakes (e.g., Lake Malawi) which were often filled by newly dammed or reversed rivers. Small lakes or swamps were also formed through volcanic activity and lava flow damming (Beadle, 1981), or from underground springs and seepages (Ashley et al., 2002, 2004). Seasonal or semipermanent pans or playas formed in shallow basins (McClanahan and Young, 1999).

These newly formed basins provided numerous colonization opportunities for animals already adapted to the river and extensive swamp (e.g. Congo, Sudd) environments which characterized Miocene Africa. Hippopotami, crocodiles, otters, and smaller mammals, fish, birds, and invertebrates colonized these lake and river margin environments with subsequent radiations through the late Cenozoic (e.g., Stewart 2003a,b, 2009; Storrs, 2003; Werdelin, 2003; Boisserie et al., 2005; Otero et al., 2009). Of particular importance were the lake and river *margin* environments, which include the littoral zone: water less than 15 feet in depth, which allows for continuous mixing and for photosynthesis to the bottom (adapted from Beadle, 1981). These margin environments are also referred to here as water/land ecotones (Shabel, 2009; see also Chapter 9 in this book). These are distinguished from the riparian zone, which is defined as "located on the *bank* (italics mine) of a natural watercourse" (Merriam-Webster). Numerous plants colonized these water/land ecotones, including emergent plants (i.e., partly submerged, including mainly sedges such as papyrus, reeds, rushes using the C4 photosynthetic pathway), floating plants (e.g., water hyacinth), and submerged plants (e.g., fennel pondweed), which spread into these new environments in the late Tertiary (Beadle, 1981; Greb et al., 2006). The Niger delta pollen sequence indicates sedges were a component of the grasslands and/or freshwater there after ca 6.6 Ma (Morley in Peters and Vogel, 2005).

Many of these plants are a food source for freshwater or freshwater-dependent animals higher in the food chain (Greb et al., 2006: 28). During more prolonged arid periods, as proposed for the Plio-Pleistocene (e.g., Trauth et al., 2007), these lake/river margin wetlands would have been refuges for hominins and other faunas dependent on water and aquatic food sources (Beadle, 1981; Greb et al., 2006; Maslin and Christensen, 2007). Aquatic sedges, and some species of mollusks, fish, birds, and mammals are adapted to severe water level fluctuations, including near-dessication, and even thrive in these conditions (Beadle, 1981; McClanahan and Young, 1999), making a dependable food source in arid periods. Would hominins have been affected by wetlands vectors associated with spread of parasitic or viral diseases? Although the vast majority of swamp-dwelling mosquito larvae do not contain malaria (Beadle, 1981), some do, and there are also many other parasitic chronic diseases which affect both humans and extant apes (e.g, Sadun et al., 1966); these probably also have affected ancestral apes and hominins. As discussed above ("Extant Apes") there is a vast literature on the consumption by extant large apes of specific aquatic and terrestrial plants; many are also consumed by local human populations.

Despite their productivity, the lake and river margin trophic niches were underexploited through the late Tertiary and Quaternary. The main consumers were hippopotamids which eat considerable amounts of lake and lakeshore grasses, and crocodiles which exploit animal prey in both the lake margin and lakeshore (riparian) ecotones. Smaller consumers include otters, turtles, cane rats, and a diversity of birds and fish. This chapter suggests that hominins too exploited this productive lake and river margin ecotone, where minimal competition existed for foods including sedges, fish, or aquatic invertebrates (other than piscivorous and/or molluscivorous fish). The only large resident predator taxon would be crocodiles (*Crocodylus*, not *Euthecodon*), certainly a formidable predator, but not, I suggest, posing the danger of the riparian (shore) zone, where hominins would be targeted by numerous terrestrial carnivore taxa, each with different prey-capture strategies (Blumenschine, 1987; Blumenschine et al., 1994: table 1). Like terrestrial carnivores, crocodiles also are at rest at certain periods of the day or season; hominins would be highly aware of the prey-capture strategies of this lone predator.

Previous studies have found that early *Homo* was most commonly associated with lake margin environments (e.g., Behrensmeyer 1975; Sikes, 1994), while *Paranthropus* was commonly associated with both lake margin and especially fluvial environments (Behrensmeyer, 1975). Early hominins may have initially filled the lake/river margin trophic niche only seasonally, as seen with extant apes (Verhaegen et al., 2002; Wrangham, 2005). As climatic conditions and food supplies worsened, they relied increasingly on these wetlands and wetland resources. Hominins undoubtedly also did utilize the riparian ecotone (lake/river shore), as indicated by location of some archaeological sites (e.g., Koobi Fora [Harris and Isaac, 1997; Pobiner et al., 2008]), in part to access trees for shelter and protection (e.g., at Olduvai [Bamford, 2005]), and for raw materials for tools.

PLIO-PLEISTOCENE CLIMATE INSTABILITY AND USE OF WETLANDS RESOURCES

Early Australopithecines

The appearance of *Australopithecus afarensis* and *A. anamensis* after 4.2 Ma is coincident with a late Cenozoic trend to drier conditions (Cerling et al., 1993; White et al., 1993; Bobe and Behrensmeyer, 2004; deMenocal, 2004; Bobe, 2006), overlain by alternating extreme wet and dry periods, including high water stands at about 4.0 and 3.4 Ma (Trauth et al., 2007). At Hadar, remains of *A. afarensis* were common (370 specimens; Kimbel et al., 1996; Reed, 2008) throughout the approximately 500,000-year sequence from about 3.4 to 2.9 Ma, indicating consistent hominin frequenting of the environments around the lake. Through this period, deposition cycled between lacustrine- and fluvial-derived sediments (Campisano and Feibel, 2007). While detailed pollen recovery and analysis from the 500,000-year section documented a diversity of terrestrial biomes associated with the hominins, attempts to associate *A. afarensis* with a particular terrestrial biome were inconclusive, as *A. afarensis* "showed no preference for any single biome" (Bonnefille et al., 2004: 12127); similarly, Reed (2008) found that "*A. afarensis* survived in a variety of habitats." Further, an ecomorphological study also indicated *A. afarensis* showed no preference for one habitat over another (Reed, 2008).

In fact, the only environmental constant throughout the 500,000-year sequence was the wetlands biome. This biome was characterized by abundant aquatic herbaceous vegetation around the paleolake and/or rivers, with Cyperaceae sedges and reeds which were consistently present – frequently in high numbers – throughout the Hadar sequence (Bonnefille et al., 2004: fig. 2). Further, the association of the highest density of *A. afarensis* remains with the submembers with large proportions of wetlands faunas (reduncines and hippos in early SH, and DD2 members/submembers [Reed, 2008]) indicates utilization of the wetlands biome by hominins.

The other constant was the unchanging microwear profile of *A. afarensis* throughout Hadar – which was very similar to that of extant gorillas, particularly mountain gorillas (*Gorilla gorilla beringei*) who eat tough leaves, grasses, and stems (Grine et al., 2006). The consistent association of *A. afarensis*' tough vegetation microwear profile with a wetlands biome of tough, fibrous aquatic vegetation throughout the Hadar sequence, suggests that *A. afarensis* was regularly feeding on aquatic sedges over time. It could be argued in fact that the consistently available wetlands and wetlands vegetation was a "buffer" for *A. afarensis* against the fluctuating terrestrial habitats and foods throughout the 500,000-year sequence at Hadar.

Similarly, at Hadar, Koobi Fora, Kanapoi, Allia Bay, and the southern Omo River, *A. anamensis* (Leakey et al., 1995) was also associated with mixed wetlands/terrestrial environments: lacustrine and fluvial floodplain deposits as well as riparian woodland and gallery forests (Feibel et al., 1991; Reed, 1997; Ward et al., 1999; Harris and Leakey, 2003a; Schoeninger et al., 2003). The faunas at Allia Bay and Kanapoi contain a large component of aquatic vertebrates – hippopotami, reduncines, fish and aquatic reptiles, several semiaquatic taxa, as well as terrestrial taxa (e.g., Harris and Leakey, 2003b). Boisserie (in Sanders, 2005) informally noted that many of the pre-2 Ma hominin remains have been associated with hippopotami and other aquatic animals, implying that they lived in the same biotopes.

Based on microwear studies, Grine et al. (2006: 301) found that "the wear fabric" seen in the teeth of *A. anamensis* and *A. afarensis* are very similar: "it is perhaps significant that this pattern appears to have remained little changed in this presumptive descendant of *'A.' anamensis*" (Grine et al., 2006: 301). In other words, *A. anamensis* was also apparently eating tough leaves and stems consistently throughout its existence; its association with wetlands environments suggests these were mainly wetlands plants.

While feeding on wetlands vegetation, *A. afarensis* and *A. anamensis* were almost certainly also inadvertently and sometimes deliberately consuming aquatic invertebrates and fish, as do bonobo chimpanzees and gorillas. Increased time in wetlands due to climatic instability meant less reliance on terrestrial foods, even seasonally. Familiar terrestrial sources of protein, micronutrients, and fats – particularly small terrestrial mammals and insects – would be less abundant and/or accessible in a wetlands environment. These hominins, especially later *A. afarensis*, would therefore be increasingly forced to incorporate freshwater sources of these macro and micronutrients – fish and invertebrates – into their diet in the dry season. I suggest this pattern of consumption of fish and aquatic invertebrates was precursive to cyclical and consistent consumption of these foods in later hominins; evidence of consumption of fish and mollusks by early *Homo* is presented below.

The hypothesis of *A. afarensis* and *A. anamensis* consuming wetlands sedges needs further testing, with stable carbon isotope studies of tooth enamel being one such test. However, the evidence cited above is strongly supportive of *A. afarensis*, *A. anamensis*, and perhaps even *Sahelanthropus*, consuming stable and abundant wetlands foods as a buffer against fluctuating terrestrial habitats and food sources (Conklin-Brittain et al., 2002; Wrangham, 2005 made a similar argument based on underground storage organs). I suggest the inadvertent inclusion of aquatic animals in the hominin diet – which occasionally was deliberate – was a precursor to deliberate and consistent consumption of these foods.

Subsequent to 2.7 Ma (or 2.8 Ma [deMenocal, 2004]), climatic instability intensified in eastern Africa, in part caused by the onset of glaciation in the northern latitudes and the suggested "coupling" of climates between northern and southern latitudes (Trauth et al., 2007). This "coupling" influenced the precessional signal in Africa (deMenocal, 2004), resulting in climate extremes (deMenocal 2004; see discussion in Maslin and Christensen, 2007; Trauth et al., 2007). Precessional forcing was primarily responsible for "extreme climate variability" (Trauth et al., 2007) every 400,000 years before 2.7 Ma, and every 800,000 years after 2.7 Ma. Large shifts in precession can significantly alter annual precipitation (Clement in Trauth et al., 2007), sometimes causing highly intensified seasonality. Long periods with highly intensified seasonality, for example, rainy seasons with very high rainfall, can result in a substantial increase in lake basins, river systems, and wetlands areas, resulting in large and/or deep lakes. Such lakes are recorded at about 2.7–2.5 Ma, 1.9–1.7 Ma, and 1.1–.9 Ma (Trauth et al., 2007). These wet periods are then followed by periods of "extreme aridity" (Trauth et al., 2007: 482). It is possible that the regressive lake cycles may occur every ca 100,000 years due to periods of eccentricity (Kingston, 2005). It should be added that while some lakes and swamps would have dried up in these arid periods – as did the Barsemoi lakes near Lake Baringo (Deino et al., 2006; Kingston et al., 2007) – many others would have existed with shallower levels, as seen with Lake Malawi during the late Pleistocene megadroughts (Cohen et al., 2007). The alternating wet and dry periods would have had a "profound influence" on the climate and vegetation of eastern Africa (Trauth et al., 2007: 483), particularly during the dry periods, when hominin populations would have been isolated around sources of potable water.

INTENSIFICATION OF WETLANDS VEGETATION EXPLOITATION

Prior to 2.7 Ma, I suggested that australopithecines utilized the consistently abundant wetlands vegetation at a minimum as a seasonal buffer against terrestrial resource instability. There are several lines of evidence indicating that several of the hominins which emerged around or after 2.7 Ma, that is, *Paranthropus* spp., early *Homo*, and *Australopithecus africanus*, were intensively exploiting wetlands and wetlands foods. First, remains of these hominins continue to be associated with lake and or river margin environments. *Paranthropus* spp. are frequently associated with abundant water and high proportions of mammals eating riparian vegetation (Reed, 1997). They are mainly associated with edaphic grasslands, near bushland and open woodland environments (Reed, 1997). Early *Homo* remains have been found together with *Paranthropus* remains in several different deposits and sites throughout eastern and southern Africa (e.g., Reed, 1997; Herries et al., 2009); several sites are reconstructed as wetlands (e.g., Reed, 1997). Some of the earliest remains of *Homo* were found at Hadar, dated at 2.3 Ma (Kimbel et al., 1996), where the paleoenvironment is interpreted as standing water with reed beds, based on associated cane rat and hippopotamus bones. Similarly, *Homo* remains are associated with well-vegetated swampy lake and river margins and aquatic/semiaquatic faunas at Olduvai Gorge and Koobi Fora (e.g., Pobiner et al., 2008; Ashley et al., 2009).

Stable carbon isotope analysis indicates that these hominins also consumed a large component of C4 foods, suggested in this chapter to be wetlands sedges. The combined results from two stable carbon isotope studies of teeth of 15 *A. africanus* individuals from Sterkfontein (ca 2.5 Ma), and nine *Paranthropus robustus* individuals (ca 1.8 Ma) from Swartkrans and Kromdraai, provide persuasive evidence that a surprisingly high proportion of the diet (ca 40% and 35%, respectively) was made up of C4 foods (van der Merwe et al., 2003; Sponheimer et al., 2005). These findings build on earlier studies showing the importance of C4 foods in australopithecine diets (e.g., Lee-Thorp et al., 1994; Sponheimer and Lee-Thorp, 1999).

A search of possible sources of C4 foods – grasses, sedges, or C4 food consumers – in fact virtually eliminated terrestrial grasses and C4 consumers as the major C4 source. Comparison of australopithecine molar wear with those of grass-eating gelada baboons show the latter have distinctive and different patterns of wear from those of australopithecines (e.g., in van der Merwe et al., 2003). Further, the low nutritional value of grasses would not sustain the protein and fat needs of early hominins (Sponheimer and Lee-Thorp, 2003; Sponheimer et al., 2005). Mammal meat is also an unlikely major source, as most of the small mammals which extant apes (chimpanzees) consume are C3 not C4 plant consumers. The few small mammals that eat C4 vegetation (cane rats, hyraxes, juvenile bovid c4 consumers) could not practically be caught and consumed in sufficient and consistent quantities to provide a strong C4 signature (e.g., Sponheimer and Lee-Thorp, 2003) in hominin fossils, although small mammals were undoubtedly eaten on occasion.

In contrast, sedges have both a higher nutritional content than terrestrial grasses, and are consistently abundant in wetlands, as was seen at Hadar (e.g., Conklin-Brittain et al., 2002). Sedges have several edible parts (rhizomes, culm bases, shoots, bulb [Copeland, 2007]), and elsewhere underground storage organs (of both aquatic and terrestrial plants) have been suggested as a potential C4 food source (Conklin-Brittain et al., 2002; Wrangham, 2005). Papyrus (*Cyperus papyrus*), an abundant giant sedge which grows chiefly around permanent swamps and lakes through central, eastern, and southern tropical Africa (Beadle, 1981), would have been an ideal hominin food source (see also Peters and Vogel, 2005). Its edible rhizome and culm have a carbohydrate and fat profile similar to a potato (van der Merwe et al., 2008). Papyrus is eaten raw by people today living around the Okavango Delta (van der Merwe et al., 2008) and elsewhere in rural Africa, as well as historically (see Tackholm and Drar, 1973), indicating that the present-day human gut can process this food. Papyrus is an occasional fallback food for common chimpanzees, and other sedges are fallback foods for gorillas and bonobos. Sedges are not as common today in southern as in eastern Africa (Peters and Vogel, 2005), but they may have grown along river banks, as with the modern Nyl, Klip, and Blood Rivers (Tooth and McCarthy, 2007). There is possible evidence of wetlands environments nearby the South African fossil cave sites (Luyt and Lee-Thorp, 2003).

The teeth particularly of *A. africanus* and to some extent *Paranthropus*, similar to *A. afarensis* and *A. anamensis*, were also adapted for consumption of sedge-like vegetation. Microwear traces on *A. africanus* teeth were similar to those of *A. afarensis* (Cartmill et al., 2009: 202); the microwear pattern on *A. afarensis* teeth compares very well with microwear traces on mountain gorilla teeth (Grine et al., 2006). Mountain gorillas consume tough, fibrous foods, similar to sedges, indicating that *A. africanus* was probably able to masticate sedges. Other dental studies also support an australopithecine diet of sedges (Puech et al., 1986). Furthermore, *P. robustus* teeth too show some microwear traces compatible with grass eating (e.g., Cartmill et al., 2009), although its very large flat molars suggest it also consumed other foods (see discussion below).

Hominins at Olduvai Gorge were also almost certainly consuming sedges. Stable carbon isotope analysis conducted on teeth of two *Paranthropus boisei* individuals from upper Bed I at Olduvai (ca 1.8 Ma) and Peninj (ca 1.5 Ma) indicate that C4 foods comprised a whopping 77% and 81%, respectively, of the diet (van der Merwe et al., 2008). Analysis of teeth of *H. habilis* from Olduvai (ca 1.8 Ma) indicated that between 23% and 49% of its diet was made up of C4 foods (van der Merwe et al., 2008). Paleolake Olduvai at about 1.8 Ma was small with evidence of level fluctuations from complete dessication to a maximum of 15 km in diameter. A lacustrine plain fringed the eastern margins of the lake, from which a 1-km^2 spring-fed wetlands area extended, with archaeological sites. The wetland developed during the dry periods, and was flooded in the wet periods (Ashley et al., 2009).

High consumption of C4 sedges at Olduvai is supported by both modern and fossil evidence of their presence. Unfortunately, sedge phytoliths preserve very poorly because of their morphologies (Albert et al., 2006: 87). Nevertheless, three species of sedge were documented, with papyrus (*Cyperus papyrus*) by far the most common (Albert et al., 2006; Bamford et al., 2006; Copeland, 2007). By analogy with modern spring and stream-fed wetlands primarily in nearby Ngorongoro Crater, sedges would have been diverse and abundant at paleolake Olduvai; trees with edible fruits would also be present (Copeland, 2007). The finding that the spring-fed wetland sites were used consistently by hominins during both wet and dry periods over the 50,000-year period (Ashley et al., 2009) further suggests that hominins were intensively accessing the lake margin ecotone, almost certainly to feed on vegetation and, I suggest, fish, and invertebrates (discussion below). Woodland and grass settings with evidence of large trees are also documented, which would have provided shelter and protection (Bamford, 2005; Ruff in Gibbons, 2008).

Similar to *A. afarensis* at Hadar, the "significant climate fluctuations" documented at Olduvai Gorge at about 1.8 Ma are *not* associated with fluctuations in hominin behavior over time, but with consistency in behavior. This is seen in consistency in tool types, consistent hominin occupation of spring-fed wetlands sites through wet and dry periods beside paleolake Olduvai (Ashley et al., 2009), and, as suggested above, in apparently frequent consumption of wetlands sedges over 50,000 years. This again supports the "buffer" hypothesis that hominins' utilization of wetlands and constantly available wetlands food resources allowed them to buffer fluctuations in surrounding terrestrial vegetation and food availability.

The hominin (*P. boisei*) from Peninj, dated to about 1.5 Ma, also showed very high C4 values (van der Merwe et al., 2008), presumed to be from consumption of C4 sedges. Sedges (*Cyperus* spp.) and bulrushes (*Typha* spp.) were a large component of the paleoenvironmental flora at Peninj (Dominguez-Rodrigo et al., 2001), also providing support that the dominant C4 signal in hominins represents consumption of sedges, and implying intensive use of wetlands by these hominins.

Stable carbon isotope studies on teeth from three early *H. erectus* specimens from Swartkrans showed that C4 foods made up about 20–25% of their diet (Lee-Thorp and van der Merwe, 1993; in van der Merwe et al., 2008), signaling that more evolved hominins were still frequenting wetlands and eating wetlands sedges. Therefore, as discussed above, there is considerable evidence – stable carbon isotope studies, hominin remains associated with lake margin wetlands, craniodental studies – which points strongly to consumption of C4 sedges by some apes and hominins since the late Miocene. This is not to say that hominins were not still consuming C3 terrestrial vegetation and fruits, but that C4 foods formed a large component of their diet. This chapter therefore suggests that the buffering effect provided by consumption of wetlands vegetation was an adaptational response to unstable climate conditions, which recurred throughout the hominin lineage, and was key to its successful survival.

THE SHIFT TO HIGH-QUALITY FOODS

Two genera of hominins – *Paranthropus* (*boisei*) and especially early *Homo* – emerge around 2.4 Ma (*P. robustus* slightly later) with a greater degree of encephalization than earlier hominins and certainly greater than apes (Aiello and Dean, 1990; McHenry and Coffing, 2000). As discussed below, a high-quality diet can act as both a releaser and a major selective force in brain growth, but there needed to be a shift from a diet of foods low in fat and protein to increased consumption of foods containing AA and particularly DHA and other important nutrients (e.g., see Chapters 2 and 3). As discussed by authors elsewhere in this volume, these nutrients are abundant in freshwater and marine fish and shellfish. There is in fact evidence that early *Homo* and *Paranthropus* were consuming freshwater fish and aquatic invertebrates.

Evidence for Aquatic Invertebrate Procurement

Australopithecine teeth show a trend over time of increasing megadontia (Teaford and Ungar, 2000), with *Paranthropus*' teeth the most megadont. Megadont teeth may initially have been an adaptation to hard, brittle fruits as fallback foods, similar to those of modern orangutan, and were therefore preadapted for crushing other hard-shelled food objects. Comparison with other mammal teeth indicates that *Paranthropus* had similar unique craniodental features as seen in durophagous carnivores which eat aquatic invertebrates, including mollusks, crabs, and other foods at the water–land ecotone (Shabel, 2009; see Chapter 9 in this book). The trend to megadonty from earlier to later australopithecines (Teaford and Ungar, 2000) may also point to some consumption of these invertebrates by earlier australopithecines.

Early *Homo*, whose remains occur in association with *Paranthropus* in several sites (e.g., Reed, 1997), may also have consumed aquatic invertebrates, although their cheek teeth relative to body size are smaller than those of the australopithecines (McHenry and Coffing, 2000), and their tooth enamel is somewhat thinner (Beynon and Wood, 1986). Dental microwear indicates that early *Homo* teeth also have less pitting than the australopithecines, suggesting that early *Homo* may have eaten fewer hard, brittle foods (Ungar et al., 2006), such as freshwater invertebrates, at least not by opening them with their teeth. The presence of numerous battering stones at early archaeological sites may reflect tools used for opening invertebrate shells (and fish crania) (see below).

Evidence for Fish Procurement

Earlier in this chapter, I have suggested that, in addition to sedges, early hominins inadvertently consumed small fish (and invertebrates) while feeding on roots of aquatic vegetation. More intense use of wetlands later resulted in more deliberate procurement of fish and freshwater invertebrates – to fill nutritional gaps – probably on a seasonal basis. After 2.7 Ma, increased climatic instability associated with fluctuating terrestrial habitats and foods, forced hominins increasingly into wetlands during longer periods of aridity. Remains of early *Homo* have been found with *Paranthropus* in several different deposits and sites throughout eastern and southern Africa (e.g., Reed, 1997; Herries et al., 2009); the two genera may have been sympatric (Reed, 1997). *Homo* may have been consuming aquatic invertebrates to some extent, but it is unlikely that *Homo* and *Paranthropus* were both heavily exploiting this same trophic niche. Instead, zooarchaeological evidence indicates that *Homo* was consuming fish, the other easily procured wetlands source of brain-selective fatty acids and other nutrients.

The most definitive evidence for fish procurement by hominins comes from lake or river margin sites near Koobi Fora, at Olduvai Gorge and in the Omo. Often these, and later sites, are associated with repeated hominin occupation over time, suggesting cyclical, seasonal patterns of capture. The Koobi Fora sites were located in the Okote Member of the Koobi Fora Formation dated to about 1.5 Ma, and probably associated with early *H. erectus* (Harris and Isaac, 1997; Pobiner, 2007; Pobiner et al., 2008; Bennett et al., 2009). The associated faunas, most of which were aquatic or semiaquatic, indicate a "significant" lacustrine and/or fluvial shallow water component, "accompanied by swampy, seasonally flooded areas with edaphic grasslands" (Pobiner et al., 2008: 107). Numerous mammal, fish, crocodile, and turtle specimens were recovered, several with cutmarks, averaging about 7% of total fauna, indicating hominin modification of the carcasses. Battering tools and other stone tools were recovered at the nearby FxJj1 site.

Fish elements with cutmarks were recovered at two of the Koobi Fora sites – FwJj 14A and GaJi 14A (Pobiner, 2007), probably the earliest known cutmarks on fish bone.

Figure 8.1 Fisherman from the Republic of the Congo holds his catch of *Clarias*, a catfish. Photograph by Bruce G. Marcot. (See color insert.)

The abundant fish bone specimens (255) were identified as *Clarias* (Fig. 8.1), a large freshwater catfish common today in rivers and lakes throughout Africa. Two spines with cutmarks were reported (Pobiner, 2007: fig. 6.41). The cutmarks may have resulted from severing the spines, a common practice among fishermen today (pers. obs.). Several other sites with fish remains in the Okote Member near Koobi Fora are associated with river channel and floodplain environments (e.g., Harris and Isaac, 1997). These sites date to around 1.5 Ma and contain stone tools and animal bones. The fish bones, mainly *Clarias*, do not have direct evidence of hominin procurement (pers. obs.). However, their association with tools and mammal remains, some with cutmarks, may represent causal association. Modified fish bones were also found at a Shungura Formation site, north of the Turkana basin. Site Omo 71 in Member E, dated to around 2.4 Ma, was probably located on a shoreline (Chavaillon, 1976). This site produced a bone (and quartz) industry, which included a worked portion of a catfish basicranium showing signs of flaking to achieve the form of a perforator, as well as bone flakes (Chavaillon, 1976: 566).

Evidence from Olduvai Gorge also strongly points to fish procurement. Dietary and archaeological evidence presented above indicates that *Paranthropus* and *Homo* spent a lot of time in the Olduvai wetlands, in part consuming sedge vegetation. It would therefore be most surprising if fish were *not* procured in these wetlands. Paleolake Olduvai had sharply seasonal water level fluctuations, probably related to 19–23 kyr precession wet/dry cycles; at times, the lake dried up (Ashley et al., 2009). Not surprisingly given these conditions, abundant remains of *Clarias* and cichlids (medium size perch-like fishes) were recovered. Two Bed I lake margin sites (FLK-22 [Zinj], FLKNN-3) and one Bed II stream channel site (BK) dated between 1.85 and 1.2 Ma (Egeland, 2008), contain large numbers of fish remains recovered in direct association with stone tools (Greenwood and Todd, 1970; Stewart, 1994). Inferring procurement of fish from these early sites requires several lines of evidence. The fish bones at these Olduvai sites show a suite of unusual features compared with naturally occurring fish assemblages: they have skewed element patterns, they are present in large numbers, repetitive procurement over time, and two bones from BK have externally derived notches which appear to be cutmarks (Stewart, 1994; fig. 6).

The bones at these three sites were heavily dominated by large *Clarias*, suggesting that, similar to the Koobi Fora sites, they represented a spawning population stranded in the widely fluctuating waters of the lake at FLK-Zinj and FLKNN-3, and in the stream channel at BK (Stewart, 1994), a situation that commonly occurs (e.g., Greenwood, 1955). Predation was hypothesized, based on skewed elements at all three sites, while hominin predation seemed most likely at BK, where evidence of possible cutmarks was found. Recent reanalysis of the Olduvai mammal bone assemblages has suggested that only the assemblages from Bed 1 site FLKN-Zinj and Bed II site BK showed more than marginal evidence of hominin activities (Egeland, 2008; see also Monahan, 1996). Hominin predation of mammal bones at the same two sites as suggested for fish predation – FLK-Zinj and the BK fish assemblages – is further support for hominin predation on fish at these sites.

The annually repetitive procurement of spawning *Clarias* populations at the Koobi Fora and Olduvai sites is mirrored in modern fishing practices, where fishermen seasonally hunt spawning populations of *Clarias* with spears (or bare hands) at the same location for generations (e.g., Greenwood, 1955; Jubb, 1967; Stewart et al., 1997; Gifford-Gonzalez et al., 1999). Similarly, many Pleistocene sites contain thousands of *Clarias* bones procured year after year during spawning runs (e.g., Van Neer, 1986; Robbins et al., 1994). *Clarias* would be particularly vulnerable to predators during "intense" arid periods in the Plio-Pleistocene (Trauth et al., 2007: 482). Lower lake/river levels and greater water level fluctuations meant large numbers of *Clarias* and other fish stranded and dying in the margins of drying lakes and rivers. Just such a situation may have occurred in the Plio-Pleistocene Rawe beds, where sand and/or silt cycles are suggested to represent annual deposition, and contain repetitive occurrences of numerous fish bones and "shells" (Frost et al., 2003). Numerous predators, including occasionally bonobos and other primates, also seasonally procure these catfish (Turnbull-Kemp, 1967; Ewer, 1973; Kruuk, 1976), although hominins, frequenting the lake/river margin ecotone, would have had early access to this source of food. *Clarias* too will burrow into mud during near-dessication of waters, to await the return of rains (Johnels,1957; Graham, 1997), and are then detected and dug out by modern fishermen, as well as, surprisingly, by bonobos (noted above). In the Sudd swamps, Dinka populations survive through the dry season living primarily on *Clarias* (Beadle, 1981). *Clarias* has no scales, which makes it a quickly procured and easily eaten meal.

Some authors have suggested that fish procurement has a low energy return (Cordain et al., 2001). In fact, procuring large numbers of *Clarias* in shallow waters with hands or spears, I would argue, involves less energy output and far greater caloric input than skinning a large mammal, severing limbs, and attempting to crack open a large femur of one animal for marrow, or as some have argued, the cranium for fats in the brain. All the while watching for and/or fending off other predators.

Clarias (and other African freshwater fish) have high fat content and also have high values of DHA and AA, the two essential long-chained PUFA necessary for brain growth and maintenance (e.g., Broadhurst et al., 1998). *Clarias* and other fish are usually in excellent condition in the dry season before spawning, because they "fatten" up to compensate for reduced feeding when spawning (e.g., Garrod, 1959; Hyslop, 1986; Lowe-McConnell, 1987), or, as with cichlids, spawning males develop a nuchal hump which apparently is a source of fat reserves (Lowe-McConnell, 1987). Most mammals are fat-depleted and are often starving near the end of the dry season, when terrestrial plant resources are also dessicated. Fish brains are also much valued and extracted by fishermen (pers. obs; Stewart et al., 1998) as they are concentrated sources of high-quality fat. The cranial bones of *Clarias* and other catfish have fused into a thin bony plate, so modified or unmodified cobbles, common at Olduvai and other sites, could have been used to crack the crania to get at the brains.

Clarias' dominance in the Koobi Fora, Olduvai, and Omo sites, its ubiquity in archaeological sites through the Pleistocene, and its popularity among modern fishermen, is directly related to its ease of capture, seasonal predictability, and high nutritional value, especially when other food sources are depleted. For hominins who frequented the lake/river margin ecotone, I suggest *Clarias* was a staple resource, procured very occasionally by earlier hominins, but more deliberately, seasonally, and consistently with increasing time spent in wetlands.

Other fish are available food sources during the arid periods, due to their adaptations to highly fluctuating water levels in wetlands. The lungfish (*Protopterus* sp.) is a large air-breathing fish which estivates in the dry season. When they sense that their water source is drying out, they burrow into mud, leaving an opening to the outside, and then producing a mucus membrane which "cocoons" them – up to several years – until wet conditions return (Perry et al., 2008). Lungfish are valued for food by fishermen today who dig out their burrows in the dry season; they are often recovered in archaeological sites. Cichlids, in particular due to their ecology and behavior, and other fish, are annually trapped in drying pools, as they construct nests in lake/river floodplains at high waters, and then are vulnerable when their nests are exposed by receding waters (e.g., Leakey, 1971).

Recent studies of Oldowan tools have found that many "battered pieces" – rocks with evidence of battering activities – are associated with Beds I and early Bed II sites at Olduvai Gorge, but may not be associated with butchering activities (Egeland, 2008), but with opening nuts or fruits (Mora and de la Torre, 2005). In fact, such battering tools would be ideal for smashing the shells of aquatic invertebrates or breaking into the crania of fish, particularly *Clarias*, for meat and brains, as modern fishermen do (Stewart, 1994).

Would fish consumption leave a craniodental footprint? Probably not. Fish meat is softer and less chewy than mammal meat. Fish locomotion is not weight bearing, so a strong and widespread network of connective tissues to maintain and support muscles is not required (Foegeding et al., 1996). Consumption of fresh fish meat also does not require the force or repetitive chewing (loading) required for fibrous vegetation or tough tissue-bound meat. If bones of small fish were consumed (adding calcium to the diet), these may leave small scratches on the enamel; in medium to larger fish, bones would probably be avoided. Robust catfish cranial shields would be smashed with rocks.

Other than cutmarks, which will always be rare in Pleistocene fish bones (e.g., Willis et al., 2008), how can we test further for hominin consumption of fish? Most promising is recent nitrogen isotope and sulfur isotope analysis of human remains (collagen) which suggest that the 40,000-year-old hominins consumed freshwater fish (Hu et al., 2009). Testing of collagen in Plio-Pleistocene hominin specimens, if enough collagen could be accessed, would confirm the consumption of freshwater fish.

PRECONDITIONS FOR ENCEPHALIZATION

Physiological Considerations

This chapter has provided evidence that *Paranthropus boisei*, *P. robustus*, and early *Homo* frequented wetlands environments and consumed wetlands vegetation, and later, consumed foods such as fish and aquatic invertebrates that were a high-quality source of essential fatty acids and other nutrients. Was a high-quality diet the major driver in the process of encephalization (e.g., Falk, 1995; Aiello, 1997)? Other processes were certainly involved. While other authors in this volume and elsewhere provide much more detail and insight into these processes, they can be summarized here. The consumption of fish and/or aquatic invertebrates by hominins would have acted both as a "releaser" (*sensu* Aiello and Wheeler, 1995; Aiello, 1997) and, I suggest, a "prime mover" in brain development. The diet of fish and/or aquatic invertebrates must have acted as a "releaser" in enacting changes in gut operation. Regular consumption of these foods would result in a decrease in hominin gut size, thereby freeing up or "releasing" energy from less efficient gut operation and channeling it to growth and maintenance of other organs, especially the brain (Aiello and Wheeler, 1995).

Other "releasers" must have been activated prior to brain growth, including mechanisms to cool the brain. Analysis of pathways by which blood drains from the hominid brain has shown that, in addition to cooling the body through evaporation of sweat which cools the arterial system, *Homo* and *Australopithecus* **but not** *Paranthropus* developed an additional mechanism to cool the brain. This was a network of tiny veins over the braincase which keeps temperatures from becoming too high. This additional cooling mechanism "released" thermal constraints in *Homo* which would have put the brakes on brain size growth (Falk, 2007). Even although *Paranthropus* may have been consuming foods (aquatic invertebrates) with high-quality fatty acids, its brain, in a broad reading of Falk's hypothesis, would have had physical limits to encephalization, so that it remained lower (relative to body size) than that of *Homo*. Examples of other "releasers" that allowed growth of the brain include climatic cooling; it is suggested that warm climates impose constraints on development of large brains (Schwartzman et al., 2009).

The Exaptive Role of High-Quality Freshwater Foods in Encephalization

I argue that a fish and aquatic invertebrate diet was also a "prime mover," by making available essential fatty acids and other nutrients with the potential to fuel development and maintenance of the hominin brain. This fish/invertebrate diet may also have been a "prime mover" in triggering brain growth through hormonal changes. The sudden, consistent, and large intake of a rarely consumed high-quality class of foods – fish and aquatic invertebrates – initiated hormonal changes in the brain. Thyroid hormones are implicated in facilitating new morphologies, including a larger brain, among early hominids (Crockford, 2003, 2008; Chapter 6 in this book). Dietary innovations – necessary because of scarcity of preferred foods – may have exaptively introduced large influxes of exogenous thyroid hormones from fish and freshwater invertebrates, which initated these new morphologies (Crockford, 2003, 2008; Chapter 6 in this book).

The suggestion that a fish and aquatic invertebrate diet was one of the "prime movers" in hominin encephalization comes with an important caveat. As discussed above, several authors have emphasized that development and maintenance of the hominin brain required access to food with specific fatty acids, particularly fish and aquatic invertebrates, on an ongoing, consistent basis (e.g., see Chapters 2 and 3). In this chapter, I have suggested that earlier hominins frequented wetlands and were inadvertently consuming tiny fish and/or invertebrates while feeding on aquatic vegetation. Later, they more deliberately procured fish seasonally when spawning or stranded in drying pools; evidence of this is in modified bones and/or fish assemblages at the Koobi Fora, Omo, and Olduvai sites. Procurement of fish and aquatic invertebrates on an ongoing seasonal basis exaptively provided the higher quality food source necessary for a more efficient gut, development of cooling mechanisms and other associated changes. However, to effect higher order changes, much greater consumption of this "different" class of foods, reinforced on a cyclical, multigenerational basis, was needed. This could only happen through climatically driven cyclical events which resulted in a consistent supply of these foods. This chapter suggests that the precessionally forced 23,000-year wet/dry cycles which resulted in transgressive and particularly regressive cycles in eastern African lakes largely drove these evolutionary events.

PRECESSIONAL FORCING, DRYING LAKES/RIVERS, AND DIE-OFFS OF AQUATIC FAUNAS

Within the moist periods at 2.7 and 1.9 Ma described above, sediments from a number of African lakes contain evidence of precessionally forced 23,000-year wet/dry cycles (Campisano and Feibel, 2007; Hopley et al., 2007; Kingston et al., 2007; Maslin and Christensen, 2007: 459; Trauth et al., 2007; Ashley et al., 2009). The clearest evidence for these cycles is found in fluviolacustrine sediments in the Barsemoi River drainage near Lake Baringo, at about 2.6 Ma (Deino et al., 2006), which record five cycles of diatomites alternating with subaerial sediments over ca. 100,000 years. Most significant is that the transitions from low levels (or dessication) to high lake, and especially high lake to dessication or low levels, could be rapid (500 years), a view supported by the absence of shallow water diatoms (Deino et al., 2006; Kingston et al., 2007). These rapid transitions would have been accompanied by climatic instability from a presumably moist to more arid period. Some lakes, such as those at Barsemoi, may have dried out completely in the dry cycles, while other larger and/or deeper lakes or spring-fed lakes in eastern Africa would have retained some water. Hominins would have inhabited the margins of these deeper or spring-fed lakes with potable water during the arid cycles.

The precessionally forced wet/dry cycles seen at Baringo are documented in other lakes across eastern Africa (Deino et al., 2006: 41), including the Gadeb sequence at about 2.7 Ma (Williams et al., 1979), Olduvai Gorge at ca 1.75 Ma (Ashley et al., 2009), the Turkana basin between about 1.9 and 1.6 Ma (Brown and Feibel, 1991; Lepre et al., 2007), and from flowstone deposits in northeastern South Africa (Hopley et al., 2007). At Lakes Albert and Edward there is also considerable evidence for wet/dry lake cycles at *about* 2.3–2 Ma (Boaz et al., 1992), although these may not have been linked to precessional forcing (see Campisano and Feibel, 2007 for a cautionary note on distinguishing global climate patterns from those caused by local environmental dynamics).

The effects of these alternating wet/dry dry cycles, and in particular the transitions from deep lakes to shallower or dessicated lakes, would have been catastrophic for aquatic faunas, particularly those in littoral ecotones. The actual transition from deep lakes to shallow or dessicated lakes on an approximately 23,000-year cycle would have been rapid, possibly within a few hundred years, with associated widely fluctuating water levels (Maslin and Christensen, 2007; Trauth et al., 2007). These wide fluctuations and/or near-dessication would have decimated the littoral biotopes and their fish and mollusk faunas. Deeper water biotopes would similarly have disappeared as the lake levels decreased, and deep lakes became increasingly shallow lakes. Wide fluctuations in lake levels, including dessication, are well documented in modern large and small eastern and southern African lakes (e.g., Beadle, 1981; Nicholson, 1998).

These catastrophic effects are recorded in numerous large-scale die-offs of fish and mollusks in eastern/central African Plio-Pleistocene deposits. Abundant fish remains are also associated with the KBS Member cycles at Koobi Fora (e.g., Schwartz, 1983; Feibel, 1988), as are numerous mollusk assemblages (Feibel, 1988; Feibel et al., 1991; Lepre et al., 2007). Fish bones and mollusk shells are documented from the precessionally forced wet/dry layers in the Baringo basin, discussed above (Deino et al., 2006). At Lakes Edward and Albert, sequential layers consisting of millions of mollusk shells and fish bones representing recurring, cyclical, and catastrophic die-offs (e.g., Stewart, 1990; Williamson, 1990; Boaz et al., 1992). While these wet/dry sedimentary cycles in Lakes Albert/Edward may not be precessionally forced, the catastrophic effects on fauna are the same. Informal observations of other eastern African lacustrine/fluvial deposits indicate that cyclical fish and mollusk death assemblages are not uncommon in Plio-Pleistocene deposits, regardless of cause.

The evidence cited above indicates that early *Homo* was consuming fish, and *Paranthropus* was consuming mollusks. The cyclical die-offs would have provided these hominins (and possibly their direct ancestors) with an ongoing windfall of fish and mollusks, which were a high-quality, cyclical source of essential fatty acids and other nutrients. This would have been reinforced consistently every 23,000 years, over a 100,000-year period, during the moist periods at 2.7 and 1.9 Ma. This recurring cyclical incorporation of high-quality foods not only fulfilled the immediate needs for energy, but had the exaptive effect of providing brain-selective nutrients, including DHA and AA, which fuelled the physiological changes necessary for encephalization. The consumption of high-quality foods on an ongoing, cyclical basis would have been the trigger for evolutionary change. These aquatic die-offs, however, would have attracted not only hominins, but also other predators. Given that the australopithecines and early *Homo* (at least, *H. habilis* [Ruff in Gibbons, 2008]) were probably still sleeping in trees, this would provide them with protection from most predators. Presumably hominins would access faunas at times of the day when other predators did not, possibly during the midday, when many predators are less active (e.g., Blumenschine et al., 1994).

Evidence of evolutionary change in hominins is found subsequent to the 2.6 Ma precessionally forced cycles, with the emergence of early *Homo* in Ethiopia at 2.33 Ma (Kimbel et al., 1996), about 2.4 Ma from Malawi (Schrenk et al., 1993) and from Baringo (originally described by Tobias, 1967, 1993 at level of family) but later classified as *Homo* (Sherwood et al., 2002); its status is still under discussion. *P. boisei* was recovered at ca 2.5 Ma from west Turkana (Walker et al., 1986), *Australopithecus garhi* at 2.5 Ma, *P. robustus* at 2 Ma (e.g., Grine, 1989) and early *H. erectus* at 1.9 Ma from east Turkana (e.g., Rightmire, 1990). As discussed above, many of these remains were recovered from lake/river margin environments. Except for *A. garhi*, all of these taxa had more encephalized brains compared to earlier hominins, with *H. habilis* having the highest, or almost highest, encephalization quotient, in part due to its small body size (Aiello and Dean, 1990; McHenry and Coffing, 2000), and also showing a different level of organization (e.g., Tobias, 1987). Specimens referred to early *H. erectus*, and possibly others referred to *H. rudolfensis*, show postcranial changes closer to those of later *Homo*, and less like the australopithecine-like postcrania of *H. habilis* (e.g., McHenry and Coffing, 2000). Other postcranial and physiological changes also occurred but are not detailed here.

MAMMAL MEAT: A LATER HOMININ ADAPTATION?

This chapter presents evidence that precessionally forced wet/dry cycles resulted in widely fluctuating lake levels and/or dessication, resulting in die-offs of aquatic faunas and a high-quality food source for hominins. Other terrestrial and semiaquatic animals –

including hippos, otters, and crocodiles – would also have died as a result of prolonged arid periods. At first glance, these animals could also have provided high-quality nutrients, protein, and fats for hominins. However, several lines of evidence indicate that these animals were not a significant source of essential fatty acids and other nutrients. First, while hominins may well have occasionally taken advantage of these other sources of food, most such animals would have been very fat-depleted and lean during arid periods. Researchers have documented that climatic deterioration – usually in the late dry season – results in a loss of protein and a consequent decline in the quality of graze and browse. The meat of herbivores is therefore of little nutritional value to consumers during this period (e.g., Speth and Spielmann, 1983; Speth, 1989). Further, DHA is less available in mammal meat than in fish and shellfish (e.g., Cordain et al., 2001). Peoples dependent on these animals are therefore eating meat with reduced fat levels; such consumption of fat-depleted meat reportedly results in severe weight-loss and other nutrition-depletion problems (Wilmsen in Speth, 1989).

Second, in contrast to obligatory aquatic animals that were directly affected by seasonal or perennial dessication of lakes, terrestrial and amphibious animals, such as hippos, otters, and crocodiles, would not have generally died *en masse* in "cycles"; they would move to find other sources of water. Therefore, mammal meat would be a more erratic source, and *not* be available as a consistent, large and long-term supply of nutrients to the brain, which was necessary to enact evolutionary change.

Third, and perhaps most critically, there is not a great deal of zooarchaeological evidence prior to the middle Pleistocene that hominins butchered mammals more than sporadically (see also, e.g., O'Connell et al., 1999; Klein et al., 2007). Prior to about 1.8 Ma, *in situ* cutmarked bones are only known from the Bouri site in Ethiopia dating to about 2.6–2.5 Ma. These are thought to indicate defleshing of animals, while limb bones were smashed apparently to obtain marrow. No artifacts were found (de Heinzelin et al., 1999). Cutmarked mammal bones from nearby Gona were not *in situ*, but surface finds (Dominguez-Rodrigo et al., 2005). At other eastern African archaeological sites in this period, mammal bones have been recovered in the same deposits as stone tools, but without evidence of causal association (e.g., Merrick and Merrick, 1976; Kibunjia, 1994; Kimbel et al., 1996).

Only a handful of Olduvai Gorge and Koobi Fora sites provide compelling evidence for butchering before the middle Pleistocene. The well-studied Bed I and II bone assemblages at Olduvai Gorge document considerable carnivore activity, inferred from marks on large mammal bones (Leakey, 1971; Blumenschine, 1995; Selvaggio and Wilder, 2001; Egeland, 2008). However, evidence of hominin modifications on mammal bones (cutmarks, smashing for marrow) or of hominin transport of bone, seems most compelling only at the Bed I site of FLK-22 (Zinj) and the Bed II site of BK (Monahan, 1996; Egeland, 2008, Faith et al., 2009), although cutmarks are documented at other sites (Blumenschine, 1995). This paucity of hominin modifications on bones is surprising given recovery of abundant stone tools at most Olduvai sites. As suggested above, stone tools may have been used for nonbutchering activities, including battering stones for cracking open freshwater invertebrate shells and catfish crania.

The excavations at three sites near Koobi Fora, as discussed earlier, have uncovered a vertebrate fauna dominated by wetlands taxa, including hippos, reduncini, cane rats, fish, and aquatic reptiles, as well as terrestrial taxa, dated between about 1.6 and 1.3 Ma (e.g., Harris and Isaac, 1997; Pobiner et al., 2008). The environments they inhabited are characterized by a significant shallow water component, accompanied by swampy areas, seasonally flooded areas with wet grasslands and a nearby gallery forest (Pobiner et al., 2008). Butchery marks were found on an average of 7% of the bones, but no stone tools were found at these sites. The sites are assumed to result from hunting or scavenging by *H. erectus*, of whom a femur was found near one of these sites (Pobiner et al., 2008). The presence of cane rats at all three sites suggests these may have been a prey item; they are commonly eaten by humans today.

Even later sites associated with *H. erectus* have little or no definitive evidence of butchery. The abundant fauna associated with Acheulian bifaces, and a partial cranium of *Homo heidelbergensis*, at Elandsfontein (ca 1.0–0.6 Ma), suggests that local Acheuleans played little role in the bone accumulation (Klein et al., 2007: 183). The very sporadic presence of mammal butchery until the middle Pleistocene has led one researcher to state: "Tool-marked bones also appear to be uncommon at other Acheulean or Acheulean-age carcass or death sites" (Klein et al., 2007: 183), suggesting that Acheulian hominins procured only a small number of large mammals; a sentiment shared by others (e.g., O'Connell et al., 1999; Pobiner et al., 2008). Further, dental microwear analysis also plays down the importance of meat: "None of these hominins (early *Homo*) probably limited their diets to either tough meat or hard USO's (underground storage organs)" (Ungar et al., 2006:90).

All in all, given the span of time that *H. habilis* and *H. erectus* existed (ca 2.4–.5 Ma) and the small number of sites in this period with definitive evidence of butchery, there is little *consistent* and irrefutable evidence of butchering mammals until the middle Pleistocene, with the advent of *H. heidelbergensis* (e.g., Rightmire, 2004), or perhaps even later with *H. sapiens*. At and after this time, some sites show cutmarked bones of large herbivores associated with large numbers of stone tools (e.g., Boxgrove [Roberts and Parfitt, 1999]) or wooden spears (e.g., Schoningen, [e.g., Thieme, 1997). However, even at well-known sites with purported evidence of large-scale hominin hunting or scavenging activities (e.g., Ambrona), re-evaluation of the data indicates a much more limited human involvement, leading Vila et al. (2005:223) to state that at Ambrona "we cannot prove hunting," plus they reject previous suggestions of marginal scavenging.

A *consistent and abundant* intake of sources of essential fatty acids and other nutrients are needed to fuel the hominin brain. There is no zooarchaeological evidence, or even a plausible scenario, to support the hypothesis that development and growth of the hominin brain occurred in association with a consistently large intake of mammal meat prior to *H. heidelbergensis*, or possibly even *H. sapiens*. Nor admittedly is there a large body of zooarchaeological evidence for early hominin procurement of fish or shellfish. However, as cited above, recent nutritional, morphological, behavioral, and isotope data increasingly shift the balance in favor of hominin consumption of aquatic foods, the only consistently accessible food source that could enable growth and maintenance of the hominin brain.

POSTSCRIPT: *H. HEIDELBERGENSIS* AND *H. SAPIENS*

The focus of this chapter has been the context with which encephalization and other associated changes occurred in early *Homo*, but I add here brief observations on later "jumps" in encephalization (see also Chapter 10). Almost all of the earliest *H. sapiens* remains are from African sites, with the earliest dating at about 195,000 Ma (McDougall et al., 2005), strongly suggesting that *H. sapiens* evolved in Africa, before subsequent dispersal event(s) (Stringer, 2000; Chapter 10 in this book). While evidence has been presented here for the use of freshwater shellfish resources by earlier hominins, marine mollusks contain more PUFA than freshwater ones (which are rich enough in any case). Based on archaeological evidence from southern and eastern Africa, *H. sapiens*, and late *H. heidelbergensis*, started exploiting marine fish and invertebrate resources (e.g., Walter et al., 2000; Marean et al., 2007) by about 200,000 Ma or slightly later. This coincides with the beginning of a long period of aridity (Crombie et al., 1997; Jahns in Finlayson, 2004).

Hominins may have been forced by the long periods of aridity to coastal locations (see Chapter 10). Coastal wetlands have C4 sedges and other vegetation, making it an easier ecological shift from a freshwater to a marine environment. While marine mollusks have different ecologies than freshwater mollusks, long-term use of the latter would make an easy transition to marine resources. Fish in the marine biome are much harder to catch than in a freshwater biome with only minimal technology, because few sizable easy-to-procure fish inhabit the marine inshore. The heavy and ongoing consumption of marine invertebrates, and later of fish, may have provided a larger incorporation of essential nutrients and thyroid hormone, leading to further encephalization. Occupation of coastal areas was also a prelude to movement out of Africa in the late Pleistocene (e.g. Stringer, 2000).

SUMMARY

The human brain has a large requirement for brain-selective nutrients, particularly two long-chained essential fatty acids, DHA and AA, to fuel normal growth of the fetal and infant human brains, and for ongoing maintenance of the adult brain. Freshwater fish and invertebrates not only have long-chain PUFA ratios most similar to that of the human brain, but also are an abundant source of other key nutrients such as iodine and iron. Therefore, logically, early hominins must have evolved in wetlands where abundant foods providing these nutrients were consistently available.

Data from both late Miocene and extant large apes indicate both consistent short-term, even obligate, use of wetlands and wetlands foods (extant apes), and long-term occupation of wetlands environments, in large part due to inaccessibility or scarcity of terrestrial foods (extant and fossil apes). Certain late Miocene African apes are associated with wetlands environments and had craniodental adaptations to tough vegetation. It is suggested that ancestral hominins were similarly adapted to wetlands and wetlands vegetation. These hominins were increasingly forced, through competition for disappearing habitats and resources, into wetlands environments or land-bordering wetlands. Of critical importance in the successful colonization of and radiation in these new habitats were the late Miocene and Pliocene formation of lake basins and lake margin biomes in eastern, central, and southern Africa, and the accompanying spread of C4 *wetlands* sedges and grasses.

An increased density of hominin remains in the Plio-Pleistocene indicates association with lake and river margin environments, not just as death sites, but as living sites. Several lines of evidence point to the consumption, often intensive, of wetlands vegetation by hominins from *Australopithecus anamensis* at least to early *Homo erectus*. At several sites, the consistent stability and apparent interconnection over time in hominin craniodental adaptations to tough and/or fibrous foods, hominin association with lake/river margin ecotones, and constant availability of wetlands vegetation, *stand in sharp contrast to the absence of responsive adaptations in hominin craniodental morphology and behavior to ongoing fluctuations in terrestrial vegetation and habitats* (e.g., Hadar, Olduvai). As with extant apes, occasional consumption of small fish and invertebrates by early hominins must have occurred while consuming aquatic vegetation. I suggest that consumption of consistently available wetlands vegetation acted as a "buffer" to increasingly unstable climate conditions and scarce terrestrial foods and habitats, an argument made by Conklin-Brittain et al. (2002) and Wrangham (2005) for underground storage organs. The inclusion of aquatic animals in the hominin diet, though at first inadvertent and occasional, was a precursor to deliberate and consistent consumption of these foods.

Increased climatic and tectonic instability in the Plio-Pleistocene created a further dependence of hominins on wetlands and their food sources. Terrestrial sources of fats, protein, and micronutrients had to be increasingly replaced by aquatic sources, leading to an abrupt shift of freshwater foods from minor to major component in the hominin diet. This intensification of consumption of fish and mollusks – sources of high-quality fats and micronutrients – exaptively triggered the role of a high-quality diet as a "releaser" for energy to be redirected toward encephalization, as well as other evolutionary changes. Precessionally forced wet and dry lake/river cycles over 100,000 years at about 2.6 and 1.9 Ma caused wide fluctuations in levels and often quite rapid "shallowing" and sometimes dessication of lakes. In these arid periods, hominin populations were probably isolated at whichever water bodies still contained potable water and edible aquatic foods. The drastic lake level fluctuations resulted in cyclical large-scale die-offs of aquatic fish and invertebrates. The consistent consumption of these unusually large amounts of freshwater fish and invertebrates, implying an anomalously high intake of "brain-selective" nutrients including specific essential nutrients and hormones, reinforced cyclically on a long-term basis, would have kick-started the encephalization process in hominins. The exaptational result of incorporating large amounts of exogenous thyroid hormones may also have played a major role in triggering this process. Evidence of evolutionary change, particularly not only encephalization, but also postcranial, physiological, and behavioral changes, is seen with the emergence of several hominin taxa, including *Paranthropus* spp. and early *Homo* after 2.7 Ma, and early *H. erectus* at around 1.9 Ma.

Hypothetical scenarios are briefly presented to account for encephalization "jumps" seen in the emergence of *H. heidelbergensis* and *H. sapiens*. These suggest that severe droughts in the middle and late Pleistocene forced hominins to the coast to exploit coastal resources, now accessible due to possession of more advanced technology. Marine foods are very rich in brain-selective nutrients. When these are consumed in anomalously high amounts as an energy source, they exaptively may have contributed to the process of encephalization and other related physiological and anatomical changes.

ACKNOWLEDGMENTS

I would like to thank Karen Chambers (Wiley) for her support and encouragement to Stephen Cunnane and myself in putting this volume together. My chapter is the culmination of many discussions with many people over time. While this chapter reflects my interpretations, any errors are solely mine. My thanks for thoughtful discussions to Kay Behrensmeyer, Rufus Churcher, Stephen Cunnane, Diane Gifford-Gonzalez, Jack Harris, Maxine Kleindienst, Yin Lam, Meave Leakey, Sam Muteti, Jim Oliver, Larry Robbins, and Nancy Sikes. I would also like to thank Bruce Marcot for his generosity in allowing me to include his photograph. My thanks to the Canadian Museum of Nature for contributing to logistical support for African fieldwork. My thanks also to the National Museums of Kenya and their staff for their help and support.

REFERENCES

Aiello, L. 1997. Brains and guts in human evolution: The expensive tissue hypothesis. *Brazilian Journal of Genetics* **20** (1):141–148.

Aiello, L. and Dean, C. 1990. *An Introduction to human evolutionary anatomy*. London: Academic Press.

Aiello, L.C. and Wheeler, P. 1995. The expensive-tissue hypothesis. The brain and the digestive system in human and primate evolution. *Current Anthropology* **36**: 199–221.

Albert, R.M., Bamford, M.K., and Cabanes, D. 2006. Taphonomy of phytoliths and macroplants in different soils from Olduvai Gorge (Tanzania) and the application to Plio-Pleistocene palaeanthropological soils. *Quaternary International* **148**:78–94.

Andrews, P., Begun, D.R., and Zylstra, M. 1997. Interrelationships between functional morphology and paleoenvironments in *Miocene hominoids*. In *Function, Phylogeny, and Fossils: Miocene Hominoid Evolution and Adaptations*, ed. D.R. Begun, C.V. Ward, and M.D. Rose, pp. 29–58. New York: Plenum Press.

Ashley, G.M., Goman, M., Hover, V.C. et al. 2002. Artesian blister wetlands, a perennial water source in the semi-arid Rift Valley of East Africa. *Wetlands* **22** (4): 686–695.

Ashley G.M., Mworia, J.M., Muasya, A.M. et al. 2004. Sedimentation and recent history of a freshwater wetland in a semi-arid environment: Loboi Swamp, Kenya, East Africa. *Sedimentology* **51**:1301–1321.

Ashley, G.M., Tactikos, J.C., and Owen, R.B. 2009. Hominin use of springs and wetlands: Paleoclimate and archaeological records from Olduvai Gorge (~1.79–1.74 Ma). *Palaeogeography, Palaeoclimatology, Palaeoecology* **272**:1–16.

Badrian, N., Badrian, A., and Susman, R.L. 1981. Preliminary observations on the feeding behavior of *Pan paniscus* in the Lomako Forest of Central Zaire. *Primates* **22**:173–181.

Badrian, N. and Malenky, R.K. 1984. Feeding ecology of *Pan paniscus* in the Lomako Forest, Zaire. In *The Pygmy Chimpanzee, Evolutionary Biology and Behaviour*, ed. R.L. Susman, pp. 275–299. New York: Plenum Press.

Bamford, M.K. 2005. Early Pleistocene fossil wood from Olduvai Gorge, Tanzania. *Quaternary International* **129**:15–22.

Bamford, M.K., Albert, R.M., and Cabanes, D. 2006. Plio-Pleistocene macroplant fossil remains and phytoliths from Lowermost Bed II in the eastern palaeolake margin of Olduvai Gorge, Tanzania. *Quaternary International* **148**:95–112.

Beadle, L.C. 1981. *The Inland Waters of Tropical Africa*. London: Longman.

Behrensmeyer, A.K. 1975. The taphonomy and paleoecology of the Plio-Pleistocene vertebrate assemblages east of Lake Rudolf, Kenya. *Bulletin of the Museum of Comparative Zoology* **146**:473–578.

Behrensmeyer, A.K. 1982. Time resolution in fluvial vertebrate assemblages. *Paleobiology* **8** (3):211–227.

Behrensmeyer, A.K. 1993. The bones of Amboseli. *National Geographic Research and Exploration* **9** (4): 402–421.

Behrensmeyer, A.K., Todd, N., Potts, R. et al. 1997. Late Pliocene faunal turnover in the Turkana Basin, Kenya and Ethiopia. *Science* **278**:1589–1594.

Bennett, M.R., Harris, J.W.K., Richmond, B.G. et al. 2009. Early hominin foot morphology based on 1.5-million-year-old footprints from Ileret, Kenya. *Science* **523** (5918):1197–1201.

Bernor, R.L. 2007. New apes fill the gap. *Proceedings of the National Academy of Sciences of the United States of America* **104** (50):19661–19662.

Beynon, A.D. and Wood, B. 1986. Variations in enamel thickness and structure in East African hominids. *American Journal of Physical Anthropology* **70**:177–193.

Blake, S., Rogers, E., Fay, J.M. et al. 1995. Swamp gorillas in northern Congo. *African Journal of Ecology* **33** (3):285–290.

Blumenschine, R.J. 1987. Characteristics of an early hominid scavenging niche. *Current Anthropology* **28** (4): 383–407.

Blumenschine, R.J. 1995. Percussion marks, tooth marks and the experimental determinations of the timing of hominid and carnivore access to long bones at FLK Zinjanthropus, Olduvai Gorge, Tanzania. *Journal of Human Evolution* **29**:21–51.

Blumenschine, R.J., Cavallo, J.A., and Capaldo, S.D. 1994. Competition for carcasses and early hominid behavioural ecology: A case study and conceptual framework. *Journal of Human Evolution* **27**:197–213.

Blumenschine, R.J., Peters, C.R., Masao, F.T. et al. 2003. Late Pliocene Homo and hominid land use from western Olduvai Gorge, Tanzania. *Science* **299**: 1217–1221.

Boaz, N.T., Bernor, R.L., Brooks, A.S. et al. 1992. A new evaluation of the significance of the late Neogene Lusso Beds, Upper Semliki Valley, Zaire. *Journal of Human Evolution* **22**:505–517.

Bobe, R. 2006. The evolution of arid ecosystems in eastern Africa. *Journal of Arid Environments* **66**:564–584.

Bobe, R and Behrensmeyer, A.K. 2004. The expansion of grassland ecosystems in Africa in relation to mammalian evolution and the origin of the genus *Homo*. *Palaeogeography, Palaeoclimatology, Palaeoecology* **207**:399–420.

Bobe, R., Behrensmeyer, A.K., and Chapman, R.E. 2002. Faunal change, environmental variability and Late Pliocene hominin evolution. *Journal of Human Evolution* **42**:475–497.

Boisserie, J.-R., Brunet, M., Andossa, L. et al. 2005. A new Late Miocene Hippopotamid from Toros-Menalla. *Journal of Vertebrate Paleontology* **25** (3):665–673.

Bonnefille, R., Potts, R., Chalié, F. et al. 2004. High-resolution vegetation and climate change associated with Pliocene *Australopithecus afarensis*. *Proceedings of the National Academy of Sciences of the United States of America* **101** (33):12125–12129.

Breuer, T., Ndoundou-Hockemba, M., and Fishlock, V. 2005. First observation of tool use in wild gorillas. *Public Library of Science* **3** (11):380.

Broadhurst, C.L., Cunnane, S.C., and Crawford, M.A. 1998. Rift Valley lake fish and shellfish provided brain-specific nutrition for early *Homo*. *British Journal of Nutrition* **79**:3–21.

Broadhurst, C. L., Wang, Y., Crawford, M.A et al. 2002. Brain-specific lipids from marine, lacustrine, or terrestrial food resources: Potential impact on early African *Homo sapiens*. *Comparative Biochemistry and Physiology Part B* **131**:653–673.

Brown, F.H. and Feibel, C.S. 1991. Stratigraphy, depositional environments, and palaeogeography of the Koobi Fora formation. In *Koobi Fora Research Project, Vol. 3: Stratigraphy, Artiodactyls and Paleoenvironments*, ed. J.M. Harris, pp. 1–30. Oxford: Clarendon Press.

Brunet, M., Guy, F., Pilbeam, D. et al. 2002. A new hominid from the Upper Miocene of Chad, Central Africa. *Nature* **418** (6894):145–151.

CAMERON, D.W. 2004. *Hominid Adaptations and Extinctions*. Seattle: University of Washington Press.

CAMPISANO, C.J. AND FEIBEL, C.S. 2007. Connecting local environmental sequences to global climate patterns: Evidence from the hominin-bearing Hadar Formation, Ethiopia. *Journal of Human Evolution* **53**:515–527.

CARNIERI, E. AND MALLEGNI, F. 2003. A new specimen and dental microwear in *Oreopithecus bambolii*. *Journal of Comparative Human Biology* **54** (1):29–35.

CARTMILL, M., SMITH, F.H., AND BROWN, K.B. 2009. *The Human Lineage*. New York: Wiley-Blackwell.

CERLING, T.E., WANG, Y., AND QUADE, J. 1993. Global ecological change in the Late Miocene: Expansion of C4 ecosystems. *Nature* **361**:344–345.

CHAPMAN, C.A., WHITE, F.J., AND WRANGHAM, R.W. 1993. Defining subgroup size in fission-fusion societies. *Folia Primatologica* **61**:31–34.

CHAVAILLON, J. 1976. Evidence for the technical practices of Early Pleistocene hominids. In *Earliest Man and Environments in the Lake Rudolf Basin*, ed. Y. Coppens, F.C. Howell, G.L. Isaac, and R.E.F. Leakey, pp. 565–574. Chicago: University of Chicago Press.

COHEN, A., STONE, J.R., BEUNING, K.R.M. ET AL. 2007. Ecological consequences of early Late Pleistocene megadroughts in tropical Africa. *Proceedings of the National Academy of Sciences of the United States of America* **104** (42):16422–16427.

CONKLIN-BRITTAIN, N.L., WRANGHAM, R.W., AND SMITH, C.C. 2002. A two-stage model of increased dietary quality in early hominid evolution: The role of fibre. In *Human Diet: Its Origins and Evolution*, ed. P.S. Ungar and M.F. Teaford, pp. 61–76. Westport, CT: Bergin and Garvey.

COPELAND, S.R. 2007. Vegetation and plant food reconstruction of lowermost Bed II, Olduvai Gorge, using modern analogs. *Journal of Human Evolution* **53**: 146–175.

CORDAIN, L., WATKINS, B.L., AND MANN, N.J. 2001. Fatty acid composition and energy density of foods available to African hominids. *World Reviews of Nutrition and Diets* **90**:144–161.

COUSINS, D. AND HUFFMAN, M.A. 2002. Medicinal properties in the diet of gorillas: An ethnopharmacological evaluation. *African Study Monographs* **23**:65–89.

CRAWFORD, M.A. 2006. Docosahexaenoic acid in neural signaling systems. *Nutrition and Health* **18**:247–260.

CRAWFORD, M.A., COSTELOE, K., DOYLE, W. ET AL. 1992. Essential fatty acids in early development. In *Polyunsaturated Fatty Acids in Human Nutrition*, ed. U. Bracco and R.J. Deckelman, pp. 93–110. New York: Raven Press.

CRAWFORD, M.A., COSTELOE, K., GHEBREMESKEL, K. ET AL. 1997. Are deficits of arachidonic and docosahexaenoic acids responsible for the neural and vascular complications of preterm babies? *American Journal of Clinical Nutrition* **S66**:S1032–S1041.

CROCKFORD, S.J. 2003. Thyroid rhythm phenotypes and hominid evolution: A new paradigm implicates pulsatile hormone secretion in speciation and adaptation changes. *Comparative Biochemistry and Physiology (Part A)* **135**:105–129.

CROCKFORD, S.J. 2008. *Rhythms of Life: Thyroid Hormone & the Origin of Species*. Victoria: Trafford.

CROMBIE, M.K., ARVIDSON, R.E., STURCHIO, N.C. ET AL. 1997. Age and isotopic constraints on Pleistocene pluvial episodes in the Western Desert, Egypt. *Palaeogeography, Palaeoclimatology, Palaeoecology* **130**:338–355.

CUNNANE, S.C. 2005. *Survival of the Fattest*. Hackensack, NJ: World Scientific.

CUNNANE, S.C. AND CRAWFORD, M.A. 2003. Survival of the fattest. Fat babies were the key to evolution of the large human brain. *Comparative Biochemistry and Physiology* **136A**:17–26.

CUNNANE, S.C., HARBIGE, L.S., AND CRAWFORD, M.A. 1993. The importance of energy and nutrient supply in human brain evolution. *Nutrition and Health* **9**:219–235.

DART, R. 1925. *Australopithecus africanus*: The man-ape from South Africa. *Nature* **115**:195–199.

DE HEINZELIN, J., CLARK, J.D., WHITE, T.D. ET AL. 1999. Environment and behavior of 2.5-million-year-old Bouri hominids. *Science* **284** (5414):625–629.

DEINO, A.L., KINGSTON, J.D., GLEN, J.M. ET AL. 2006. Precessional forcing of lacustrine sedimentation in the late Cenozoic Chemeron Basin, Central Kenya Rift, and calibration of the Gauss/Matuyama boundary. *Earth and Planetary Science Letters* **247**:41–60.

DEMENOCAL, P. 1995. Plio-Pleistocene African climate. *Science* **270**:53–59.

DEMENOCAL, P.B. 2004. African climate change and faunal evolution during the Pliocene-Pleistocene. *Earth and Planetary Science Letters* **220**:3–24.

DE WAAL, F.B.M. 1990. *Peacemaking among Primates*. Boston: Harvard University Press.

DE WAAL, F.B.M. AND LANTING, F. 1997. *Bonobo: The Forgotten Ape*. Berkeley: University of California Press.

DOMINGUEZ-RODRIGO, M., LOPEZ-SAEZ, J.A., VINCENS, A. ET AL. 2001. Fossil pollen from the Upper Humbu formation of Peninj (Tanzania): Hominid adaptation to a dry, open Plio-Pleistocene savanna environment. *Journal of Human Evolution* **40**:151–157.

DOMINGUEZ-RODRIGO, M., PICKERING, T.R., AND SEMAW, S. 2005. Cutmarked bones from Pliocene archaeological sites at Gona, Afar, Ethiopia: Implications for the function of the world's oldest stone tools. *Journal of Human Evolution* **48**:109–121.

DORAN, D.M., MCNEILAGE, A., GREER, D. ET AL. 2002. Western lowland gorilla diet and resource availability: New evidence, cross-site comparisons, and reflections on indirect sampling methods. *American Journal of Primatology* **58**:91–116.

DORAN-SHEEHY, D.M, GREER, D., MONGO, P. ET AL. 2004. Impact of ecological and social factors on ranging in western gorillas. *American Journal of Primatology* **64**:207–222.

EGELAND, C. 2008. Patterns of early hominid site use at Olduvai Gorge. *Mitteilungen der Gesellschaft für Urgeschichte*. **17**:9–37.

EWER, R.F. 1973. *The Carnivores*. Ithaca, NY: Cornell University Press.

FAITH, J.T., DOMINGO-RODRIGUEZ, M., AND GORDON, A.D. 2009. Long-distance carcass transport at Olduvai Gorge?

A quantitative examination of Bed I skeletal element abundances. *Journal of Human Evolution* **56**:247–256.

FALK, D. 1995. Comment on: The expensive-tissue hypothesis: The brain and the digestive system in human and primate evolution. *Current Anthropology* **36**:212–213.

FALK, D. 2007. Evolution of the primate brain. In *Handbook of Paleaoanthropology Vol. 2: Primate Evolution and Human Origins*, ed. W. Henke, H. Rothe, AND I. Tattersall, pp. 1133–1162. New York: Springer-Verlag.

FARQUHARSON, J., COCKBURN, F., PATRICK, W.A. ET AL. 1992. Infant cerebral cortex phospholipid fatty-acid composition and diet. *Lancet* **340**:810–813.

FAY, J. AND AGNAGNA, M. 1992. Census of gorillas in northern Republic of Congo. *American Journal of Primatology* **27**:275–284.

FAY, J.M., AGNAGNA, M., MOORE, J. ET AL. 1989. Gorillas *(Gorilla gorilla gorilla)* in the Likouala swamp forests of North Central Congo. *International Journal of Primatology* **10**:477–486.

FEIBEL, C.S. 1988. *Paleoenvironments of the Koobi Fora Formation, Turkana Basin, Northern Kenya*. Unpublished PhD dissertation, University of Utah.

FEIBEL, C.S., HARRIS, J.M., AND BROWN, F.H. 1991. Palaeoenvironmental context for the late Neogene of the Turkana basin. In *Koobi Fora Research Project, Vol. 3*, ed. J.M Harris, pp. 321–346. Oxford: Clarendon Press.

FINLAYSON, C. 2004. *Neanderthals and Modern Humans: An Ecological and Evolutionary Perspective*. Cambridge: Cambridge University Press.

FOEGEDING, E.A., LANIER, T.C., AND HULTIN, H.O. 1996. Characteristics of edible muscle tissues. In *Food Chemistry*, ed. O.R. Fennema, pp. 879–943. San Francisco: CRC Press.

FOLEY, R. 1987. *Another Unique Species*. London: Longman.

FROST, S.R., PLUMMER, T., BISHOP, L.C. ET AL. 2003. Partial cranium of *Cercopithecoides kimeui* Leakey, 1982 from Rawi Gully, Southwestern Kenya. *American Journal of Physical Anthropology* **122**:191–199.

GALBANY, J., MOYÀ-SOLÀ, S., AND Pérez- PÉREZ, A. 2005. Dental microwear variability on buccal tooth enamel surfaces of extant catarrhini and the Miocene fossil *Dryopithecus laietanus* (Hominoidea). *Folia Primatologica* **76**:325–341.

GARROD, D.J. 1959. The growth of *Tilapia esculenta* Graham in Lake Victoria. *Hydrobiologia* **36**:268–298.

GIBBONS, A. 2008. Snapshots from the meeting (American Association of Physical Anthropologists, April 9–12; Columbus). *Science* **320**:608–609.

GIFFORD-GONZALEZ, D.P., STEWART, K.M., AND RYBCZYNSKI, N. 1999. Human activities and site formation at modern lake margin foraging camps in Kenya. *Journal of Anthropological Archaeology* **18**:397–440.

GOODALL, J. 1986. *The Chimpanzees of Gombe*. Boston: Houghton Mifflin Publishing.

GRAHAM, J.B. 1997. *Air-Breathing Fishes*. New York: Academic Press.

GREB, S.F., DIMICHELE, W.A., AND GASTALDO, R.A. 2006. Evolution and importance of wetlands in earth history. In *Wetlands Through Time*, ed. S.F. Greb, and W.A. DiMichele. *Geological Society of America Special Paper* **399**:1–41.

GREENWOOD, P.H. 1955. Reproduction in the Cat-fish *Clarias mossambicus* Peters. *Nature* **176** (4480):516–517.

GREENWOOD, P.H. AND TODD, E.J. 1970. Fish remains from Olduvai. In *Fossil Vertebrates of Africa, Vol. 2*, ed. L.S.B. Leakey and R.J.G. Savage, pp. 225–241. London: Academic Press.

GRINE, F.E. 1989. New hominid fossils from the Swartkrans Formation (1979–1986) excavations: Craniodental specimens. *American Journal of Physical Anthropology* **79**: 409–449.

GRINE, F.E., UNGAR, P.S., TEAFORD, M.F. ET AL. 2006. Molar microwear in *Praeanthropus afarensis*: Evidence for dietary stasis through time and under diverse paleoecological conditions. *Journal of Human Evolution* **51**: 297–319.

HAILE-SELASSIE, Y., SUWA, G., AND WHITE, T.D. 2004a. Late Miocene teeth from Middle Awash, Ethiopia, and early hominid dental evolution. *Science* **303** (5663): 1503–1505.

HAILE-SELASSIE, Y., WOLDEGABRIEL, G., WHITE, T.D. ET AL. 2004b. Mio-Pliocene mammals from the Middle Awash, Ethiopia. *Geobios* **37** (4):536–552.

HARRIS, J.M. AND LEAKEY, M.G., EDS. 2003a. Geology and vertebrate paleontology of the Early Pliocene site of Kanapoi, Northern Kenya. *Contributions in Science 498*. Los Angeles: Natural History Museum of Los Angeles County.

HARRIS, J.M. AND LEAKEY, M.G. 2003b. Introduction. In *Geology and Vertebrate Paleontology of the Early Pliocene Site of Kanapoi, Northern Kenya*, eds. J.M. Harris and M.G. Leakey, pp. 1–9. Los Angeles, CA: Natural History Museum of Los Angeles County.

HARRIS, J.W.K. and ISAAC, Ll. 1997. Sites in the upper KBS, Okote, and Chari Members: Reports. In *Koobi Fora Research Project, Vol. 5: Plio-Pleistocene Archaeology*, eds. G.Ll. Isaac and B. Isaac, pp. 115–236. Oxford: Clarendon Press.

HARRISON, T. AND ROOK, L. 1997. Enigmatic anthropoid or misunderstood ape? The phylogenetic status of *Oreopithecus bambolii* reconsidered. In *Function, Phylogeny and Fossils: Miocene Hominoid Evolution and Adaptation*, ed. D.R. Begun, C.V. Ward, and M.D. Rose, pp. 327–362. New York: Plenum.

HASHIMOTO, C., TASHIRO, Y., KIMURA, D. ET AL. 1998. Habitat use and ranging of wild bonobos (*Pan paniscus*) at Wamba. *International Journal of Primatology* **19** (6): 1045–1060.

HERRIES, A.I.R., CURNOE, D., AND ADAMS, J.W. 2009. A multi-disciplinary seriation of early *Homo* and *Paranthropus* bearing palaeocaves in southern Africa. *Quaternary International* **202**:14–28.

HOPLEY, P.J., WEEDEN, G.P., MARSHALL, J.D. ET AL. 2007. High- and low-latitude orbital forcing of early hominin habitats in South Africa. *Earth and Planetary Letters* **256**:419–432.

HU, Y., SHANG, H., TONG, H. ET AL. 2009. Stable isotope dietary analysis of the Tianyuan 1 early modern human.

Proceedings of the National Academy of Sciences of the United States of America **106** (27):10971–10974.

Huffman, M.A. 1994. The medicinal use of plants by chimps in the wild. *Pan Africa News* **1**:1.

Hyslop, E.J. 1986. The growth and feeding habits of *Clarias anguillaris* during their first season in the floodplain pools of the Sokoto-Rima river basin, Nigeria. *Journal of Fish Biology* **30** (2):183–193.

Idani, G., Kuroda, S., Kano, T. et al. 1994. Flora and vegetation of Wamba forest, central Zaire with reference to bonobo (*Pan paniscus*) foods. *Tropics* **3**:309–332.

Johnels, A.G. 1957. The mode of terrestrial locomotion in *Clarias*. *Oikos* **8** (2):122–129.

Jubb, R.A. 1967. *Freshwater Fishes of Southern Africa*. Capetown: Balkema.

Kano, T. 1979. Pilot study of pygmy chimpanzees (*Pan paniscus*). In *The Great Apes*, ed. D. Hamburg, and E. McCown, pp. 122–135. Palo Alto, CA: Benjamin Cummings.

Kaplan, G. and Rogers, L. 1994. *Orang-Utans in Borneo*. Armidale, NSW: University of New England Press.

Kempf, E. 2009. Patterns of water use in primates. *Folia Primatologica* (in press).

Kibunjia, M. 1994. Pliocene archaeological occurrences in the Lake Turkana basin. *Journal of Human Evolution* **27**:159–171.

Kimbel, W.H., Walter, R.C., Johanson, D.C. et al. 1996. Late Pliocene *Homo* and Olduwan tools from the Hadar Formation (Kada Hadar Member) Ethiopia. *Journal of Human Evolution* **31**:549–561.

Kingdon, J. 2003. *Lowly Origin: Where, When and Why Our Ancestors First Stood Up*. Princeton, NJ: Princeton University Press.

Kingston, J.D. 2005. Orbital controls on seasonality. In *Seasonality in Primates*, ed. D.K. Brockman and C.P. van Schaik, pp. 519–543. Cambridge: Cambridge University Press.

Kingston, J.D., Deino, A.L., Edgar, R.K. et al. 2007. Astronomically forced climate change in the Kenyan Rift Valley 2.7–2.55 Ma: Implications for the evolution of early hominin ecosystems. *Journal of Human Evolution* **53**:487–503.

Kinzey, W.G. 1984. The dentition of the pygmy chimpanzee, *Pan paniscus*. In *The Pygmy Chimpanzee: Evolutionary Biology and Behavior*, ed. R.L Susman, pp. 65–88. New York: Plenum Press.

Klein, R.G., Avery, G., Cruz-Uribe, K. et al. 2007. The mammalian fauna associated with an archaic hominin skullcap and later Acheulean artifacts at Elandsfontein, Western Cape Province, South Africa. *Journal of Human Evolution* **52**:164–186.

Knott, C.D. 2005. Energetic responses to food availability in the great apes: Implications for hominin evolution. In *Seasonality in Primates*, ed. D.K. Brockman and C.P. van Schaik, pp. 351–379. Cambridge: Cambridge University Press.

Köhler, M., Moyà-Solà, S., and Alba, D.M. 2001. Eurasian hominoid evolution in the light of recent *Dryopithecus* findings. In *Hominoid Evolution and Climate Change in Europe*, ed. L. de Bonis, G.D. Koufos, and P. Andrews, pp. 192–210. Cambridge: Cambridge University Press.

Kono, R.T. and Suwa, G. 2008. Enamel distribution patterns of extant human and hominoid molars: Occlusal versus lateral enamel thickness. *Bulletin of the National Museum of Natural Sciences Series D* **34**:1–9.

Kordos, L. and Begun, D.R. 2002. Rudabánya: A Late Miocene subtropical swamp deposit with evidence of the origin of the African apes and humans. *Evolutionary Anthropology* **11**:45–47.

Kruuk, H. 1976. *The Spotted Hyena: A Study of Predation and Social Behaviour*. Chicago: University of Chicago Press.

Kunimatsu, Y., Nakatsukasa, M., Sawada Y. et al. 2007. A new Late Miocene great ape from Kenya and its implications for the origins of African great apes and humans. *Proceedings of the National Academy of Sciences of the United States of America* **104** (49): 19220–19225.

Kuroda, S., Nishihara, T., Suzuki, S. et al. 1996. Sympatric chimpanzees and gorillas in the Ndoki Forest, Congo. In *Great Ape Societies*, ed. W.C. McGrew, L.F. Marchant, T. Nishida, pp. 71–80. Cambridge: Cambridge University Press.

Lam, Y.M. 2008. What have taphonomic studies taught us about early hominin behaviour? *Evolutionary Anthropology* **17**:158–161.

Lambert, J.E. 2007. Seasonality, fallback strategies, and natural selection: A chimpanzee versus cercopithecoid model for interpreting the evolution of hominin diet. In *Evolution of Human Diet: The Known, the Unknown, and the Unknowable*, ed. P. Ungar, pp. 324–343. Oxford: University of Oxford Press.

Leakey, M.D. 1971. *Olduvai Gorge, Vol. 3: Excavations in Beds I and II, 1960–1963*. Cambridge: Cambridge University Press.

Leakey, M.G., Feibel, C.S., MacDougall, I. et al. 1995. New four-million-year-old hominid species from Kanapoi and Allia Bay, Kenya. *Nature* **376** (6541):565–571.

Lebatard, A.-E., Bourlès, D.L., Duringer, P. et al. 2008. Cosmogenic nuclide dating of *Sahelanthropus tchadensis* and *Australopithecus bahrelghazali*: Mio-Pliocene hominids from Chad. *Proceedings of the National Academy of Sciences of the United States of America* **105** (9):3226–3231.

Lee-Thorp, J.A. and van der Merwe, N.J. 1993. Stable carbon isotope studies of Swartkrans fossils. In *Swartkrans: A Cave's Chronicle of Early Man*, ed. C.K. Brain, *Transvaal Museum Monograph* **8**, pp. 251–256. Pretoria.

Lee-Thorp, J.A., van der Merwe, N.J., and Brain, C.K. 1994. Diet of *Australopithecus robustus* at Swartkrans from stable carbon isotopic analysis. *Journal of Human Evolution* **27**:361–372.

Lepre, C.J., Quinn, R.L., Joordens, J.C.A. et al. 2007. Plio-Pleistocene facies environments from the KBS Member, Koobi Fora Formation: implications for climate controls on the development of lake-margin hominin habitats in the north-east Turkana Basin (northwest Kenya). *Journal of Human Evolution* **53**: 504–514.

Liu, W. and Zheng, L. 2005. Tooth wear difference between the Yuanmou hominoid and *Lufengpithecus*. *International Journal of Primatology* **26** (2):491–506.

López-Antoñanzas, R., Sen, S., and Mein, P. 2004. Systematics and phylogeny of the cane rat (Rodentia: Thryonomyidae). *Zoological Journal of the Linnaean Society* **142** (3):423–444.

Lowe-McConnell, R.H. 1987. *Ecological Studies in Tropical Fish Communities*. Cambridge: Cambridge University Press.

Luyt, J. and Lee-Thorp, J.A. 2003. Carbon isotope ratios of Sterkfontein fossils indicate a marked shift to open environments ca. 1.7 Ma. *South African Journal of Science* **99**:271–273.

Lyman, L. 1994. *Vertebrate Taphonomy*. Cambridge: Cambridge Manuals in Archaeology.

Magliocca, F. and Gautier-Hion, A. 2002. Mineral content as a basis for food selection by western lowland gorillas in a forest clearing. *American Journal of Primatology* **57**:67–77.

Malenky, R.K., Kuroda, S., Vineberg, E.O. et al. 1996. The significance of terrestrial herbaceous foods for bonobos, chimpanzees and gorillas. In *Chimpanzee Cultures*, ed. R.W. Wrangham, W.C. McGrew, F.B.M. de Waal, and P. Heltne, pp. 59–75. Boston: Harvard University Press.

Marean, C.W., Bar-Matthews, M., Bernatchez, J. et al. 2007. Early human use of marine resources and pigment in South Africa during the Middle Pleistocene. *Nature* **449**:905–909.

Marshall, A.J. and Wrangham, R.W. 2007. Evolutionary consequences of fallback foods. *International Journal of Primatology* **28**:1219–1235.

Maslin, M.A. and Christensen, B. 2007. Tectonics, orbital forcing, global climate change, and human evolution in Africa: Introduction to the African paleoclimate special volume. *Journal of Human Evolution* **53**: 443–464.

McBrearty, S. and Jablonski, N.G. 2005. First fossil chimpanzee. *Nature* **437**:105–108.

McCauley J.F., Schaber G.G., Breed C.S. et al. 1982. Subsurface valleys and geoarchaeology of the Eastern Sahara revealed by Shuttle Radar. *Science* **218** (4516): 1004–1020.

McCauley J.F., Breed C.S., Issawi B. et al. 1998. Spaceborne Imaging Radar (SIR) geologic results in Egypt (a review: 1982–1997). *Proceedings of the Egyptian Geological Survey Centennial Conference 1996, Special Publication, Geological Survey Egypt*, ed. G.M. Naim, **75**:489–527.

McClanahan, T. and Young, T.P. 1999. *East African Ecosystems and Their Conservation*. Oxford: Oxford University Press.

McDougall I., Brown F.H., and Fleagle J.G. 2005. Stratigraphic placement and age of modern humans from Kibish, Ethiopia. *Nature* **433**:733–736.

McGrew, W.C. 1977. Socialization and object manipulation of wild chimpanzees. In *Primate Bio-Social Development*, ed. S. Chevalier-Skolnikoff and F.E. Poirier, pp. 261–288. New York: Garland.

McGrew, W.C., Marchant, L.F., and Nishida, T. 1996. *Great Ape Societies*. Cambridge: Cambridge University Press.

McHenry, H.M. and Coffing, K. 2000. *Australopithecus* to *Homo*: Transformations in body and mind. *Annual Review of Anthropology* **29**:125–146.

Merceron, G., Blondel, C., de Bonis, et al. 2004. Dental microwear analysis on *Ouranopithecus* and bovids from the Vallesian (Late Miocene) of Macedonia, Greece: Paleoenvironmental implications. In *5th International Symposium on Eastern Mediterranean Geology, Thessaloniki, Greece, April 14–20, 2004. Proceedings, Vol. 1*, ed. A. Chatzipetros and A. Pavlides, pp. 335–336.

Merrick, H.V. and Merrick, J.P.S. 1976. Archaeological occurrences of earlier Pleistocene age from the Shungura Formation. In *Earliest Man and Environments in the Lake Rudolf Basin*, ed. Y. Coppens, F.C. Howell, G. Ll. Isaac, and R.E.F. Leakey, pp. 574–584. Chicago: University of Chicago Press.

Milton, K. 1999. A hypothesis to explain the role of meat-eating in human evolution. *Evolutionary Anthropology* **8**:11–21.

Monahan, C.M. 1996. New zooarchaeological data from Bed II, Olduvai Gorge, Tanzania: Implications for hominid behavior in the Early Pleistocene. *Journal of Human Evolution* **31**:93–128.

Mora, R. and de la Torre, I. 2005. Percussion tools in Olduvai Beds I and II (Tanzania): Implications for early human activities. *Journal of Anthropological Archaeology* **24**:179–192.

Nicholson, S.E. 1998. Historical fluctuations of Lake Victoria and other lakes in the northern Rift Valley of East Africa. In *Environmental Change and Response in East African Lakes*, ed. J.T. Lehman, pp. 7–37. New York: Springer.

Nishida, T. 1972. Preliminary information of the pygmy chimpanzees of the Congo Basin. *Primates* **13** (4): 415–425.

Nishida, T. 1980. Local differences in responses to water among wild chimpanzees. *Folia Primatologica* **33**: 189–209.

Nishida, T. and Uehara, S. 1983. Natural diet of chimpanzees (*Pan troglodytes schweinfurthii*): Long-term record from the Mahale Mountains, Tanzania. *African Study Monographs* **3**:109–130.

Nishihara, T. 1995. Feeding ecology of western lowland gorillas in the Nouabal'e-Ndoki National Park, Congo. *Primates* **36** (2):151–168.

O'Connell, J.F., Hawks, K., and Blurton-Jones, N.G. 1999. Grandmothering and the evolution of *Homo erectus*. *Journal of Human Evolution* **36**:461–485.

Ohigashi, H., Jisaka, M., Takagaki, T. et al. 1991. Bitter principle and a related steroid glucoside from *Vernonia amygdalina*, a possible medicinal plant for wild chimpanzees. *Agricultural and Biological Chemistry* **55**: 1201–1203.

Otero, O., Pinton, A., Mackaye, H.T. et al. 2009. Fishes and palaeogeography of the African drainage basins: Relationships between Chad and neighbouring basins throughout the Mio-Pliocene. *Palaeogeography, Palaeoclimatology, Palaeoecology* **274**:134–139.

Patterson, N., Richter, D.J., Gnerre, S. et al. 2006. Genetic evidence for complex speciation of humans and chimpanzees. *Nature* **441**:1103–1108.

Perry, S.F., Euverman, R., Wang, T. et al. 2008. Control of breathing in African lungfish (*Protopterus dolloi*): A comparison of aquatic and cocooned (terrestrialized) animals. *Respiratory Physiology and Neurobiology* **160**: 8–17.

Peters, C.R. and Vogel, J.C. 2005. Africa's wild C_4 plant foods and possible early hominid diets. *Journal of Human Evolution* **48**:219–236.

Plummer, T., Bishop, L.C., Ditchfield, P. et al. 1999. Research on Late Pliocene Oldowan sites at Kanjera South, Kenya. *Journal of Human Evolution* **36**:151–170.

Pobiner, B.L. 2007. *Hominin-Carnivore Interactions: Evidence from Modern Carnivore Bone Modification and Early Pleistocene Archaeofaunas (Koobi Fora, Kenya; Olduvai Gorge, Tanzania)*. Unpublished PhD dissertation, Rutgers University.

Pobiner, B.L., Rogers, M.J., Monahan C.M, et al. 2008. New evidence for hominin carcass processing strategies at 1.5 Ma, Koobi Fora, Kenya. *Journal of Human Evolution* **55**:103–130.

Potts, R. 1998. Environmental hypotheses of hominin evolution. *Yearbook of Physical Anthropology* **41**:93–136.

Potts, R. 2004. Paleoenvironmental basis of cognitive evolution in great apes. *American Journal of Primatology* **62**:209–228.

Poulsen, J.R. and Clark, C.J. 2004. Densities, distributions, and seasonal movements of gorillas and chimpanzees in swamp forest in Northern Congo. *International Journal of Primatology* **25** (2):285–306.

Puech, P.F., Cianfarani, F., and Albertini, H. 1986. Dental microwear features as an indicator for plant food in early hominids: A preliminary study of enamel. *Human Evolution* **1**:507–515.

Reed, K.E. 1997. Early hominid evolution and ecological change through the African Plio-Pleistocene. *Journal of Human Evolution* **32**:289–322.

Reed, K.E. 2008. Paleoecological patterns at the Hadar hominin site, Afar Regional State, Ethiopia. *Journal of Human Evolution* **54**:743–768.

Remis, M.J. 1997. Ranging and grouping patterns of a western lowland gorilla group at Bai Hokou, Central African Republic. *American Journal of Primatology* **43**:110–130.

Remis, M.J., Dierenfeld, E.S., Mowry, C.B et al. 2001. Nutritional aspects of western lowland gorilla (*Gorilla gorilla gorilla*) diet during seasons of fruit scarcity at Bai Hokou, Central African Republic. *International Journal of Primatology* **23**:231–250.

Reynolds, V., Plumptre, A.J., Greenham, J. et al. 1998. Condensed tannins and sugars in the diet of chimpanzees (*Pan troglodytes schweinfurthii*) in the Budongo Forest, Uganda. *Oecologia* **115** (3):331–336.

Rightmire, G.P. 1990. *The Evolution of Homo erectus: Comparative Anatomical Studies of an Extinct Human Species*. Cambridge: Cambridge University Press.

Rightmire, G.P. 2004. Brain size and encephalization in early to mid-Pleistocene *Homo*. *American Journal of Physical Anthropology* **124**:109–123.

Rizkalla, C., Blanco-Silva, F., and Gruver, S. 2007. Modeling the impact of Ebola and bushmeat hunting on western lowland gorillas. *Ecohealth* **4** (2):151–155.

Robbins, L.H., Murphy, M.L., Stewart, K.M. et al. 1994. Barbed bone points, paleoenvironment, and the prehistory of fish exploitation in the Western Kalahari Desert, Botswana. *Journal of Field Archaeology* **21** (2):257–264.

Roberts, M.G. and Parfitt, S.A. 1999. Boxgrove: A Middle Pleistocene hominid site at Eartham Quarry, Boxgrove, West Sussex. *English Heritage Archaeological Report 17*. London: English Heritage.

Rode, K.D., Chapman, C.A., Chapman, L.J. et al. 2002. Mineral resource availability and consumption by *Colobus* in Kibale National Park, Uganda. *International Journal of Primatology* **24** (3):541–573.

Rodman, P.S. 1988. Diversity and consistency in ecology and behaviour. In *Orang-utan Biology*, ed. J.H. Schwartz, pp. 31–52. Oxford: Oxford University Press.

Rogers, M.E., Abernethy, K., Bermejo, M. et al. 2004. Western Gorilla Diet: A Synthesis from Six Sites. *American Journal of Primatology* **64**:173–192.

Rook, L., Bondioli, L., Köhler, M. et al. 1999. *Oreopithecus* was a bipedal ape after all: Evidence from the iliac cancellous architecture. *Proceedings of the National Academy of Sciences of the United States of America* **96** (15):8795–8799.

Sadun, E.H., von Lichtenberg, F., Hickman, R.L. et al. 1966. *Schistosomiasis mansoni* in the Chimpanzee: Parasitologic, clinical, serologic, pathologic and radiologic observations. *American Journal of Tropical Medicine and Hygiene* **15** (4):496–506.

Sakamaki, T. 1998. First record of algae-feeding by a female chimpanzee at Mahale. *Pan-Africa News* **5** (1):1.

Sanders, R. 2005. UC Berkeley, French scientists find missing link between the whale and its closest relative, the hippo. University of California; UC Newsroom, January 24, 2005. http://www.universityofcalifornia.edu/news/article/6875.

Schoeninger, M.F., Reeser, H. and Hallin, K. 2003. Paleoenvironment of *Australopithecus anamensis* at Allia Bay, East Turkana, Kenya: Evidence from mammalian herbivore enamel stable isotopes. *Journal of Anthropological Archaeology* **22**:200–207.

Schrenk, F., Bromage, T.G., Betzler, C.G. et al. 1993. Oldest *Homo* and Pliocene biogeography of the Malawi Rift. *Nature* **365**:833–836.

Schwartz, H.E. 1983. *Paleoecology of Late Cenozoic Fishes from the Turakan Basin, Northern Kenya.*

Unpublished PhD dissertation, University of California-Santa Cruz.

Schwartzman, D., Middendorf, G. and Armour-Chelu, M. 2009. Was climate the prime releaser for encephalization? *Climate Change* **95** (3–4):439–447.

Ségalen, L., Lee-Thorp, J.A., and Cerling, T. 2007. Timing of C_4 grass expansion across sub-Saharan Africa. *Journal of Human Evolution* **53**:549–559.

Selvaggio, M.M. and J. Wilder, J. 2001. Identifying the involvement of multiple carnivore taxa with archaeological bone assemblages. *Journal of Archaeological Science* **28**:465–470.

Senut, B. and Pickford, M. 2001. The geological and faunal context of Late Miocene hominid remains from Lukeino, Kenya. *Comptes Rendus de l'Academie des Sciences, Series IIA – Earth and Planetary Science* **332** (2):145–152.

Shabel, A.B. 2009. *Craniodental morphology and biogeochemistry of African carnivorans: toward a new model of Plio-Pleistocene hominin evolution.* Unpublished PhD thesis. University of California.

Sherwood, R.J., Ward, S.C., and Hill, A. 2002. The taxonomic status of the Chemeron temporal (KNM-BC 1). *Journal of Human Evolution* **42**:153–184.

Sikes, N.E. 1994. Early hominid habitat preferences in East Africa: Paleosol carbon isotopic evidence. *Journal of Human Evolution* **27**:25–45.

Speth, J.D. 1989. Early hominid hunting and scavenging: The role of meat as an energy source. *Journal of Human Evolution* **18**:329–343.

Speth, J.D. and Spielmann, K.A. 1983. Energy source, protein metabolism and hunter-gatherer subsistence strategies. *Journal of Anthropological Archaeology* **2**:1–31.

Sponheimer, M. and Lee-Thorp, J.A. 1999. Isotopic evidence for the diet of an early hominid, *Australopithecus africanus*. *Science* **283** (5400):368–370.

Sponheimer, M. and Lee-Thorp, J.A. 2003. Differential resource utilization by extant great apes and australopithecines: Towards solving the C4 conundrum. *Comparative Biochemistry and Physiology Part A* **136**:27–34.

Sponheimer, M., Lee-Thorp, J.A., de Ruiter, D.J. et al. 2005. Hominins, sedges and termites: New carbon isotope data from the Sterkfontein Valley and Kruger National Park. *Journal of Human Evolution* **48**:301–312.

Sponheimer, M., Passey, B.H., de Ruiter, D.J. et al. 2006. Isotopic evidence for dietary variability in the early hominin *Paranthropus robustus*. *Science* **314**: 980–982.

Stewart, K.M. 1990. Fossil fish from the Upper Semliki. In *Evolution of Environments and Hominidae in the African Western Rift Valley*, ed. N.T. Boaz. *Virginia Museum of Natural History, Memoir* **1**:141–162.

Stewart, K.M. 1994. Early hominid utilisation of fish resources and implications for seasonality and behaviour. *Journal of Human Evolution* **27**:229–245.

Stewart, K.M. 2003a. Fossil fish remains from Mio-Pliocene deposits at Lothagam, Kenya. In *Lothagam: The Dawn of Humanity in Eastern Africa*, ed. M.G. Leakey and J.M. Harris, pp. 75–115. New York: Columbia University Press.

Stewart, K.M. 2003b. Fossil fish remains from the Pliocene Kanapoi site, Kenya. *Contributions in Science* **498**:21–39.

Stewart, K.M. 2009. Fossil fish from the Nile and its Southern Basins. In *The Nile: Origin, Environments, Limnology and Human Use*, ed. H. Dumont, pp. 677–705. New York: Springer.

Stewart, K.M., Gifford-Gonzalez, D.P., and Rybczynski, N. 1997. Characteristics of modern foraging camps and their faunas from Lake Turkana, Kenya. *Anthropozoologia* **25–26**:763–766.

Stokes, E., Malonga R., and Rainey, H. 2008. New gorilla population estimates in Northern Congo Republic. *Gorilla Journal* **37**:1.

Storrs, G.W. 2003. Late Miocene-Early Pliocene crocodilian fauna of Lothagam, southwest Turkana Basin, Kenya. In *Lothagam: The Dawn of Humanity in Eastern Africa*, ed. M.G. Leakey and J.M. Harris, pp. 137–159. New York: Columbia University Press.

Stringer, C. 2000. Coasting out of Africa. *Nature* **405**:24–27.

Susman, R.L., Badrian, N., Badrian, A. et al. 1985. Positional behavior and feeding ecology of the pygmy chimpanzee (*Pan paniscus*): First year results of the Lomako Forest Pygmy Chimpanzee Project. *National Geographic Research* **20**:725–739.

Suwa, G., Reiko, T.K., Shigehiro, K. et al. 2007. A new species of great ape from the Late Miocene epoch in Ethiopia. *Nature* **448** (7156):921–924.

Tackholm, V. and Drar, M. 1973. *Flora of Egypt, Vol. II*. Königstein: Otto Koeltz Antiquariat.

Teaford, M. F. and Ungar, P.S. 2000. Diet and the evolution of the earliest human ancestors. *Proceedings of the National Academy of Sciences of the United States of America* **97** (25):13506–13511.

Thieme, H. 1997. Lower Paleolithic hunting spears from Germany. *Nature* **385**:807.

Tobias, P.V. 1967. Pleistocene deposits and new fossil localities in Kenya. *Nature* **215**:476–480.

Tobias, P.V. 1987. The brain of *Homo habilis*: A new level of organization in cerebral evolution. *Journal of Human Evolution* **16**:741–761.

Tobias, P.V. 1993. Earliest *Homo* not proven. *Nature* **361**:307.

Tooth, S. and McCarthy T.S. 2007. Wetlands in drylands: Geomorphological and sedimentological characteristics, with emphasis on examples from southern Africa. *Progress in Physical Geography* **31** (1):3–41.

Trauth, M.H., Maslin, M.A., Deino, A.L. et al. 2007. High- and low-latitude forcing of Plio-Pleistocene East African climate and human evolution. *Journal of Human Evolution* **53**:475–486.

Tsujikawa, H. 2005. The palaeoenvironment of *Samburupithecus kiptalami* based on its associated fauna. *African Study Monographs, Supplement* **32**:51–62.

Turnbull-Kemp. P. 1967. *The Leopard*. Capetown: Howard Timmins.

Tutin, C.E.G. and Fernandez, M. 1992. Insect-eating by sympatric lowland gorillas (*Gorilla gorilla gorilla*) and chimpanzees (*Pan troglodytes troglodytes*) in the Lopé

Reserve, Gabon. *American Journal of Primatology* **28** (1): pp. 29–40.

TUTIN, C.E.G. AND FERNANDEZ, M. 2009. Composition of the diet of chimpanzees and comparisons with that of sympatric lowland gorillas in the Lopé Reserve, Gabon. *American Journal of Primatology* **30** (3):195–211.

UEHARA, S. 1990. Utilization patterns of a marsh grassland within the tropical rain forest by the bonobos (*Pan paniscus*) of Yalosidi, Republic of Zaire. *Primates* **31** (3): 311–322.

UNGAR, P.S. 1996. Dental microwear of European Miocene catarrhines: Evidence for diets and tooth use. *Journal of Human Evolution* **31**:335–366.

UNGAR, P.S. 2004. Dental topography and diets of *Australopithecus afarensis* and early *Homo*. *Journal of Human Evolution* **46**:605–622.

UNGAR P.S. AND KAY R.F. 1995. The dietary adaptations of European Miocene catarrhines. *Proceedings of the National Academy of Sciences of the United States of America* **92**:5479–5481.

UNGAR, P.S., GRINE, F.E., TEAFORD, M.F. ET AL. 2006. Dental microwear and diets of African early *Homo*. *Journal of Human Evolution* **50**:78–95.

VAN DER MERWE, N.J., MASAO, F.T., AND BAMFORD, M.K. 2008. Isotopic evidence for contrasting diets of early hominins *Homo habilis* and *Australopithecus boisei* of Tanzania. *South African Journal of Science* **104**: 153–155.

VAN DER MERWE, N.J., THACKERAY, J.F., LEE-THORP, J.A. ET AL. 2003. The carbon isotope ecology and diet of *Australopithecus africanus* at Sterkfontein, South Africa. *Journal of Human Evolution* **44**:581–597.

VAN NEER, W. 1986. Some notes on the fish remains from Wadi Kubbaniya (Upper Egypt: Late Palaeolithic). *Fish and Archaeology*, BAR International Series **294**: 103–113.

VERHAEGEN, M., PUECH, P.-F., AND MUNRO, S. 2002. Aquarboreal Ancestors? *Trends in Ecology and Evolution* **17** (5):212–217.

VIGNAUD, P., DURINGER, P., MACKAYE, H.T. ET AL. 2002. Geology and palaeontology of the Upper Miocene Toros-Menalla hominid locality, Chad. *Nature* **418**: 152–155.

VILA, P., SOTO, E., SANTONJA, M. ET AL. 2005. New data from Ambrona: Closing the hunting *versus* scavenging debate. *Quaternary International* **126-128**:223–250.

VOGEL, E.R., VAN WOERDEN, J.T., LUCAS, P.W. ET AL. 2008. Functional ecology and evolution of hominoid molar enamel thickness: *Pan troglodytes schweinfurthii* and *Pongo pygmaeus wurmbii*. *Journal of Human Evolution* **55**:60–74.

VRBA, E.S. 1985. Ecological and adaptive changes associated with early hominid evolution. In *Ancestors: The Hard Evidence*, ed. E. Delson, pp. 63–71. New York: Alan R. Liss.

VRBA, E.S. 1988. Late Pliocene climatic events and hominid evolution. In *Evolutionary History of the "Robust Australopithecines"*, ed. F. Grine, pp. 405–426. New York: Aldine de Gruyter.

WALKER, A., LEAKEY, R.E., HARRIS, J.M. ET AL. 1986. 2.5-Myr *Australopithecus boisei* from West of Lake Turkana, Kenya. *Nature* **322** (7):517–522.

WALSH, P.D., ABERNATHY, K. A., BERMEJO, M. ET AL. 2003. Catastrophic ape decline in western Equatorial Africa. *Nature* **1566**:1–3.

WALTER, R.C., BUFFLER, R.T., BRUGGEMANN, J.H. ET AL. 2000. Early human occupation of the Red Sea coast of Eritrea during the last interglacial. *Nature* **405**:65–69.

WARD, C., LEAKEY, M., AND WALKER, A. 1999. The new hominid species *Australopithecus anamensis*. *Evolutionary Anthropology* **7**:197–205.

WELLS, J.C.K. AND STOCK, J.T. 2007. The biology of the colonizing ape. *Yearbook of Physical Anthropology* **50**:191–222.

WERDELIN, L. 2003. Carnivora from the Kanapoi hominid site, Turkana Basin, Northern Kenya. *Contributions in Science* **498**:115–132.

WHITE, T.D., SUWA, G., AND ASFAW, B. 1994. *Australopithecus ramidus*, a new species of early hominid from Aramis, Ethiopia. *Nature* **371**:306–312.

WHITE, T.D., SUWA, G., AND ASFAW, B. 1995. Corrigendum: *Australopithecus ramidus*, a new species of early hominid from Aramis, Ethiopia. *Nature* **375**:88.

WHITE, T.D., SUWA, G., HART, W.K. ET AL. 1993. New discoveries of *Australopithecus* at Maka in Ethiopia. *Nature* **366**:261–265.

WHITE, T.D., ASFAW, B., BEYENE, Y. ET AL. 2009. *Ardipithecus ramidus* and the paleobiology of early hominids. *Science* **326**:75–86.

WILLIAMS, M.A.J., WILLIAMS, F.M., GASSE, F. ET AL. 1979. Plio-Pleistocene environments at Gadeb prehistoric site, Ethiopia. *Nature* **282** (1):29–39.

WILLIAMSON, L. AND USONGO, L. 1996. Survey of Gorillas *Gorilla gorilla* and Chimpanzees *Pan troglodytes in* the Reserve de Faune du Dja, Cameroun. *African Primates* **2** (2):67–72.

WILLIAMSON, P. 1990. Late Cenozoic mollusk faunas from the North Western African Rift (Uganda-Zaire). In *Evolution of Environments and Hominidae in the African Western Rift Valley*, ed. N.T. Boaz. *Virginia Museum of Natural History, Memoir* **1**:125–141.

WILLIS, L.M., EREN, M.I., RICK, T.C. 2008. Does butchering fish leave cutmarks? *Journal of Archaeological Science* **35**:1438–1444.

WOLDEGABRIEL, G., HAILE-SELASSIE, Y., RENNE, P.R. ET AL. 2001. Geology and palaeontology of the Late Miocene Middle Awash Valley, Afar rift, Ethiopia. *Nature* **412**:175–178.

WOLDEGABRIEL, G., AMBROSE, S.A., BARBONI, D., ET AL. 2009. The geological, isotopic, botanical, invertebrate, and lower vertebrate surroundings of *Ardipithecus ramidus*. *Science* **326** (65):65e1–65e5.

WRANGHAM, R. 1977. Feeding behaviour of chimpanzees in Gombe National Park, Tanzania. In *Primate Ecology: Studies of Feeding and Ranging Behaviour in Lemurs, Monkeys and Apes*, ed. T.H. Clutton-Brock, pp. 503–538. New York: Academic Press.

WRANGHAM, R.W., CHAPMAN, C.A., AND CLARK, A.P. 1996a. Social ecology of Kanyawara chimpanzees: Implications for understanding the costs of great ape groups. In *Great Ape Societies*, ed. W.C. McGrew, L.F. Marchant, and T. Nishida, pp. 45–57. Cambridge: Cambridge University Press.

Wrangham, R.W. 2005. The Delta Hypothesis. In *Interpreting the Past*, eds. D.E. Lieberman, R.J. Smith, and J. Kelley. pp. 231–243. Boston: Brill Academic Publishers, Inc.

Wrangham, R.W., McGrew, W.C., de Waal, F.B.M. et al., eds. 1996b. *Chimpanzee Cultures*. Boston: Harvard University Press.

CHAPTER 9

BRAIN SIZE IN CARNIVORAN MAMMALS THAT FORAGE AT THE LAND–WATER ECOTONE, WITH IMPLICATIONS FOR ROBUST AUSTRALOPITHECINE PALEOBIOLOGY

Alan B. Shabel

INTRODUCTION

This chapter provides a preliminary analysis of brain size in the extant carnivoran mammals in order to explore the relationship between neuroanatomy, manual dexterity, and durophagy (hard-object feeding). The primary goal is to encourage the study of the Carnivora as a comparative context for the investigation of paleoanthropological hypotheses. Previously, I studied the craniodental morphology of the extant Carnivora as the basis for a new paleoecological reconstruction of the robust australopithecines (*Paranthropus*), and I analyzed the biogeochemistry of the extant African Carnivora to test the new model against the existing alternative models (Shabel, 2009).

The durophage–ecotone model reconstructs the robust australopithecines as opportunistic consumers of hard-shelled food objects at the land–water interface, including eggs, crabs, and mollusks, as well as nuts (Shabel, 2009). In the course of testing the new model, I made several observations on brain size in carnivoran species that forage at the land–water ecotone, and these are the basis of additional investigation in this chapter. Studies of brain size are an important part of mammalogy, anthropology, and biology in general, and studies of carnivoran brains have been an important focus at least since Mivart (1885). The carnivoran brain represents a crossroads of natural historical processes, and it is influenced by ecological, ethological, historical, and genetic variables. However, the separate influences of natural history variables on brain size are difficult to disentangle (Gittleman, 1986a,b, 1993).

The mammalian clade Carnivora is taxonomically and ecologically diverse, and it provides a powerful comparative context for testing models of primate ecology and evolution. The Carnivora includes 282 extant species that consume meat, bone, fish, mollusks, crabs, insects, earthworms, bamboo, and fruit. Some carnivoran species are omnivorous – mixing vegetation, invertebrates, and vertebrates, in some combination – and many are

Human Brain Evolution, Edited by Stephen C. Cunnane and Kathlyn M. Stewart

Figure 9.1 *Atilax paludinosus*, the marsh mongoose, is shown here using the downward hurling technique to crack a freshwater bivalve mollusk on a stone anvil. This figure is adapted from still images from the film, "Crocodiles: Here Be Dragons" (Survival Anglia Ltd., 1990). Illustration by Noel Sirivansanti.

opportunistic. As is true for vertebrates in general, the morphology of the carnivoran skull is closely correlated with the physical properties of major food items (Radinsky, 1981; Christiansen and Wroe, 2007), and the carnivoran species that forage for hard-shelled prey at the land–water ecotone are craniodentally robust. The suite of features that characterizes the skulls of the wetland durophages is also exhibited by the robust australopithecines, and the durophage–ecotone model is grounded in this anatomical analogy.

Wetland Durophagy

Three lineages of durophagous wetland carnivorans are the focus of this chapter: *Atilax* (marsh mongooses), *Aonyx* (clawless otters), and *Procyon cancrivorus* (crab-eating raccoons). *Atilax* is a relatively basal member of the family of true mongooses (Herpestidae), and this solitary mongoose is widespread in sub-Saharan Africa where it forages for crabs, mollusks, frogs, rodents, eggs, and insects in waterside environments (Rowe-Rowe, 1978; Baker, 1989, 1992; Fig. 9.1). *Atilax* overlaps across much of its range with *Aonyx*, a member of the weasel family (Mustelidae), and *Aonyx* is another major crab consumer that also eats fish, frogs, other small vertebrates, and, in some regions, earthworms (Larivière, 2001a,b; Kruuk, 2006; Fig. 9.2). The afrotropical *Atilax* and *Aonyx* clades are compared with the neotropical *P. cancrivorus*, an omnivorous member of the Procyonidae that preys on crustaceans and mollusks in mangrove, riverine, and some upland systems (Wainwright, 2007). The ecology of the crab-eating raccoons remains relatively little studied.

Atilax, *Aonyx*, and *P. cancrivorus* are not closely related in the phylogeny of the Carnivora (Flynn et al., 2005). The convergent evolution of craniodental features associated with durophagy in these lineages is striking: overall massive skull, wide zygomatic arches, prominent sagittal crest, robust dentary, high ascending ramus, expanded postcanine dentition, and reduced anterior dentition. At the level of the occlusal surface of the crushing teeth, the wetland durophages exhibit bulbous dental cusps, and in the extreme condition, these teeth are grossly inflated and the teeth are "puffy" in appearance (Shabel, 2009). All of these craniodental features are also characteristic of the robust australopithecines.

Another apparent morphological convergence exhibited by *Atilax*, *Aonyx*, and *Procyon* is in the structure of their forepaws. Pocock (1916, 1921a,b) was the first author

Figure 9.2 *Aonyx capensis*, the African clawless otter, is a relatively large lutrine (12–19 kg) with a blunt head, powerful neck, and hypertrophied masticatory muscles. The digits of the hands are largely unwebbed, and those of the hindfeet are partly webbed. The distal phalangeal bones of *Aonyx* are reduced, and the keratinous claws are rudimentary or completely lost (*a-onyx* means "clawless"). *A. capensis* relies on its dexterous hands when foraging for crabs, frogs, and slow-moving fish in shallow water and semiterrestrial environments. Photo taken by Alan B. Shabel. (See color insert.)

to describe the hand (manus) structure of these three taxa, and he specifically related the absence of webbing between the digits in *Atilax* and *Procyon* to their emphasis on manual foraging. The loss of interdigital webbing probably increases the mobility of the digits and permits more sensitive control of the fingertips in these taxa (Ewer, 1973; Estes, 1991). The majority of carnivoran species capture their food with their mouths, whereas the wetland durophages actively seek and probe for food with their hands. In *Aonyx*, the complete loss of the foreclaws and the development of fleshy excrescences on the apexes of the fingers are additional specializations related to manual foraging (Kingdon, 1977).

Independently of the studies of manus structure, a series of analyses showed that the brains of *Atilax*, *Procyon*, and *Aonyx* are relatively enlarged and complex in the regions associated with the sensorimotor activity of the forepaws (Welker and Seidenstein, 1959; Radinsky, 1968, 1975; Willemsen, 1992). In a separate analysis, Sheppey and Bernard (1984) found that *Atilax* and *Aonyx* had the largest brains of their respective families in southern Africa, and *Aonyx* had the highest encephalization quotient of any of the 30 species of southern African carnivoran that the authors examined. Sheppey and Bernard (1984) argued that the large brains of *Aonyx* and *Atilax* could be explained in part by their high "feeding efficiency." I predict that the large brain size of the wetland durophages is correlated with their increased brain complexity, and that this is related to the habit of extractive foraging for nutritious, hard-shelled prey at the land–water ecotone. In this chapter, I measure cranial capacity in a large number of carnivoran species as a first step in the study of wetland durophagy as an integrated natural history syndrome.

Measurement of Brain Volume

One method to measure brain volume (BV) is to make a cast of the endocranium of an intact skull and then to measure the volume of the endocast. Although accurate, this method is laborious, and it can be damaging to intact skulls. Another method to estimate cranial capacity is to fill the endocranium with a packing material such as glass beads or plant seeds. The packing material is then decanted and its volume measured, and this quantity is used to estimate BV. A third method to estimate cranial capacity, and the one followed here, is based on the measurement of three chords on the external surface of the cranium (the ectocranial surface), as demonstrated by Young (1959) for rats, Martin (1990) and Elton et al. (2001) for primates, and Finarelli (2006) for carnivorans. Finarelli (2006) reported that estimates of BV based on ectocranial proxy are closely correlated with estimates based on endocranial packing ($R^2 = 0.983$), as well as with estimates based on endocranial casts. Estimates of brain size based on ectocranial proxy correlate strongly with actual brain size because carnivoran brains are closely appressed to the cranial bones (Fig. 9.3).

The natural history variable most closely correlated with brain size is whole body size, and the latter term is most commonly estimated as whole body mass. Because of the strong correlation between brain mass and whole body mass, many studies statistically correct brain mass for body mass to estimate relative "encephalization" (Jerison, 1973). In the absence of actual brain mass data, it is necessary to estimate brain mass from BV.

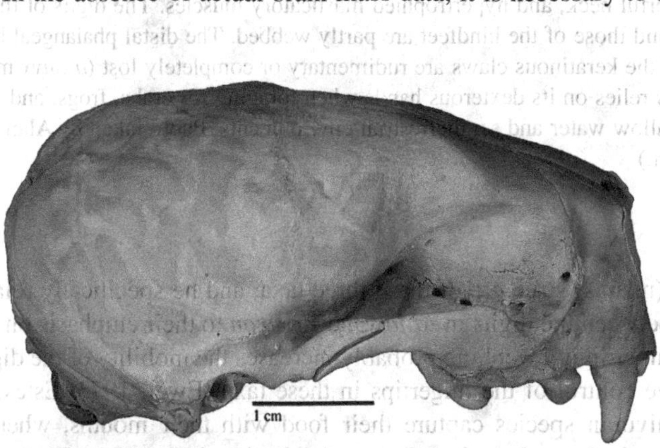

Figure 9.3 This specimen of *Enhydra lutris* (USNM 285430), a young sea otter from Amchitka Island, Alaska, illustrates the close connection between the brain and the cranial bones in a carnivoran. The purple-red stain on the braincase is probably related to the consumption of heavily pigmented sea urchins by these animals. Photograph by Alan B. Shabel. (See color insert.)

BV can be equated with brain mass under the assumption that 1 mL = 1 g of brain matter (e.g., Bininda-Emonds and Gittleman, 2000). However, the conversion of BV to brain mass is not trivial (Hemmer, 2007), and a better understanding of the relationship between these variables is clearly needed. As a contribution to the study of carnivoran neuroanatomy, and in order to shed light on trends in australopithecine evolution, I measured BV by ectocranial proxy on 977 carnivoran individuals from 144 extant species (representing all 13 extant families, excluding pinnipeds). The relative degree of encephalization was analyzed at the species level in the context of the Carnivora as a whole, as well as in local subclades, in order to explore the trends in brain size evolution in a phylogenetic context.

METHODS

Neurocranial length (NCL), neurocranial width (NCW), and neurocranial height (NCH) were measured with digital calipers on specimens in museum collections in North America, Europe, and Africa (Fig. 9.4). The sex and collection history of all specimens are listed in appendix 2.2 of Shabel (2009), except for 70 specimens that are included here for the first time. BV was estimated for each individual according to the formula of Finarelli (2006), $\ln(BV) = 1.06 \times \ln(NCH) + 0.28 \times \ln(NCL) + 1.27 \times \ln(NCW) - 6.23$ ($R^2 = 0.98$), where the ectocranial chords are in millimeters and BV is in cubic centimeters. The individual results were averaged across sexes to calculate species means.

Body mass estimates for each species were taken from Smith et al. (2003, updated MOM v3.6.1), except in a small number of cases where no data were given in that compilation, or where the given data were clearly inconsistent with other studies. In these cases, it was necessary to rely on additional sources, as indicated in the footnotes in Table 9.1. The body mass estimates chosen for *Aonyx* require a more complete explanation. There is confusion in the literature over the body mass of the two *Aonyx* species, with most published estimates derived from the work of Kingdon (1977, 1997). I reviewed all field studies on *Aonyx*, and I compiled data from all available museum specimens with associated information, in order to estimate body mass in this genus. It was found that 14.0 kg was a good working average for *Aonyx capensis*, but there were no data at all for the body mass of *Aonyx congicus*. Most studies have listed *A. congicus* as heavier than *A. capensis* (e.g., Smith et al., 2003; Meiri et al., 2005), and for that reason, I assigned a body mass value of 15.0 kg to *A. congicus*.

Encephalization indexes for each species were calculated as the natural logarithm of observed BV (as measured by ectocranial proxy) divided by the natural log of expected BV. The expected BVs were determined by the lines of best fit of ordinary least squares (OLS) and reduced major axis (RMA) regressions of ln(BV) against ln(body mass). All species were analyzed in relation to the Carnivora as a whole ($n = 144$); *Atilax* and the other herpestids were analyzed in relation to clade Herpestidae; and *Procyon* and *Aonyx* were analyzed in relation to superfamily Musteloidea (this taxon includes Ailuridae, Mephitidae, Procyonidae, and Mustelidae). Encephalization indexes greater than 1.0 represent relatively large brains, and indexes less than 1.0 represent relatively small brains. Space limitations restrict this presentation to families Herpestidae, Procyonidae, and Mustelidae.

RESULTS

BV estimates (cm^3) for herpestid, procyonid, and mustelid species are given in Table 9.1. The new data are compared with the results based on endocranial packing from Sheppey and Bernard (1984) and Gittleman (1986a,b, 1991), as well as with previous data based on the ectocranial proxy technique (Finarelli and Flynn, 2009); see Fig. 9.5.

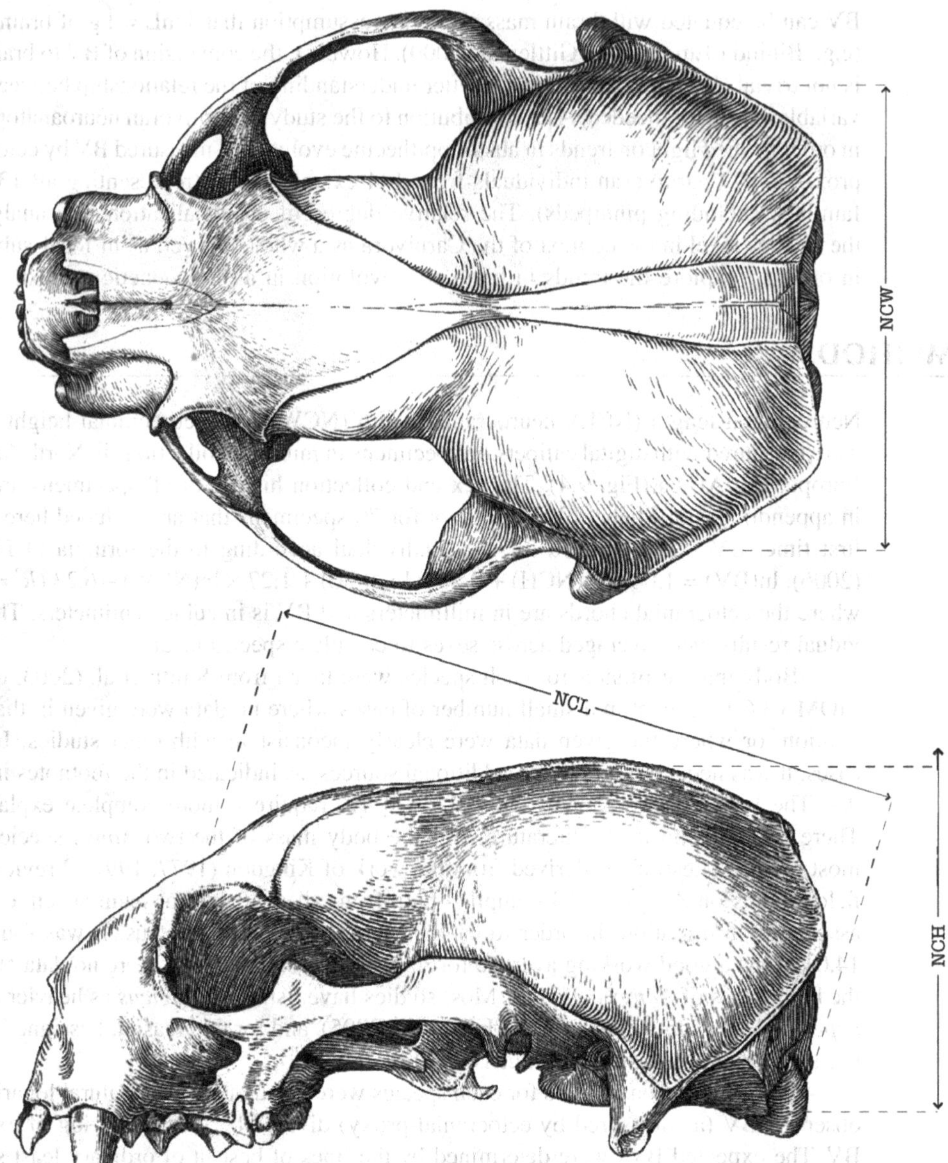

Figure 9.4 The three ectocranial chords used to estimate carnivoran brain volume. Neurocranial length (NCL) is the distance from nasion to opisthion, neurocranial width (NCW) is the maximum biparietal (bisquamosal) distance, and neurocranial height (NCH) is the distance from the basicranium to the superiormost point on the parietals, not including the sagittal crest (if present). Modified from Allen, J.A. 1924. Carnivora collected by the American Museum Congo Expedition. *Bulletin of the American Museum of Natural History* **47** (3):73–281 + Plates.

TABLE 9.1 Brain size estimates for carnivoran species of the families Herpestidae, Procyonidae, and Mustelidae

Carnivoran clade	n	Mass (g)	Brain size (cm^3)	Gittleman	Finarelli and Flynn (2009)	Sheppey and Bernard (1984)	All Carnivora		Local subclade	
							EI OLS	EI RMA	EI OLS	EI RMA
Herpestidae										
Atilax paludinosus	29	3300	29.05 ± 2.15	28.50	25.95	29.0	1.02	1.01	1.07	1.05
Herpestes naso	3	3000	22.55 ± 0.84	25.53	22.92		0.96	0.97	1.01	1.00
Herpestes urva	17	1863	22.06 ± 1.05	20.91	21.69		1.06	1.04	1.10	1.07
Herpestes brachyurus	4	2000	21.84 ± 1.41				1.03	1.03	1.08	1.05
Herpestes semitorquatus	1	3162[a]	21.42 ± n.a.				0.93	0.95	0.98	0.98
Herpestes smithii	1	1861	15.18 ± n.a.	13.74	12.99		0.93	0.95	0.96	0.97
Herpestes vitticollis	1	2995	28.11 ± n.a.	25.79			1.03	1.02	1.08	1.06
Herpestes fuscus	3	1360[a]	12.93 ± 0.22	12.55			0.94	0.96	0.97	0.97
Herpestes auropunctatus	11	434[b]	7.46 ± 0.64				1.02	1.02	1.00	1.01
Herpestes edwardsi	5	1324	11.33 ± 0.99	10.49	10.97		0.90	0.93	0.92	0.94
Herpestes javanicus	6	750	9.68 ± 0.75				0.97	0.99	0.98	0.99
Herpestes ichneumon	11	2850	22.75 ± 1.06	23.34	21.78	24.0	0.97	0.98	1.02	1.01
Galerella pulverulenta	13	797	11.16 ± 0.61	11.02		11.0	1.02	1.02	1.03	1.02
Galerella sanguinea	12	550	8.71 ± 0.83	8.76		7.0	1.02	1.02	1.01	1.01
Ichneumia albicauda	14	3500	24.07 ± 2.20	24.29	21.36	23.0	0.95	0.96	1.00	1.00
Cynictis penicillata	14	836	11.41 ± 1.76	10.49	9.54	10.0	1.01	1.01	1.02	1.02
Paracynictis selousi	11	1640	15.89 ± 0.76	15.80			0.97	0.98	1.00	1.00
Bdeogale crassicauda	3	1550	15.48 ± 1.50	16.95	15.55		0.97	0.98	1.01	1.00
Bdeogale jacksoni	9	2500	21.39 ± 0.95				0.98	0.99	1.03	1.02
Bdeogale nigripes	3	2500	23.88 ± 1.88				1.01	1.01	1.06	1.04
Rhynchogale melleri	5	2500	16.51 ± 0.61	16.95			0.90	0.93	0.94	0.95
Suricata suricatta	13	726	11.83 ± 1.19	10.28	11.80	10.0	1.07	1.05	1.07	1.06
Mungos gambianus	3	1500	11.34 ± 0.73				0.87	0.91	0.90	0.92
Mungos mungo	13	1925	11.66 ± 0.94	10.49	10.86	10.0	0.83	0.88	0.86	0.89
Crossarchus alexandri	12	1500	13.18 ± 0.97				0.92	0.95	0.95	0.96
Crossarchus platycephalus	1	1250	11.75 ± n.a.				0.92	0.95	0.95	0.96
Crossarchus obscurus	5	1250	10.32 ± 0.61	9.78	10.81		0.87	0.91	0.90	0.92
Dologale dybowskii	2	350	8.17 ± 0.27				1.15	1.11	1.11	1.09

TABLE 9.1 Continued

Carnivoran clade	*n*	Mass (g)	Brain size (cm^3)	Gittleman	Finarelli and Flynn (2009)	Sheppey and Bernard (1984)	All Carnivora		Local subclade	
							EI OLS	EI RMA	EI OLS	EI RMA
Helogale hirtula	6	289	6.79 ± 0.26				1.13	1.10	1.07	1.07
Helogale parvula	11	300	5.65 ± 0.53	4.76	6.07	5.0	1.00	1.01	0.96	0.98
Procyonidae										
Bassariscus astutus	4	1130	15.59 ± 1.22	16.44	16.45		1.05	1.04	1.04	1.03
Bassariscus sumichrasti	4	900	20.54 ± 1.54	19.30	18.72		1.23	1.15	1.21	1.14
Nasua narica	4	4030	39.65 ± 3.09	37.00[f]	36.46		1.07	1.05	1.04	1.02
Nasua nasua	3	3794	34.44 ± 1.48	29.96			1.04	1.03	1.01	1.01
Potos flavus	4	3000	30.43 ± 2.54	25.53	29.76		1.05	1.04	1.02	1.01
Procyon cancrivorus	21	6950	57.95 ± 4.83	59.45			1.07	1.04	1.03	1.02
Procyon lotor	17	5525	47.51 ± 5.68	40.04	43.52		1.06	1.04	1.02	1.01
Mustelidae										
Taxidea taxus	12	7108	56.55 ± 4.05	48.91	57.90		1.06	1.04	1.02	1.01
Mellivora capensis	13	9050[c]	86.38 ± 9.38	72.97	73.89	88.0	1.12	1.08	1.08	1.05
Meles meles	7	9530[a]	46.63 ± 3.98	50.40	36.41		0.96	0.97	0.93	0.95
Arctonyx collaris	3	6356	44.54 ± 7.33	49.40	51.67		1.01	1.01	0.98	0.98
Eira barbara	9	3910	42.89 ± 6.01	35.87	42.59		1.10	1.06	1.07	1.04
Martes pennanti	9	4000	30.35 ± 3.69	31.82	30.25		0.99	0.99	0.96	0.97
Gulo gulo	12	14,525	76.22 ± 9.07	78.26	78.00		1.01	1.00	0.97	0.98
Martes flavigula	4	2500	34.90 ± 2.08	34.12	26.52		1.14	1.09	1.11	1.07
Martes foina	4	1541	18.76 ± 2.52	20.91	24.66		1.04	1.03	1.02	1.02
Martes melampus	1	1000	18.41 ± n.a.				1.16	1.11	1.14	1.09
Martes americana	14	1250	16.40 ± 1.74	15.80	17.98		1.05	1.03	1.03	1.02
Martes martes	1	1300	17.74 ± n.a.	20.09			1.07	1.05	1.05	1.03
Martes zibellina	4	1130	19.91 ± 2.17	18.54	19.59		1.15	1.10	1.13	1.08
Melogale moschata	3	939	15.18 ± 1.21	15.07			1.10	1.07	1.08	1.05
Lyncodon patagonica	1	225	6.61 ± n.a.				1.23	1.16	1.25	1.17
Galictis cuja	2	1000	16.85 ± 3.77	15.03	16.29		1.12	1.08	1.10	1.07
Galictis vittata	4	3200	25.43 ± 3.29	24.29	24.22		0.98	0.99	0.96	0.97
Vormela peregusna	3	450	8.13 ± 2.02	4.76	7.07		1.04	1.04	1.04	1.03
Ictonyx libyca	4	625	6.41 ± 0.33	4.48[g]	6.27		0.84	0.89	0.83	0.89

Ictonyx striatus	14	1300	12.16 ± 1.30	9.78	10.16	10.0	0.93	0.95	0.91	0.94
Poecilogale albinucha	6	340	5.33 ± 0.81	4.76	6.08	5.0	0.92	0.95	0.92	0.96
Mustela erminea	11	120	3.09 ± 1.15	4.01	3.58		0.94	0.98	0.99	1.02
Mustela nivalis	11	100	2.29 ± 0.65	1.99	2.32		0.79	0.87	0.84	0.92
Mustela lutreola	2	440	8.81 ± 1.16	8.50			0.95	0.97	0.94	0.96
Mustela lutreolina	1	707	7.44 ± n.a.				0.83	0.88	0.82	0.88
Mustela nigripes	4	850	9.97 ± 0.86	8.50	9.32		0.95	0.97	0.94	0.96
Mustela frenata	12	147	4.16 ± 0.70	4.01	3.62		1.12	1.10	1.17	1.13
Mustela vison	14	945	9.36 ± 1.04	8.50	8.59		0.90	0.93	0.88	0.92
Pteronura brasiliensis	7	24,000	114.50 ± 5.88	85.63	99.71		1.03	1.01	0.99	0.99
Lontra canadensis	14	8087	53.68 ± 5.86	52.98	46.56		1.02	1.01	0.99	0.99
Lontra felina	6	4500[d]	38.46 ± 1.87	38.86	35.84		1.04	1.03	1.01	1.00
Lontra longicaudis	12	6555	51.63 ± 6.53				1.05	1.03	1.01	1.01
Lontra provocax	2	7500	54.60 ± 4.72				1.04	1.02	1.01	1.00
Enhydra lutris	41	23,500	136.30 ± 9.03	125.21	113.43		1.07	1.04	1.03	1.01
Hydrictis maculicollis	18	4000	48.60 ± 5.56	40.04	39.54	49.0	1.13	1.08	1.10	1.06
Lutra lutra	7	11,000	45.76 ± 7.21	42.10	35.36		0.93	0.95	0.90	0.93
Lutra sumatrana	3	5500	43.78 ± 8.42				1.03	1.02	1.00	1.00
Amblonyx cinereus	18	3990	35.18 ± 2.65	38.09	31.71		1.04	1.02	1.01	1.00
Lutrogale perspicillata	4	9967	67.50 ± 2.24	64.72	50.50		1.04	1.03	1.01	1.00
Aonyx capensis	50	15,000[e]	101.12 ± 10.79	94.63	76.63	108.0	1.07	1.04	1.03	1.02
Aonyx congicus	34	16,000[e]	103.54 ± 10.29				1.07	1.04	1.03	1.01

Species means (± SD) are compared with several earlier publications. All body mass values were taken from Smith et al. (2003, updated MOM v3.6.1), except as indicated by footnote. All values for Gittleman were based on Gittleman (1986a), except as indicated in the footnotes. Encephalization indexes (EI) were calculated based on ordinary least squares (OLS) and reduced major axis (RMA) regressions for the Carnivora as a whole, as well as in relation to local subclades (Herpestidae for the herpestid species, and Musteloidea for the procyonids and mustelids).

Mass references:

[a] Meiri et al. (2005);

[b] Nellis (1989);

[c] Vanderhaar and Hwang (2003);

[d] Larivière (1998);

[e] this study.

Gittleman references:

[f] Gittleman (1986b);

[g] Gittleman (1991).

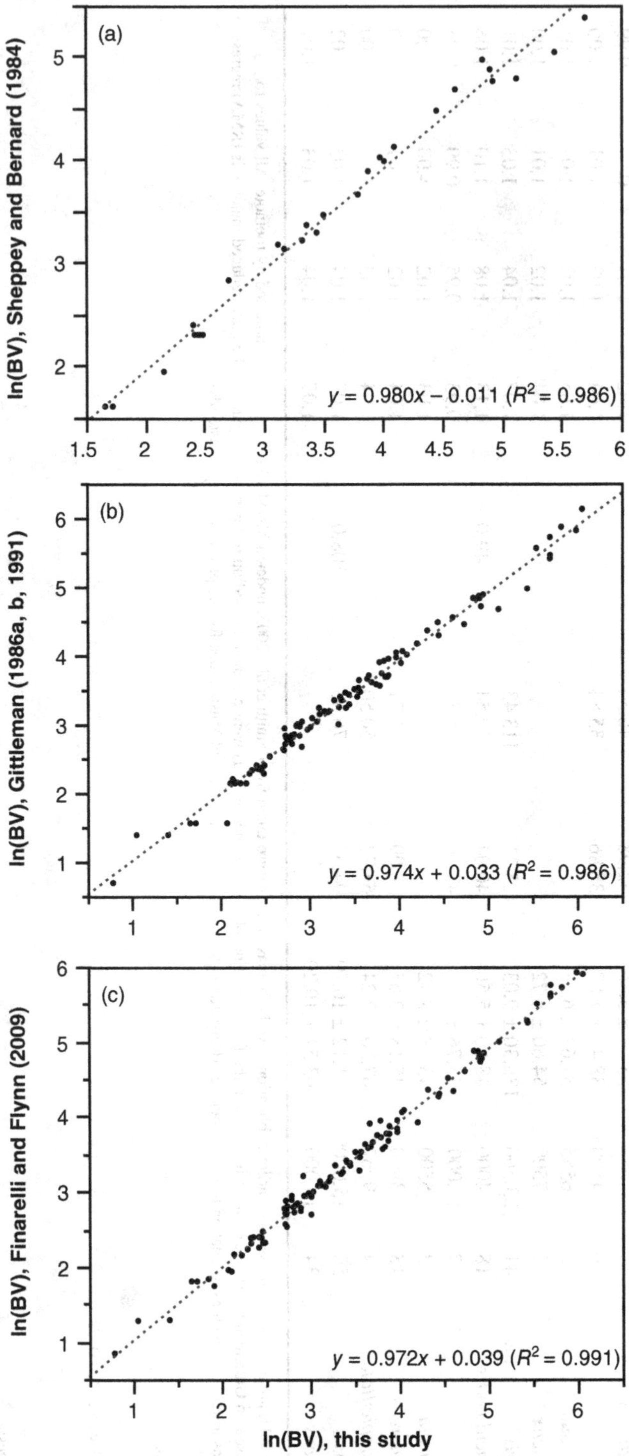

Figure 9.5 Comparison of the estimates of carnivoran brain volume (BV) obtained in this study with the estimates made by previous investigators. Values are species means. Specimens in (a) and (b) were measured by the decanting technique, and specimens in (c) were measured by ectocranial proxy (the Finarelli method that was followed here).

DISCUSSION

At present, we have a good working database on brain size at the species level for extant Carnivora, and this chapter includes BV estimates for >50% of all extant species, including 88% of the herpestids, 69% of the mustelids, and 50% of the procyonids. At this stage of research, we arguably have a better understanding of carnivoran BV than we do of carnivoran body mass. Indexes of encephalization based on the regression of brain size on body mass incorporate the error associated with the latter term, and for this reason, encephalization indexes must be interpreted with caution.

My estimates of brain size based on ectocranial proxy correlated strongly ($R^2 > 0.98$) with estimates from studies based on the decanting technique, reinforcing the earlier finding of Finarelli (2006) that the two methods produce consistent results. The strong correlation of my results with those of Finarelli and Flynn (2009) demonstrates the repeatability of the ectocranial proxy technique. The latter technique is considerably faster, cheaper, and less destructive than alternative methods, and it is suitable for estimating brain size in a large number of individuals.

Atilax and the Durophagous Mongooses

Atilax paludinosus, the marsh mongoose, has the absolutely largest brain of any extant mongoose species, and the encephalization index of this species relative to other herpestids is also high (1.05–1.07), in spite of its overall large body size (3.3 kg). It is likely that the absolutely and relatively large brain of *Atilax* is related, in some way, to its foraging mode (manual dexterity), its diet (high quality, easy to digest food), and the behavior of its prey (relatively slow moving).

Although none of the other mongooses matches *Atilax* in absolute brain size, some consistent patterns emerge from the examination of species with high encephalization indexes. The three *Herpestes* species with encephalization indexes >1.05 are all wetland durophages from southeast Asia (*Herpestes urva*, *Herpestes brachyurus*, and *Herpestes viticollis*), and they are all craniodentally robust. In this way, these species generally resemble the condition seen in *Atilax*.

Two of the other three species with encephalization indexes >1.05 are dwarfed forms (*Helogale hirtula* and *Dologale dybowskii*), and the extremely low body mass of these species seems to have a disproportionate influence on their index of encephalization. The third species, *Suricata suricatta*, the meerkat, is also a small-bodied species, but this alone cannot account for its relatively high encephalization, and brain structure in meerkats deserves additional study.

Aonyx and Other Large-Brained Otters

The clawless otters of Africa (*A. capensis* and *A. congicus*) exhibit larger absolute brain sizes than any other African mustelid species. The mustelids with the largest brains in the world are also otters, viz, the giant otter (*Pteronura*) and sea otter (*Enhydra*). However, all of these otters are also heavy bodied overall, and for this reason, their encephalization indexes are not exceptionally high.

Anomalously high encephalization indexes in the Mustelidae are exhibited by two of the dwarfed forms (*Mustela frenata* and *Lyncodon patagonica*), as was the case in the Herpestidae. The omnivorous African honey badger (*Meles meles*) exhibits a high encephalization index in my analysis, as do several of the martens (*Martes* spp.), and each of these cases deserves additional study.

The African spotted-necked otter (*Hydrictis maculicollis*) exhibits a high encephalization index in relation to other musteloids (1.06–1.10) and in relation to the Carnivora as a whole (1.08–1.13). This species is smaller and more aquatic than *Aonyx*, with which it shares its range in the Afrotropics. *Hydrictis* is an open-water piscivore that relies on vision, foreclaws, and teeth in the capture of prey (the digits are fully webbed for swimming).

The *Procyon* Raccoons

The crab-eating raccoons (*P. cancrivorus*) have the largest absolute brain size of any living procyonid by a wide margin. However, because of the large estimated body size of this species, the encephalization index is not exceptionally high (1.02–1.03 in relation to other musteloids, 1.04–1.07 in relation to the Carnivora as a whole). *P. cancrivorus* does appear to be slightly more encephalized than the common raccoon, *Procyon lotor*, and this deserves closer study. The anomalously high encephalization index of *Bassariscus sumichrasti* is probably related to dwarfing.

Comparison of Herpestidae, Mustelidae, and Procyonidae

The clearest result from this analysis of brain size data in the herpestids, mustelids, and procyonids is that the extant species that forage for wetland and aquatic prey have the largest absolute brain sizes of their respective families. The significance of this trend is not entirely clear, however, because the wetland durophages also tend to be heavy bodied overall. For this reason, although the encephalization indexes of the wetland durophages are consistently high (>1.0), both with respect to the Carnivora as a whole as well as in relation to local subclade, these values do not tend to be exceptionally high.

The finding of large absolute brain size in the wetland durophages is more interesting in light of the published neuroanatomical studies on these taxa. Radinsky (1968) found that the brains of *Aonyx* and *Enhydra* were complex in the regions associated with the projection area of the forelimb, and he argued that the two lineages had independently derived this condition. Radinsky (1975) found that *Atilax* had the most complex sulcate pattern of all mongoose species that he examined, and he linked this feature to increased tactile sensitivity, finer motor control, or both.

The connection between neuroanatomy and manual dexterity has been most closely studied in the common raccoon, *P. lotor* (Welker and Seidenstein, 1959; Welker et al., 1964). Many authors have commented on the exceptional manual dexterity of this species (e.g., Lyall-Watson, 1963). Mivart (1885: 10) observed that the brain convolutions of the crab-eating raccoon, *P. cancrivorus*, were more complex than those of *P. lotor*, but this observation does not appear to have been investigated further.

Gittleman (1986a) tentatively suggested that specializations related to "forepaw manipulation" could be linked to increased brain size in carnivorans. Iwaniuk et al. (1999) investigated this suggestion but did not find a correlation between brain size and their index of forelimb dexterity. However, Iwaniuk et al. (1999) cautioned that the mustelids and herpestids were underrepresented in their analysis, and they made no mention of *Atilax*, *Aonyx*, and *P. cancrivorus*. Iwaniuk et al. (2000: 1121) did observe "a tendency for riparian species (those species inhabiting watercourses) to have relatively high degrees of forelimb dexterity," but that observation was also not pursued further.

Taken together, the neuranatomical analyses of Herpestidae, Mustelidae, and Procyonidae suggest that the members of these families with the absolutely largest brains

(*Atilax*, *Aonyx*, and *P. cancrivorus*, respectively) also tend to have relatively complex brains, particularly in the neural regions associated with the sensorimotor dexterity of the forepaws. All three of these lineages inhabit the ecotone between terrestrial and aquatic systems, where they forage for prey primarily by hand. These same species show clear adaptations to the mastication of hard-shelled food objects in the robusticity of their jaws and teeth (Shabel, 2009). Given our knowledge of carnivoran phylogeny, it is clear that the wetland durophage syndrome evolved independently in these three groups. The evolution of wetland durophagy in Carnivora reveals a deep connection between neuroanatomy, craniodental morphology, manus structure, diet, and habitat preference in this clade.

Implications for Robust Australopithecine Natural History

Fossils of both *Atilax* and *Aonyx* have been found at the classic robust australopithecine sites of Olduvai and Swartkrans (Petter, 1973; Turner, 1993; Watson, 1993), and it is clear that these wetland durophages overlapped in space and time with *Paranthropus*. The durophage–ecotone model reconstructs *Atilax*, *Aonyx*, and *Paranthropus* as comembers of a wetland durophage foraging guild (*sensu* Root, 1967). As is the case in many vertebrate guilds, body size differentiation among the guild members can help explain their ecological coexistence. *Atilax*, *Aonyx*, and *Paranthropus* differ in average body mass by more than threefold (3.3, 12–18, and 32–49 kg, respectively), and their BVs differ by a similar factor (29, 100, and 450 cm^3, respectively). For a review of australopithecine brain size, see Falk et al. (2000); for a review of australopithecine body mass, see McHenry (1992). However, it must be made clear that the carnivorans *Atilax* and *Aonyx* are only considered rough analogs for the primate *Paranthropus*. *Atilax* and *Aonyx* consume little or no plant resources, whereas *Paranthropus* is likely to have exploited fruits, nuts, and belowground plant storage organs when available.

The durophage–ecotone model of the robust australopithecines provides a clear mechanism, based on habitat and trophic preference, to explain the long-term coexistence of *Paranthropus* and early *Homo*: the robust australopithecines gathered hard-shelled foods at the land–water ecotone when faced with competition from *Homo*. According to this view, the hands of *Paranthropus* were exapted to perform the dexterous manipulation behaviors necessary to collect and extract hard-shelled food objects, and the massive skull was well suited for oral processing of hard and brittle shells (it is assumed that the heaviest shells would have been cracked extraorally). The nutritiousness of ecotonal hard-shelled food items, and possibly their relative ease of capture, can help explain the mechanism by which the large brain of *Paranthropus* was evolved and sustained. Biogeochemical analyses (trace elements and stable light isotopes) of a wide range of African vertebrates and invertebrates did not falsify the durophage–ecotone model, and the new model fits the morphological evidence better than any of the alternatives (Shabel, 2009).

The durophage–ecotone model is focused on the robust australopithecines and the mechanism for their coexistence with other mammals, including early *Homo*. The new model does not, however, specify the niche of early *Homo* (see Chapter 8), and the model is consistent with a full range of ecological hypotheses related to the evolution of *Homo sapiens* (*Paranthropus* went extinct more than a million years before the emergence of *H. sapiens*). It bears repeating that all paleoanthropological hypotheses should be tested as working alternatives using the most rigorous available methods. Questions pertaining to diet and habitat preference remain at the heart of paleoanthropological inquiry, and we are wise to approach these problems empirically and objectively. Although the vast majority of early hominins were fossilized at the land–water interface, the extent to which they actually foraged in such systems remains poorly understood.

ACKNOWLEDGMENTS

The specimens relied on for this study are housed in museums in North America, Europe, and Africa, and I gratefully acknowledge the assistance that I received from the curators and staff at all of these institutions.

REFERENCES

BAKER, C.M. 1989. Feeding habits of the water mongoose (*Atilax paludinosus*). *Zeitschrift für Säugetierkunde* **54**:31–39.

BAKER, C.M. 1992. Atilax paludinosus. *Mammalian Species* **408**:1–6.

BININDA-EMONDS, O.R.P. AND GITTLEMAN, J.L. 2000. Are pinnipeds functionally different from fissiped carnivores? The importance of phylogenetic comparative analyses. *Evolution* **54** (3):1011–1023.

CHRISTIANSEN, P. AND WROE, S. 2007. Bite forces and evolutionary adaptations to feeding ecology in carnivores. *Ecology* **88** (2):347–358.

ELTON, S., BISHOP, L.C., AND WOOD, B.A. 2001. Comparative context of Plio-Pleistocene hominin brain evolution. *Journal of Human Evolution* **41**:1–27.

ESTES, R.D. 1991. *The Behavior Guide to African Mammals: Including Hoofed Mammals, Carnivores, Primates*. Berkeley: The University of California Press.

EWER, R.F. 1973. *The Carnivores*. Ithaca: Cornell University Press.

FALK, D., REDMOND, J.C., GUYER, J. ET AL. 2000. Early hominid brain evolution: A new look at old endocasts. *Journal of Human Evolution* **38** (5):695–717.

FINARELLI, J.A. 2006. Estimation of endocranial volume through the use of external skull measures in the Carnivora (Mammalia). *Journal of Mammalogy* **87** (5): 1027–1036.

FINARELLI, J.A. AND FLYNN, J.J. 2009. Brain-size evolution and sociality in Carnivora. *Proceedings of the National Academy of Sciences of the United States of America* **106** (23):9345–9349.

FLYNN, J.J., FINARELLI, J.A., ZEHR, S. ET AL. 2005. Molecular phylogeny of the Carnivora (Mammalia): Assessing the impact of increased sampling on resolving enigmatic relationships. *Systematic Biology* **54** (2): 317–337.

GITTLEMAN, J.L. 1986a. Carnivore brain size, behavioral ecology, and phylogeny. *Journal of Mammalogy* **67** (1): 23–36.

GITTLEMAN, J.L. 1986b. Carnivore life history patterns: Allometric, phylogenetic, and ecological associations. *The American Naturalist* **127** (6):744–771.

GITTLEMAN, J.L. 1991. Carnivore olfactory bulb size: Allometry, phylogeny and ecology. *Journal of Zoology* **225**:253–272.

GITTLEMAN, J.L. 1993. Carnivore life histories: A reanalysis in the light of new models. In *Mammals as Predators*. Symposia of the Zoological Society of London, ed. N. Dunstone and M.L. Gorman, pp. 65–86. Oxford: Clarendon Press.

HEMMER, H. 2007. Estimation of basic life history data of fossil hominoids. In *Handbook of Paleoanthropology*, ed. W. Henke and I. Tattersall, pp. 587–619. Berlin: Springer-Verlag.

IWANIUK, A.N., PELLIS, S.M., AND WHISHAW, I.Q. 1999. Brain size is not correlated with forelimb dexterity in fissiped carnivores (Carnivora): A comparative test of the principle of proper mass. *Brain, Behavior and Evolution* **54** (3):167–180.

IWANIUK, A.N., PELLIS, S.M., AND WHISHAW, I.Q. 2000. The relative importance of body size, phylogeny, locomotion, and diet in the evolution of forelimb dexterity in fissiped carnivores (Carnivora). *Canadian Journal of Zoology* **78** (7):1110–1125.

JERISON, H.J. 1973. *Evolution of the Human Brain and Intelligence*. London: Academic Press.

KINGDON, J. 1977. *East African Mammals: An Atlas of Evolution in Africa. Volume III, Part A (Carnivores)*. Chicago: University of Chicago Press.

KINGDON, J. 1997. *The Kingdon Field Guide to African Mammals*. San Diego: Academic Press.

KRUUK, H. 2006. *Otters: Ecology, Behaviour, and Conservation*. New York: Oxford University Press.

LARIVIÈRE, S. 1998. Lontra felina. *Mammalian Species* **575**:1–5.

LARIVIÈRE, S. 2001a. Aonyx congicus. *Mammalian Species* **650**:1–3.

LARIVIÈRE, S. 2001b. Aonyx capensis. *Mammalian Species* **671**:1–6.

LYALL-WATSON, M. 1963. A critical re-examination of food "washing" behaviour in the raccoon (*Procyon lotor* Linn.). *Proceedings of the Zoological Society of London* **141**:371–393.

MARTIN, R. 1990. *Primate Origins and Evolution: A Phylogenetic Reconstruction*. Princeton: Princeton University Press.

MCHENRY, H.M. 1992. How big were early hominids. *Evolutionary Anthropology* **1** (1):15–20.

MEIRI, S., SIMBERLOFF, D.S., AND DAYAN, T. 2005. Insular carnivore biogeography: Island area and mammalian optimal body size. *The American Naturalist* **165** (4): 505–514.

MIVART, S.G.J. 1885. Notes on the cerebral convolutions of the Carnivora. *Journal of the Linnaean Society of London* **19**:1–25.

NELLIS, D.W. 1989. Herpestes auropunctatus. *Mammalian Species* **342**:1–6.

PETTER, G. 1973. Carnivores Pleistocènes du ravin d'Olduvai (Tanzanie). In *Fossil Vertebrates of Africa. Volume* **3**, ed. L.S.B. Leakey, R.J.G. Savage, and S.C. Coryndon, pp. 43–104 + Plates 101–104. London: Academic Press.

POCOCK, R.I. 1916. On the external characters of the mongooses (Mungotidae). *Proceedings of the Zoological Society of London*, pp. 349–374.

POCOCK, R.I. 1921a. The external characters and classification of the Procyonidae. *Proceedings of the Zoological Society of London*, pp. 389–422.

POCOCK, R.I. 1921b. On the external characters of some species of Lutrinae (otters). *Proceedings of the Zoological Society of London*, pp. 535–546.

RADINSKY, L.B. 1968. Evolution of somatic sensory specialization in otter brains. *Journal of Comparative Neurology* **134**:495–506.

RADINSKY, L.B. 1975. Viverrid neuroanatomy: Phylogenetic and behavioral implications. *Journal of Mammalogy* **56** (1):130–150.

RADINSKY, L.B. 1981. Evolution of skull shape in carnivores. 1. Representative modern carnivores. *Biological Journal of the Linnean Society* **15**:369–388.

ROOT, R.B. 1967. The niche exploitation pattern of the blue-gray gnatcatcher. *Ecological Monographs* **37** (4):317–350.

ROWE-ROWE, D.T. 1978. The small carnivores of Natal. *The Lammergeyer* **25**:1–48.

SHABEL, A.B. 2009. Craniodental morphology and biogeochemistry of African carnivorans: Toward a new model of Plio-Pleistocene hominin evolution. PhD thesis, University of California.

SHEPPEY, K. AND BERNARD, R.T.F. 1984. Relative brain size in the mammalian carnivores of the Cape Province of South Africa. *South African Journal of Zoology* **19** (1): 305–308.

SMITH, F.A., LYONS, S.K., ERNEST, S.K.M. ET AL. 2003. Body mass of late quaternary mammals. *Ecology* **84** (12): 3403 [also *Ecological Archives* E3084–E3094].

TURNER, A. 1993. New fossil carnivore remains from Swartkrans. In *Swartkrans: A Cave's Chronicle of Early Man*, Transvaal Museum Monograph No. 8, ed. C.K. Brain, pp. 151–165. Pretoria: Transvaal Museum.

VANDERHAAR, J.M. AND HWANG, Y.T. 2003. *Mellivora capensis. Mammalian Species* **721**:1–8.

WAINWRIGHT, M. 2007. *The Mammals of Costa Rica: A Natural History and Field Guide*. Ithaca: Cornell University Press.

WATSON, V. 1993. Composition of the Swartkrans bone accumulations, in terms of skeletal parts and animals represented. In *Swartkrans: A Cave's Chronicle of Early Man*, Transvaal Museum Monograph No. 8, ed. C.K. Brain, pp. 35–73. Pretoria: Transvaal Museum.

WELKER, W.I., JOHNSON, J.I. JR., AND PUBOLS, B.H. JR. 1964. Some morphological and physiological characteristics of the somatic sensory system in raccoons. *American Zoologist* **4** (1):75–94.

WELKER, W.I. AND SEIDENSTEIN, S. 1959. Somatic sensory representation in the cerebral cortex of the raccoon (*Procyon lotor*). *Journal of Comparative Neurology* **111** (3):469–501.

WILLEMSEN, G.F. 1992. A revision of the Pliocene and Quaternary Lutrinae from Europe. *Scripta Geologica* **101**:1–115.

YOUNG, R.W. 1959. The influence of cranial contents on postnatal growth of the skull in the rat. *American Journal of Anatomy* **105** (3):383–415.

CHAPTER 10

COASTAL DIET, ENCEPHALIZATION, AND INNOVATIVE BEHAVIORS IN THE LATE MIDDLE STONE AGE OF SOUTHERN AFRICA

John Parkington

INTRODUCTION

Some 35 years ago, Desmond Clark (1975) predicted that Africa was in the process of moving from peripheral to paramount in the narrative of human evolution. Nowhere has this been more dramatically vindicated than in the arena of modern human origins and in the region of southern Africa. It is now quite clear that in terms of the appearance of our species, Africa is no laggard but inarguably leader. How has this rethinking come about and what is the current understanding of the role of Africa, specifically southern Africa, in the topical debates around the emergence of anatomically and behaviorally modern people? What are the emergent patterns in the late mid-Pleistocene and early Late Pleistocene archaeology of this area? More specifically, how are dietary choices, technological innovations and arguably symbolic manifestations seen as related in the appearance of people modern behaviorally as well as anatomically?

As recently as the 1970s, the hominin cranium and postcranial fragments from Kabwe (formerly Broken Hill) in Zambia were considered Neanderthal-like (see Klein's 1973 account). They were unprofessionally excavated, had little context, and may have been associated with Middle Stone Age (MSA) stone tools. At that time, the transition from MSA to Later Stone Age (LSA) artifacts in southern Africa was regarded as late, perhaps as late as 15,000 years ago, far more recent than the supposedly similar transition from Middle to Upper Paleolithic in Europe. These now discredited views allowed non-African archaeologists to present southern Africa as a backwater housing late-surviving archaic hominins and their stone tools long after the creative explosion of *Homo sapiens*, paleolithic art, and advanced blade and burin assemblages in Europe. Few expressed alternative views, though Peter Beaumont and John Vogel pioneered the position that *H. sapiens* had, in fact, first appeared on the plains of southern Africa, another prophetic if slightly unfocused view (Beaumont and Vogel, 1972a,b).

It is perhaps not surprising that so soon after the first shift toward political independence in sub-Saharan Africa, European achievements were expected to precede similar developments elsewhere. The technological and artistic flowering of Upper Paleolithic hunter-gatherer innovations was taken to reflect the sudden emergence of "people like us."

Human Brain Evolution, Edited by Stephen C. Cunnane and Kathlyn M. Stewart

Often presented as a package of biologic and cultural features, this manifestation included the earliest *H. sapiens* in the form of Cro-Magnon skeletons; sophisticated blade and burin stone tool assemblages made with innovative punch struck techniques, later known as Mode 4 in the Graham Clark scheme (Clark, 1989); an array of decorative and decorated objects in stone or, more often, organic raw materials, including the earliest unquestioned portable and parietal art; and an abundance of shaped and perforated objects clearly intended as ornaments. Despite the recognition among some authorities that this apparent package appeared too suddenly to have arisen in the region, the external origin of these innovative behaviors was not emphasized. Nor was the narrative of the accretion of individual traits within the package systematically attempted, so poor was the evidence in potential source areas.

With the advantage of a much better archaeological record, I develop here a synthesis that identifies a range of innovative behaviors from southern Africa that far predate the European Upper Paleolithic, link them to the early appearance of fully modern people and associate both with a shift toward more regular and systematic exploitation of coastal resources. Our chronology for these phenomena is improving rapidly and will increasingly allow us to test the viability of the connections I propose.

CHANGES

Things began to look somewhat different soon after this, prompted in large part by the excavations at the sites near Klasies River (for sites mentioned in the text, see Fig. 10.1) in the late 1960s and subsequent publication of the analyses in 1982 (Singer and Wymer, 1982). This fieldwork showed a deep sequence of about 20 m of MSA assemblages housed in a shellfish-rich depositional matrix with interesting technological patterns and extremely important human skeletal associations. With a substantial number of radiocarbon dates near the top of this stack of shell middens, many of them infinite, it became clear that the whole of the MSA, at least here, was beyond the range of this technique. The apparent temporal dislocations of the Early Stone Age (ESA) to MSA and MSA to LSA transitions from the European equivalents of Lower to Middle and Middle to Upper Paleolithic were revealed as the result of reliance on early and spuriously finite radiocarbon dates. In particular, it was recognized that the MSA was contemporary with the Middle Paleolithic, which it closely resembles, and that Africa was at least as developed as Europe throughout the first half of the Late Pleistocene. The termination of MSA and Middle Paleolithic was more or less contemporary at approximately 35,000 years ago.

A new technique would be needed to date both Middle Paleolithic and MSA sites. John Vogel had the vision to invite to southern Africa in the mid-1980s a group of luminescence dating experts, an initiative that has resulted in a much improved chronology. After some technical innovations, the method of single-grain optically stimulated luminescence (OSL) has become one of the industry standards for dating MSA assemblages (Jacobs et al., 2008). By this procedure, anomalous grains that do not reflect the depositional events contemporary with MSA toolmaking can be rejected. Another method, not yet showing the tight distribution of age estimates claimed for OSL (see Tribolo et al., 2008), is the dating of burnt siliceous stone artifacts by thermoluminescence (TL). The transitions of ESA to MSA and Lower to Middle Paleolithic are still somewhat loosely dated and even more poorly understood, but the MSA now stands contemporary with the Middle Paleolithic, lasting from about 250,000 years ago to about 35,000 years ago, and southern Africa no longer appears technologically behind. Quite the contrary, as we will see.

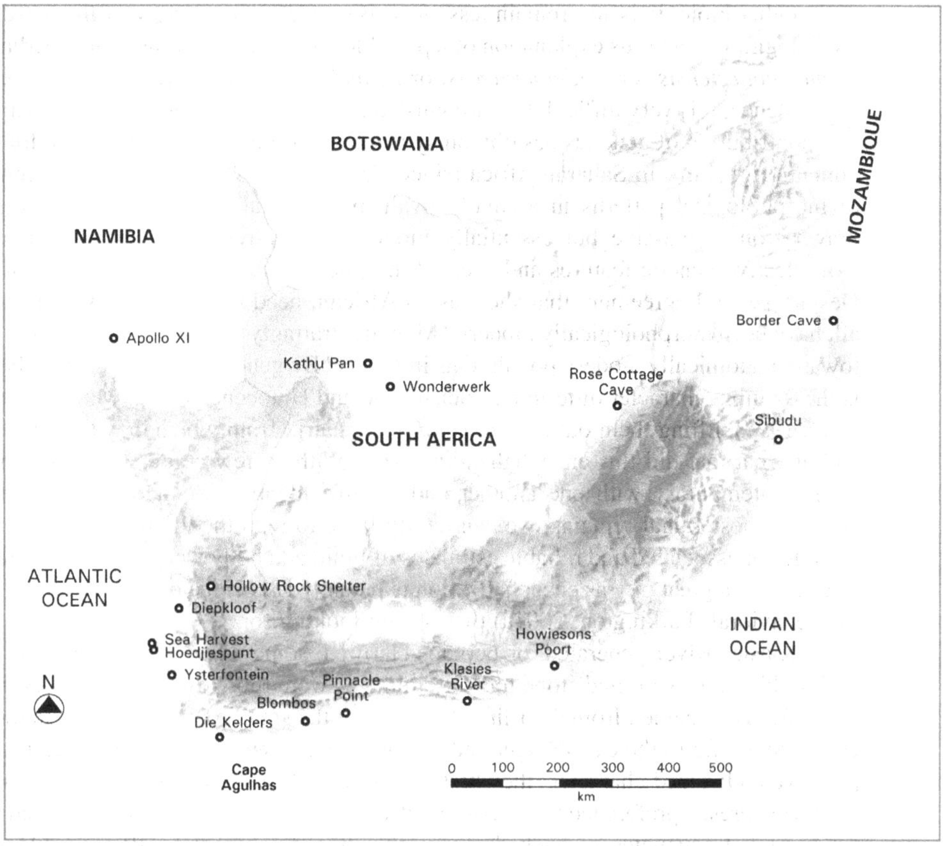

Figure 10.1 The distribution of key Middle Stone Age sites mentioned in the text. Illustration is created by and a copyright of Neil Rusch.

Alongside this has been the reevaluation of the sparse but valuable human skeletal record for the Middle and Late Pleistocene of southern Africa. All reference to "Neanderthal-like" characteristics have been expunged, and the record from the early remains at Kabwe, Bodo, and Elandsfontein through Florisbad and Hoedjiespunt and on to the fragments from Klasies River, Die Kelders, Border Cave, Sea Harvest, and Blombos, all of these latter associated with MSA stone tools, is now described as a clear trend toward anatomically modern *H. sapiens* (Klein, 1999). Although the paucity of specimens leaves much to be desired and the contemporary morphological variability remains to be explained, the transition through the late Middle and early Late Pleistocene toward "people like us" stands in stark contrast to the contemporary shift toward Neanderthal morphology in Europe. What also stands out is the fact that anatomical modernity, however poorly defined, emerged between the early MSA at Florisbad, described as knocking on the door of modernity (Grun et al., 1996), and the late MSA at Klasies River, almost universally termed modern (Rightmire and Deacon, 1991, 2001). The accepted date for the Florisbad skeletal material is about 260,000 years (Grun et al., 1996) and for the bulk of the Klasies River skeletal remains about 110,000 years (Deacon, 2008). We apparently became us during the MSA, but not during the Middle Paleolithic, in Africa and not in Europe, almost certainly between these dates.

Other biologic issues remain less clear. Because of the poverty of the fossil record, the recognition and thus explanation of a possible speciation event in Africa, perhaps from *Homo rhodesiensis* (or *heidelbergensis*, or *helmei*) to *H. sapiens* remains elusive. Given the evidence, it is very difficult to distinguish a long slow transition from any abrupt event. The contribution of genetics has obviously had a great influence on the search for modern human origins in sub-Saharan Africa (since Cann et al., 1987). However, the integration of morphological patterns in a small, invaluable skeletal sample with genetic patterns drawn from a massive but essentially modern survey remains somewhat unresolved. Nonselective genetic features and skeletal morphologies are telling two different stories. Despite general agreement that she was an African, need the putative Eve, mother of us all, have been morphologically modern? Most frustratingly, what was driving the transition toward anatomically modern populations in Africa? By general consensus, encephalization is the defining character of terminal Pleistocene and Holocene people (Martin, 1983; Ruff et al., 1997), lifting them out of the mammalian matrix from which they derive; but when and at what rate did this encephalization arise? With so few complete crania, so few of them contemporary with one another and so infrequently associated with substantially complete postcranial remains, we can hardly hope to map the march to encephalization. Samples of associated and reliable estimates of brain and body weights are embarrassingly rare. This is a great pity as encephalization is patently the bridge between the biologic and the behavioral (Parkington, 2001a) that we are looking for.

Klasies River generated, or perhaps clarified, some other key realizations. Housed within this set of stratified stone tool assemblages were those known as Howiesons Poort (hereafter HP), named from their first discovery at the small rock shelter also in the southern Cape not far to the east (Singer and Wymer, 1982; Deacon, J., 1995). Because of their perceived advanced character (they had ribbon-like blades struck from single or opposed platform cores, a preference for fine-grained stone raw materials and were marked by relatively high frequencies of backed, geometric-shaped segments), HP assemblages were expected to form a transition to the LSA assemblages of the Holocene. Some late LSA assemblages also have backed segments, though they are much smaller and struck from different kinds of cores. Interestingly, burins were noted at the type site, at Klasies River (Singer and Wymer, 1982) and in subsequent descriptions of HP assemblages (e.g., Vogelsang, 1998). The Klasies River stratigraphy showed, however, that HP assemblages were followed as well as preceded by "normal" MSA assemblages with a preference for triangular flake production based on radial or, occasionally, Levallois cores, and a lesser interest in the finer-grained raw materials (Wurz, 2005, 2008). At about the same time, Peter Beaumont's excavations at Border Cave in the northeastern part of southern Africa confirmed the position of the HP within and not atop the MSA sequence (Beaumont et al., 1978). The HP, in other words, appeared to be an innovative blip in the MSA. The nature of this stone tool innovativeness is interesting as it highlights the production of blades and, albeit infrequently, burins.

Another innovative manifestation in the MSA, surprisingly not reflected in the Klasies River sites, is the Still Bay assemblage type (hereafter SB). Like the HP, it seemed often to occur as the only assemblage type at a site, making it difficult to place within the range of MSA variability. The innovation of the SB, at least in the domain of stone tool technology, lies in the production of bifacially flaked leaf points. With an increase in excavated sites, it has emerged that the SB precedes the HP but not by very long (Rigaud et al., 2006; Wadley, 2007). A recent subcontinent wide survey using single-grain OSL (Jacobs et al., 2008) suggested that HP assemblages date between 60,000 and 65,000 years and that SB assemblages date between 70,000 and 75,000 years, though the claim is that the durations may in fact be less than 5000 years each. If, as many believe, the MSA began at least 250,000 years ago (Klein, 1999), then these innovative assemblage types mark the late but not, intriguingly, the latest phase of a long period of stone tool production based on flake, not blade manufacture. What attracted many European archaeologists to the HP in particular was the recognition of their Mode 4-like character at a time well before the earliest date of the Upper Paleolithic (see, e.g., Mellars, 2006).

It might be helpful to distinguish here the terms innovative and precocious. By precocious we would imply that artifact-making habits appear early and preempt or prefigure later developments. The fact is that neither the HP nor the SB is precocious in the sense of anticipating local early LSA stone toolmaking patterns. The earliest assemblages generally ascribed to the LSA in southern Africa are very informal collections dominated by the use of bipolar cores with extremely low frequencies of formally retouched artifacts, most of them made on tiny bladelets (Mitchell, 2002). In the Cape, at least, they are heavily dominated by quartz and seem difficult to relate to any MSA assemblage types. Ironically, if the HP anticipates anything, it is the Mode 4 assemblages of the European Upper Paleolithic. Innovative, in contrast, implies something new, something rarely if ever seen before. This would not be an unreasonable description of the segments and other backed tools of the HP or the bifacial points of the SB. It might also be a term realistically applied to the soft hammer produced blades of the HP, which differ in dimensions and production character from earlier blades and flake blades of the MSA and even ESA (Rigaud et al., 2006).

Even more innovative than the decisions about stone tool production are the non-stone associations of HP and SB assemblages. Although not so clearly expressed at Klasies River, these innovations have rightly made headlines at Blombos (Henshilwood et al., 2002, 2004), Diepkloof (Parkington et al., 2005; Rigaud et al., 2006; Porraz et al., 2008; Tribolo et al., 2008), and Sibudu (Wadley, 2007, 2008) where they have further distinguished the HP and SB from underlying and overlying MSA assemblages. Blombos has no HP assemblages but does have a series of SB assemblages with no MSA above them but plenty below (Jacobs et al., 2006). All are older than 70,000 years. Despite initial reservations by some, the bifacial points here are clearly associated with large numbers of ochre fragments – several of them marked, substantial numbers of perforated tick shell beads and a few, quite persuasively shaped, bone tools. All of these are innovative in the sense defined above and arguably precocious in a global context. Diepkloof has produced, along with a good deal of beveled and striated ochre, a large number of marked fragments of ostrich eggshell, some of them perforated as water-containing and storage flasks, definitively associated with HP artifacts. All are older than 63,000 years. Diepkloof and Sibudu are significant in demonstrating unequivocally the stratification of HP on top of SB assemblages, a sequence previously suspected at other sites in the subregion. Sibudu has produced worked pieces of ochre and perforated tick shell beads from SB assemblages. These are older than 60,000 years. Most of these finds are the earliest widely accepted occurrences of such things in archaeological assemblages anywhere.

Apart from a possible burial of an infant at Border Cave (Beaumont et al., 1978), possible because of some doubts as to its age (Sillen and Morris, 1996), there are no claims for regular intentional burials among later MSA people. Quite the contrary, but no less interesting, the scattered fragments from Klasies River have been interpreted as reflecting cannibalism, arguably a ritual rather than dietary practice as so often among much later societies (White, 1987; Deacon and Deacon, 1999). Symbolism is often cited as the underlying characteristic of these innovations, and most authorities are tempted to implicate language as an implicit but unconfirmed enabling framework. Without language, it is said, how could common readings of symbols be guaranteed? Thus, HP and SB people make extensive use of pigments made from ochre to denote the concept of blood, or even life, to distinguish individuals by the wearing of personal ornaments, to mark ownership of objects by adding designs that differentiate mine from yours, and to reflect beliefs in death and what follows by deliberate interment. In short, in Donald's (1991) terms, they are communicating by making and marking the external world with mutually intelligible signs and symbols.

A useful connecting framework may be Wurz's (2008) suggestion of memory. All the innovative behaviors listed above leave tangible and lasting traces in the immediate surroundings of the makers, users, and their kin, which serve to materialize common experiences and underline common values. The surroundings, from immediate to regional, have now become landscapes marked with material items linking times, places, and people. We have no way of being sure that earlier MSA, even ESA, groups did not also do this with more subtle, less enigmatically modified objects. The increased incidence of artifacts charged with meaning, however, apparently beyond the functional and utilitarian realms, is persuasive. This is particularly important, perhaps, because this arguably symbolic innovativeness appears to coincide with the stone tool innovations of the SB and HP and, to judge by the records at Diepkloof and Sibudu, to disappear along with them before the end of the MSA. Indeed, in the emerging picture of MSA traditions in southern Africa, it is the disappearance as much as the appearance of innovations that surprises us.

These behaviors are embedded, of course, in a hunting and gathering context that needs to be understood. For some time, Richard Klein (Klein and Cruz-Uribe, 1996) commented on differences in hunting and gathering practices he sees between MSA and LSA people, as reflected in the faunal records of southern African sites. Constrained by the availability of evidence, his comparisons usually contrast Holocene LSA with much earlier MSA, sometimes including HP or SB assemblages. His assertions that Holocene and Marine Isotope Stage 5 (MIS5) practices differ are strong, but not undisputed. There may very well be robust differences in hunting choices or gathering decisions from 120,000 years ago to the recent past, and these are quite likely to be related to technological advances, increased population densities, and resultant prey selection, as some have argued (Marean, in press). What does emerge, however, is the key role played by coastal resource use in the diets of MSA people in southern Africa, especially in the later MSA.

To situate coastal MSA sites such as Klasies River, Die Kelders, Blombos, and the Pinnacle Point caves, and even the near coastal Diepkloof and Sibudu, we need to understand MSA site distributions across the subcontinent. There are thousands of known MSA sites, almost all of them open sites, all except a miniscule fragment undated and barely described. With extremely few exceptions (Kathu Pan and Wonderwerk in Beaumont and Morris, 1990, come to mind), every one of the HP and almost all SB sites are located on or near the present coastline or in the near coastal mountain chains. A very significant proportion of MSA sites that are not HP or SB, on the other hand, all undated, is located well away from the coast. Although some SB and, less clearly, HP sites are in the open, the majority is located in caves and rock shelters. In this sense, HP and SB groups were precocious in that they anticipated the local LSA preference for rocky outcrops and some kind of sheltered overhang within them. ESA and, we might add, earlier MSA sites are usually located along pan edges or stream courses, whereas LSA people certainly, and later MSA people arguably, were more drawn to rocky terrain. Clearly, we are disadvantaged by the loss of land and sites to the Late Pleistocene rise in sea level (Fig. 10.2), but it is not simply the archaeologists' preferences that have determined a near coastal distribution for SB, HP, and early LSA occupation.

The abundance of shell midden sites from MIS5 and, perhaps, earlier (Parkington, 2001a; Marean et al., 2007) is a key element in understanding innovation in the late MSA. Since the publication of the Klasies River excavations and that of the more ephemeral

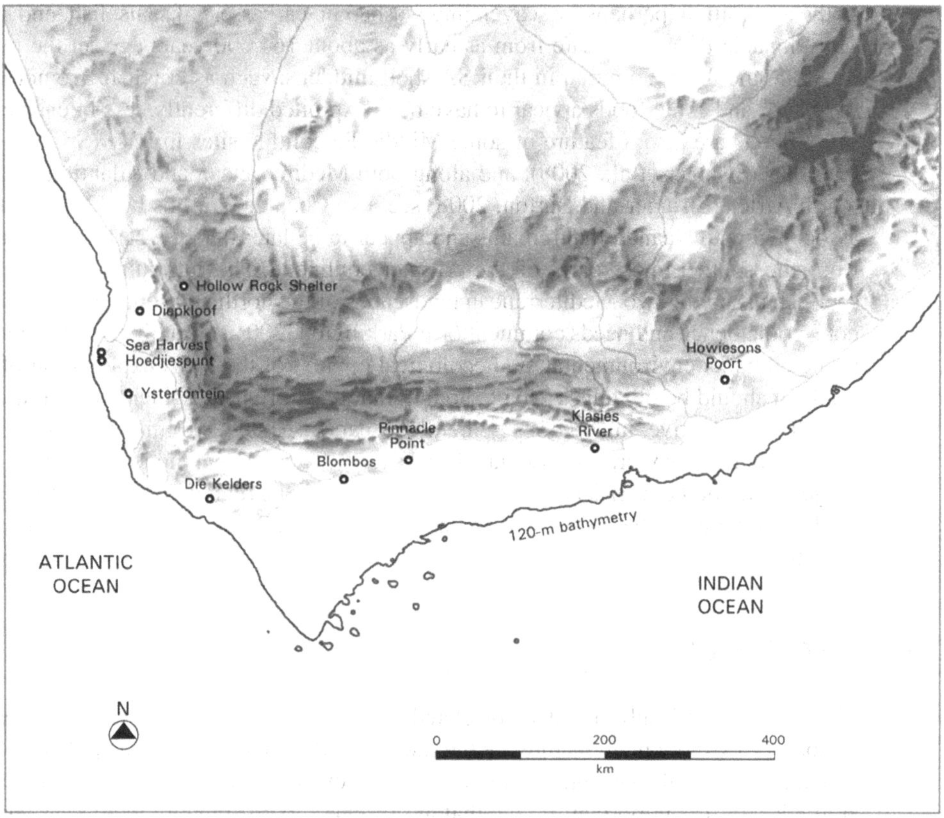

Figure 10.2 Extension of the southern African coastline to the 120-m offshore bathymetric contour. Illustration is created by and a copyright of Neil Rusch. Source data courtesy of the Council for Geoscience.

research at Sea Harvest near Saldanha (Volman, 1978), it has been clear that accumulations of marine food waste dating from Pleistocene high sea level episodes are common along both the Atlantic and Indian Ocean coastlines of southern Africa (Fig. 10.2; see also Parkington et al., 2004). Single-grain OSL and TL dating of these MSA shell middens has shown that they may date from a range of time periods when the coastline was within reach of people located along the present shore, that is, when sea levels were as they are now or a little lower. Uranium series disequilibrium and electron spin resonance (ESR) dating support the assertion that these sites may be as little as 70,000 or as much as 164,000 years old although much more work is needed to obtain accurate as well as precise ages (Klein et al., 2004; Marean et al., 2007; Avery et al., 2008; Jacobs et al., 2008). It is likely that many date from the generally high sea levels of MIS5, which makes them earlier than almost all other shell middens from anywhere else in the world.

So far, no ESA shell middens have been found in southern Africa. It is not clear yet whether this results from poor preservation of early organic materials, the destruction of early shell middens by rising and falling sea level, the failure of archaeologists to detect deeply buried sites, or a real shift in behavioral patterns, an innovation in food gathering habits. There is no lack of activity among archaeologists of the region, and we should soon whittle down these possibilities. Recorded marine food exploitation begins in the MSA well before the appearance of the innovative SB and HP and includes shellfish gathering as well as the acquisition, perhaps by scavenging washed up carcasses, of seals, fish, and seabirds. The Pinnacle Point sites date from as early as about 164,000 years ago, in the middle of MIS6. Some foods common in the LSA shell middens, such as crayfish, are not found in the MSA, and some foods appear to have been exploited differently, less intensively.

Shells are also a feature of some Middle Paleolithic sites in Italy (Stiner, 1994), at Gibraltar (Stringer et al., 2008), and along both Mediterranean and Atlantic coastlines of North Africa (Stringer and Barton, 2008; see also Chapter 7 in this book). Here, they are associated with Neanderthal remains and appear to be later than the earliest shell middens from the southernmost parts of the African continent. The intriguing pattern of early shellfish use in the two Mediterranean landscapes of the north and south of Africa is surely not a coincidence and needs comment (see Parkington, 2001a). But first, we need to sketch in the climatic, environmental, and ecological context of the MSA exploitation of marine foods. It should be obvious by now, though, that there are near-coincident, perhaps meaningfully consecutive, innovations in diet, landscape use, stone toolmaking practices, arguably symbolic behavior, and associated shifts toward anatomical modernity within the time range of the southern African MSA. Far from laggard, the region is clearly in the vanguard of "becoming us." The task now is to describe and understand the process and investigate its extensiveness.

CLIMATE CHANGE

The two key and substantially correlated environmental changes that characterize the period of MSA tool manufacture are the rise and fall of sea level and the pulsing of aridity that marks the "glacial" phases of the climate cycle (Stokes et al., 1997, 1998; Carto et al., 2009). It is important to note that at present, and presumably in all warmer ("interglacial") episodes of the last 200,000 years such as the Holocene, only a very small fraction of southern Africa, essentially the southwestern tip, is influenced by winter rainfall. For the most part, the region is under the influence of summer rainfall drawn in from the warm Indian Ocean and decreasing in amount from east to west. Even the extremely arid western half of the subcontinent north of the winter rainfall zone receives its moisture from the east. During warmer episodes, southern Africa would have resembled the landscape of the recent past – arid in the west, moist in the east. The climate history scenarios show, however, that such episodes are much shorter than the cooler times, when much greater areas would have been distinctly unattractive to hunters and gatherers and when the premium on behavioral innovation would have been much greater.

During colder periods, the Indian Ocean was cooler, less moisture was advected from it, the passage of moisture across the eastern continental ranges was impeded, and an extremely extensive aridity set in over almost all of southern Africa. Effectively at these times, the Cape Mediterranean ecosystem, although more extensive, is detached from subtropical Africa, linked only by tenuous connections along the eastern side and offered the chance to drift apart, culturally as well as genetically, from the bulk of the African continent. This is presumably how (and when) Cape Africans became recognizably different linguistically and genetically from their subtropical neighbors. The Cape is, thus, distinguished geologically as the core of the Folded Belt of mountains with characteristically poor soils, but diverse topographic habitats; climatically as the somewhat unrepresentative winter rainfall zone; biologically as the home of the Fynbos Biome, a Mediterranean-type ecosystem, the smallest of the Earth's six plant kingdoms; and, in human terms, as the origin and core of the San hunter-gatherers with their distinctive physiology, language, and lifestyle.

The sea level history of southern Africa conforms to the global pattern of oscillations in response to the accumulation and release of ice in the polar regions. At the Cape, there is an enormous expanse of continental shelf covered by less than 120m of seawater, meaning that rising and falling sea levels repetitively submerge and expose some tens of thousands of square kilometers of land at what would arguably be the most attractive latitude, substantially enlarging the fynbos landscape at times when the interior further north would have been all but uninhabitable (Fig. 10.2). What is of significance is the timing of the juxtapositions of marine and terrestrial resources and the relationship of these to the behaviors reflected in the archaeological record (Parkington, 2001a, 2005; Marean, 2009).

The Holocene record from excavations around the Cape makes it very clear that the key resources in the Fynbos Biome are the geophytes, underground regenerative organs that are easily the most attractive carbohydrate sources for local hunter-gatherers (Parkington, 2001b). Especially in the west, but wherever rainfall is seasonally uneven, these food parcels are generally small, more or less deeply buried, variably nutritious, and cyclically available in response to the growth regimes of the plants themselves. The edible geophytes are limited to periods after the plants have flowered, when corms develop and then remain dormant until the next growing season. Because the window of edibility is defined by the different, staggered growth cycles of individual species, there is some challenge to gatherers whose carbohydrate energy consumption is largely determined by these hidden but attractive foods. There is a premium on intelligence. The fynbos vegetation is rich in geophytes but very poor in productivity and so has a very low animal carrying capacity. Protein sources along the coastlines would have been especially attractive to gatherers as complementary nutritional contributions to a balanced diet.

The scenario envisaged is one of expansion and contraction of the Fynbos Biome in response to the falling and rising of sea levels. The exposed southern Cape can be seen as an island of some size providing a resource landscape never far from the coast, generating dietary opportunities open to those intellectually capable of understanding the fluctuations in availability and somewhat isolated from regions to the north. The interior in colder periods would have been seasonally unpleasant, though topographically diverse, and perhaps an area visited only in summers from home bases along the contemporary coastline.

Shellfish and geophytes, protein and carbohydrate food parcels characterized by different but equally challenging cyclic fluctuations in availability, provide the selective framework within which gatherers needed to operate and evolve. Hilary Deacon (1995) argued that the well-defined hearths associated with shellfish and what he believed to be humified geophyte remains at Klasies River, are the archaeological residue of this integrated diet, one that survived in the Cape to be well documented by literate travelers of the last three centuries (Parkington, 1984). This pattern, quintessentially San in terms of technology, landscape use, and social relations has lasted, not without modification, from some part of the later MSA. The challenges of buried geophytes and tidally submerged shellfish constitute the dietary reward for capable, strategic gatherers (Parkington, 2001a), an arena in which increased intelligence and planning would have been at a premium.

One critical reward is the acquisition of abundant long-chain fatty acid supplies by the exploitation of shellfish and other marine foods. As described elsewhere (Crawford et al., 1999; Broadhurst et al., 2002; Chapters 2 and 3 in this book), encephalization is an expensive investment and requires adequate amounts of nutrients of many kinds. We need to recognize here that encephalization is not simply, or even primarily, the selection for a *larger* brain. Human brains are far more connected, allowing far more simultaneous synaptic linkages, than those of any other mammalian species. Such increases in connectivity have been a long-term feature of primate evolution but have reached their peak in our own species. The notion of EQ, the encephalization quotient (a measure of the extent to which actual brain size differs from that predicted from the species' body size) of Harry Jerison (1973), is a proxy, barely adequate but at least measurable, reflection of the gains in processing capacity that have been made in only the last few hundred thousand years of hominin evolution. In this sense, encephalization is the bridge between biology and behavior, the skeletal structure that reflects a massive increase in planning, forethought, mental agility, and, we might say, the capacity to innovate.

Prominent among requirements for encephalization are the lipids that constitute 60% of all brain matter, more especially docosahexaenoic acid (DHA), the long-chain derivative of the omega-3 chain that is required on all synaptic junctions and all retinal receptors (Broadhurst et al., 2002; Chapters 2–4 in this book). The effects of deprivation of this critical and essential ingredient on mental and visual acuity are well documented clinically (Cunnane et al., 2007; Chapter 2 in this book). Without adequate supplies, populations would not be able to respond to evolutionary imperatives to select for larger, more intricately wired brains. DHA is far more accessible through the marine food chain than either the freshwater or the terrestrial food chains. Particularly because the need for DHA is greatest in the third trimester of fetal development or in the first year of postnatal life, the capacity of mothers to access intertidal marine foods with little effort was probably crucial. Accessing the brains of game killed in the veld is less easily envisaged, and there are arguments as to the sufficiency of DHA supplies in game muscle meat (Cordain et al., 2001, 2002). Near coastal populations would have been at a distinct advantage in generating the fatty acid base to respond to an evolutionary call for larger, more connected brains by consuming preformed DHA.

A NEW NARRATIVE

We are now in a position to generate a new narrative of later human evolution in Africa very different from that addressed by Desmond Clark in the mid-1970s. Unlike him, though, I concentrate on the likely scenario at the Cape. I omit here any discussion of the claims of East Africa as the likely source area of modern people, an idea favored by many (see, e.g., Klein, 2008). It is remarkable how different this claim is to a rival one from southern Africa, being almost completely based on the apparently early dates for rather poorly provenanced skeletal remains from Herto (White et al., 2003), Omo (McDougall et al., 2005), and Laetoli (Hay, 1987). Unlike the somewhat later remains from the Cape, these are marginally modern, tenuously dated, and completely dissociated from any signs of innovative behaviors. We can expect some attempts to integrate the essentially anatomical case for East Africa with the essentially behavioral case for South Africa.

There are some serious challenges, however, to the construction of a plausible narrative for any subregion, especially in the arena of explanation and the identification of cause and effect. How might the threads of diet, technology, and brain function be connected in a way that is testable, better refutable, from the archaeological record? The first challenge is the chronological precision available to us. Even with OSL dating, our knowledge of the age of an assemblage might carry an error margin of 5000 years or more. If key changes in climate happened abruptly and reversed just as abruptly (see Carto et al., 2009), we might not be able to distinguish a time of increased aridity from a time of increasing moisture. How then could we hope to link these into a narrative implicating climate change? The second challenge is even more serious. Although some archaeologists expect and try to find links between climate, environment, and stone tool assemblages, few are capable of generating necessary and directional linkages. How, we might ask, would making more blades, or inventing the bifacial leaf point, help in times of complex, only broadly understood climate or vegetation shifts?

I suggest that a factor worth considering is the pulsing subcontinental loss of viable territory associated with the combination of sea level change and aridity, specifically the difficult times within MIS6 between 190,000 and 130,000 years ago, and later, MIS4 between 80,000 and 60,000 years ago. Such repetitive challenges forced populations into constrained and quite specific resource landscapes, requiring initiative and rewarding innovation. I am not suggesting any specific connections between resources and artifact types, although these may well have existed, but rather the general impact of climatically related pulsing. The repeated isolation of Cape populations may well have led to cultural as well as genetic drift, with the expansion of any new forms generated into the rest of the continent, when connections increased again. The key features of the restricted core at the southern extremity of Africa were arguably seasonality and juxtaposition, with the seasonal and tidal challenges of the Fynbos Biome and Benguela upwelling structuring responses (Parkington, 2001a). Something rather similar may have led to early shellfish exploitation at another Mediterranean landscape at the northern extremity of Africa 7000 km away, with similar innovative manifestations (Bouzouggar et al., 2007).

Like their emergent Neanderthal cousins in Europe, emergent *H. sapiens* in southern Africa would have been under persistent pressure to detect new food sources. In particular, these include foods not previously utilized because they are relatively inaccessible or require some processing – the embedded foods of Foley and Lee (1991). Through much of MIS6, especially the cold spikes associated with Heinrich Events (Stokes et al., 1997; Stokes et al., 1998; Carto et al., 2009), groups of hunters and gatherers were encapsulated in an island of fynbos landscape without a high animal biomass and with few sizable aboveground plant foods. The shoreline constituted a linear patch with easily the best returns for hunting and gathering, with guaranteed returns from sedentary, predictable items such as shellfish and washed-up seals, whales, and seabirds. These foods also contribute essential marine fatty acids as well as iodine (Crockford, 2003) to all collectors including pregnant and nursing women, the (ethnographically) traditional gatherers of shellfish. In the interior, the carbohydrate deficit was made up by developing an understanding of geophyte life cycles and corm availabilities. Gatherers smart enough to understand both of these and the ways they can be integrated, were rewarded with the potential to afford a brain that made them and their children even smarter.

Are abundant intertidal sources of DHA strictly essential to meet the costs of encephalization (Cunnane et al., 2007), or could hunter-gatherers have accessed enough from the more limited amounts in the terrestrial food web (Cordain et al., 2002)? This is hard to determine, especially as there is evidence that pregnant women may be able to improve the efficiency of their chain elongation of shorter fatty acid chains in response to the demands of supporting a third trimester fetus (Carlson and Kingston, 2007). Whatever the resolution of this debate, later MSA shellfish-consuming hunters and gatherers in the Cape transformed their toolmaking behavior, developed innovative networks of shared memory, and were the first unquestionably anatomically modern, fully encephalized people of whom we know. A testable prediction of this idea is that regular shellfish consumption contributes to encephalization, which in turn facilitates innovation in many spheres. As the resolution of these events in time and place becomes clear and as the interconnections between diet, encephalization, and innovation emerge, we should understand the origins of our species much better. It seems likely that the southernmost region of Africa played a key, *paramount* role.

Dedicated to the memory of J. Desmond and Betty Clark.

REFERENCES

AVERY, G., HALKETT, D., ORTON, J., STEELE, T., TUSENHUIS, M., AND KLEIN, R. 2008. The Ysterfontein 1 Middle Stone Age rock shelter and the evolution of coastal foraging. *South African Archaeological Society Goodwin Series* **10**:66–89.

BEAUMONT, P.B., DE VILLIERS, H., AND VOGEL, J.C. 1978. Modern man in sub-Saharan Africa prior to 49000 years BP: A review and evaluation with particular reference to Border Cave. *South African Journal of Science* **74**: 409–419.

BEAUMONT, P.B. AND MORRIS, D. 1990. *Guide to Archaeological Sites in the Northern Cape*. Kimberley: Southern African Association of Archaeologists.

BEAUMONT, P.B. AND VOGEL, J.C. 1972a. On a new radio carbon chronology for Africa south of the equator, part I. *African Studies* **31**:66–89.

BEAUMONT, P.B. AND VOGEL, J.C. 1972b. On a new radio carbon chronology for Africa south of the equator, part II. *African Studies* **31**:155–182.

BOUZOUGGAR, A., BARTON, N., VANHAEREN, M., D'ERRICO, F., COLLCUTT, S., HIGHAM, T., HODGE, E., PARFITT, S., RHODES, E., SCHWENNINGER, J., STRINGER, C., TURNER, E., WARD, S., MOUTMIR, A., AND STAMBOULI, A. 2007. 82,000-year-old shell beads from North Africa and implications for the origins of modern human behaviour. *Proceedings of the National Academy of Sciences of the United States of America* **104**:9964–9969.

BROADHURST, C.L., WANG, Y., CRAWFORD, M.A., CUNNANE, S.C., PARKINGTON, J.E., AND SCHMIDT, W. 2002. Brain-specific lipids from marine, lacustrine or terrestrial food resources: Potential impact on early African *Homo sapiens*. *Comparative Biochemistry and Physiology. Part B* **131**:653–673.

CANN, R.L., STONEKING, M., AND WILSON, A.C. 1987. Mitochondrial DNA and human evolution. *Nature* **325**: 31–36.

CARLSON, B.A. AND KINGSTON, J.D. 2007. Docosahexaenoic acid biosynthesis and dietary contingency: Encephalization without aquatic constraint. *American Journal of Human Biology* **19**:585–588.

CARTO, S.L., WEAVER, A.J., HETHERINGTON, R., LAM, Y., AND WIEBE, E.C. 2009. Out of Africa and into an ice age: On the role of global climate change in the late Pleistocene migration of early modern humans out of Africa. *Journal of Human Evolution* **56**:139–151.

CLARK, J.D. 1975. Africa in prehistory: Peripheral or paramount? *Man* **10**:175–198.

CLARK, J.G.D. 1989. *World Prehistory: A New Outline*. Cambridge: Cambridge University Press.

CORDAIN, L., WATKINS, B.A., FLORANT, G.L., KELHER, M., ROGERS, L., AND LI, Y. 2002. Fatty acid analysis of wild ruminant tissues: Evolutionary implications for reducing diet-related chronic disease. *European Journal of Clinical Nutrition* **56**:1–11.

CORDAIN, L., WATKINS, B.A., AND MANN, N.J. 2001. Fatty acid composition and energy density of foods available to African hominids. *World Reviews of Nutrition and Diets* **90**:144–161.

CRAWFORD, M.A., BLOOM, M., BROADHURST, C.L., SCHMIDT, W.F., CUNNANE, S.C., GEHBREMESKEL, K., LINSEISEN, F., LLOYD-SMITH, J., AND PARKINGTON, J. 1999. Evidence for the unique function of docosahexaenoic acid during the evolution of the modern human brain. *Lipids* **34** (Suppl.):S39–S47.

CROCKFORD, S.J. 2003. Thyroid rhythm phenotypes and hominid evolution: A new paradigm implicates pulsatile hormone secretion in speciation and adaptation changes. *Comparative Biochemistry and Physiology. Part A* **135**: 105–129.

CUNNANE, S.C., PLOURDE, M., STEWART, K., AND CRAWFORD, M.A. 2007. Docosahexaenoic acid and shore-based diets in hominin encephalization: A rebuttal. *American Journal of Human Biology* **19**:578–581.

DEACON, H.J. 1995. Two late Pleistocene-Holocene archaeological depositories from the southern Cape, South Africa. *South African Archaeological Bulletin* **50**: 121–131.

DEACON, H.J. 2008. The context of the 1967-8 sample of human remains from Cave 1 Klasies River Main Site. *South African Archaeological Goodwin Series* **10**: 143–149.

DEACON, H.J. AND DEACON, J. 1999. *Human Beginnings in South Africa: Uncovering the Secrets of the Stone Age*. Cape Town: David Philip.

DEACON, J. 1995. An unsolved mystery at the Howieson's Poort name site. *South African Archaeological Bulletin* **50**:110–120.

DONALD, M. 1991. *Origins of the Modern Mind: Three Stages in the Evolution of Culture and Cognition*. Cambridge: Harvard University Press.

FOLEY, R.A. AND LEE, P.C. 1991. Ecology and energetics of encaphalization in hominid evolution. *Philosophical Transactions of the Royal Society of London* **334**: 223–232.

GRUN, R., BRINK, J.S., TAYLOR, L., STRINGER, C.B., FRANCISCUS, R.G., AND MURRAY, A.S. 1996. Direct dating of Florisbad hominid. *Nature* **382**:500–501.

HAY, R.L. 1987. Geology of the Laetoli area. In *Laetoli: A Pliocene Site in Northern Tanzania*, ed. M.D. Leakey and J.M. Harris, pp. 23–47. Oxford: Oxford University Press.

HENSHILWOOD, C.S., D'ERRICO, F., VANHAEREN, M., VAN NIEKERK, K., AND JACOBS, Z. 2004. Middle Stone Age shell beads from South Africa. *Science* **384**: 404.

HENSHILWOOD, C.S., D'ERRICO, F., YATES, R., JACOBS, Z., TRIBOLO, C., DULLER, G.A.T., MERCIER, N., SEALY, J.C., VALLADAS, H., WATTS, I., AND WINTLE, A.G. 2002. Emergence of modern human behaviour: Middle Stone Age engravings from South Africa. *Science* **295**: 1278–1280.

JACOBS, Z., DULLER, G.A.T., WINTLE, A.G., AND HENSHILWOOD, C.S. 2006. Extending the chronology of deposits at Blombos Cave, South Africa, back to 140 ka using optical dating of single and multiple grains of quartz. *Journal of Human Evolution* **20**:1–19.

Jacobs, Z., Roberts, R.G., Galbraith, R.F., Deacon, H.J., Grün, R., Mackay, A., Mitchell, P., Vogelsang, R., and Wadley, L. 2008. Ages for the Middle Stone Age of Southern Africa: Implications for human behaviour and dispersal. *Science* **322**:733–735.

Jerison, H.J. 1973. *Evolution of the Human Brain and Intelligence*. London: Academic Press.

Klein, R.G. 1973. Geological antiquity of Rhodesian man. *Nature* **244**:311–312.

Klein, R.G. 1999. *The Human Career: Human Biological and Cultural Origins*. 2nd ed. Chicago: The University of Chicago Press.

Klein, R.G. 2008. Out of Africa and the evolution of human behaviour. *Evolutionary Anthropology* **17**:267–281.

Klein, R.G., Avery, G., Cruz-Uribe, K., Halkett, D., Parkington, J.E., Steele, T., Volman, T.P., and Yates, R. 2004. The Ysterfontein I Middle Stone Age site, South Africa, and early human exploitation of coastal resources. *Proceedings of the National Academy of Sciences of the United States of America* **101**: 5708–5715.

Klein, R.G. and Cruz-Uribe, K. 1996. Exploitation of large bovids and seals at middle and later Stone Age sites in South Africa. *Journal of Human Evolution* **31**: 315–334.

Marean, C.W. (ed.). In press. Shellfish gathering, marine palaeoecology and modern human behaviour: Perspectives from Cave PP13b, Pinnacle Point, South Africa. *Journal of Human Evolution*. Special Issue: Pinnacle Point.

Marean, C.W., Bar-Matthews, M., Bernatchez, J., Fisher, E., Goldberg, P., Herries, A.I.R., Jacobs, Z., Jerardino, A., Karkanas, P., Minichillo, T., Nilssen, P.J., Thompson, E., Watts, I., and Williams, H.M. 2007. Early human use of marine resources and pigment in South Africa during the Middle Pleistocene. *Nature* **449**:905–908.

Martin, R.D. 1983. Human brain evolution in an ecological context. Fifty-second James Arthur Lecture on the evolution of the human brain. New York: American Museum of Natural History.

McDougall, I., Brown, F.H., and Fleagle, J.G. 2005. Stratigraphic placement and age of modern humans from Kibish, Ethiopia. *Nature* **433**:733–736.

Mellars, P. 2006. Going east: New genetic and archaeological perspectives on the modern human colonization of Eurasia. *Science* **313**:796–800.

Mitchell, P. 2002. *The Archaeology of Southern Africa*. Cambridge: Cambridge University Press.

Parkington, J.E. 1984. Soaqua and Bushmen: Hunters and robbers. In *Past and Present in Hunter Gatherer Studies*, ed. C. Schrire, pp. 151–174. Orlando: Academic Press.

Parkington, J.E. 2001a. Milestones: The impact of the systematic exploitation of marine foods on human evolution. In *Humanity from African Naissance to Coming Millennia*, ed. P.V. Tobias, M.A. Raath, J. Moggi-Cecchi, and G.A. Doyle, pp. 327–336. Johannesburg: Firenze University Press and Witwatersrand University Press.

Parkington, J.E. 2001b. Mobility, seasonality and Southern African hunter gatherers. *South African Archaeological Bulletin* **56**:1–7.

Parkington, J.E. 2005. Middens and moderns: Shellfishing and the Middle Stone Age of the Western Cape, South Africa. *South African Journal of Science* **99**:243–247.

Parkington, J., Poggenpoel, C.A., Halkett, D., and Hart, T.J.G. 2004. Initial observations on the Middle Stone Age coastal settlement in Western Cape, South Africa. In *Settlement Dynamics of the Middle Palaeolithic and Middle Stone Age*, Vol. **2**, ed. N. Conard, pp. 5–21. Tuebingen: Kerns Verlag.

Parkington, J.E., Poggenpoel, C., Rigaud, J.-P., and Texier, P.-J. 2005. From tool to symbol: The behavioural context of intentionally marked ostrich eggshell from Diepkloof, Western Cape. In *From Tools to Symbols: From Early Hominids to Modern Humans*, ed. F. d'Errico and L. Blackwell, pp. 475–492. Johannesburg: Witwatersrand University Press.

Porraz, G., Texier, P.-J., Rigaud, J.-P., Parkington, J., Poggenpoel, C., and Roberts, D.L. 2008. Preliminary characterization of a Middle Stone Age lithic assemblage preceding the "classis" Howieson's Poort complex at Diepkloof Rock Shelter, Western Cape Province, South Africa. *South African Archaeological Society Goodwin Series* **10**:105–121.

Rigaud, J.-P., Poggenpoel, C., and Parkington, J.E. 2006. Le mobilier Stillbay et Howiesons Poort de l'abri Diepkloof. La Chronologie du Middle Stone Age subafricain et ses implications. *Comptes Rendus Palévol* **5**:839–849.

Rightmire, G.P. and Deacon, H.J. 1991. Comparative studies of Late Pleistocene human remains from Klasies River Mouth, South Africa. *Journal of Human Evolution* **20**:131–156.

Rightmire, G.P. and Deacon, H.J. 2001. New human teeth from Middle Stone Age deposits at Klasies River, South Africa. *Journal of Human Evolution* **41**:535–544.

Ruff, C.B., Trinkhaus, E., and Holliday, T.W. 1997. Body mass and encephalization in Pleistocene *Homo*. *Nature* **387**:173–176.

Sillen, A. and Morris, A. 1996. Diagenesis of bone from Border Cave: Implications for the age of the Border Cave hominids. *Journal of Human Evolution* **31**:499–506.

Singer, R. and Wymer, J. 1982. *The Middle Stone Age at Klasies River Mouth in South Africa*. Chicago: Chicago University Press.

Stiner, M.C. 1994. *Honor among Thieves: A Zoological Study of Neanderthal Ecology*. Princeton: Princeton University Press.

Stokes, S., Haynes, G., Thomas, D.S.G., Horrocks, J.L., Higginson, M., and Malifa, M. 1998. Punctuated aridity in southern Africa during the last glacial cycle: The chronology of linear dune construction in the northeastern Kalahari. *Palaeogeography, Palaeoclimatology, Palaeoecology* **137**:305–322.

Stokes, S., Thomas, D.S.G., and Washington, R. 1997. Multiple episodes of aridity in southern Africa since the last interglacial period. *Nature* **388**:154–158.

Stringer, C.B. and Barton, R.N.E. 2008. Putting North Africa on the map of modern human origins. *Evolutionary Anthropology* **17**:5–7.

Stringer, C.B., Finlayson, J.C., Barton, R.N.E., Fernandez-Jalvo, Y., Caceres, I., Sabin, R.C., Rhodes, E.J., Currant, A.P., Rodriguez-Vidal, J., Giles-Pacheco, F., and Riquelme-Cantal, J.A. 2008. Neanderthal exploitation of marine mammals in Gibraltar. *Proceedings of the National Academy of Sciences of the United States of America* **105**:14319–14324.

Tribolo, C., Mercier, N., Valladas, H., Joron, J.L., Guibert, P., Lefrais, Y., Selo, M., Texier, P.-J., Rigaud, J.-Ph., Porraz, G., Poggenpoel, C., Parkington, J.E., Texier, J.-P., and Lenoble, A. 2008. Thermoluminescence dating of a Stillbay-Howiesons Poort sequence at Diepkloof Rock Shelter (Western Cape, South Africa). *Journal of Archaeological Science* **30**:1–10.

Vogelsang, R. 1998. *Middle Stone Age Fundstellen in Sudwest-Namibia*. Koln: Heinrich-Barth Institut.

Volman, T.P. 1978. The early archaeological evidence for shellfish collecting. *Science* **201**:911–913.

Wadley, L. 2007. Announcing a Still Bay Industry at Sibudu Cave. *Journal of Human Evolution* **52**:681–689.

Wadley, L. 2008. The Howieson's Poort industry of Sibudu Cave. *South African Archaeological Society Goodwin Series* **10**:122–132.

White, T.D. 1987. Cannibalism at Klasies River? *Sagittarius* **2**:7–9.

White, T.D., Asfaw, B., DeGusta, D., Gilbert, H., Richards, G.D., Suwa, G., and Howell, F.C. 2003. Pleistocene *Homo sapiens* from Middle Awash, Ethiopia. *Nature* **423**:742–747.

Wurz, S. 2005. Exploring and quantifying technological differences between the MSA I, MSA II and Howiesons Poort at Klasies River. In *From Tools to Symbols: From Early Hominids to Modern Humans*, ed. F. d'Errico and L. Blackwell, pp. 418–440. Johannesburg: Witwatersrand University Press.

Wurz, S. 2008. Modern behaviour at Klasies River. *South African Archaeological Goodwin Series* **10**:150–156.

CHAPTER 11

HUMAN BRAIN EVOLUTION: A NEW WETLANDS SCENARIO

Stephen C. Cunnane and Kathlyn M. Stewart

HUMAN BRAIN EVOLUTION

This book aimed to integrate knowledge about nutritional and metabolic constraints on human brain development and function, together with fossil and other archaeological evidence to build a plausible basis for explaining the unique evolution of the human brain. Humans today need a higher quality diet than other primates because we do have larger, more highly developed brains. Indeed, the evidence from the World Health Organization is painfully clear that as much as a quarter of the world's population is not able to meet the brain's requirement for several nutrients, most notably iodine and iron. Nevertheless, "need" cannot have been the original reason our distant hominin ancestors started to evolve larger brains. The issue then was to fortuitously consume foods that met the nutritional and metabolic requirements of a somewhat larger brain *before* brain growth could increase the requirement for those nutrients.

Anatomically modern humans have only been present for about the past ca. 200,000 of the past 2 million years, which is around 10% of the total time during which hominins have existed. Even if evolution of a larger brain and higher intelligence were desirable features, let us recall that natural selection does not take orders – what happens, happens. The term *exaptation* (see Chapter 1) succinctly makes this crucial Darwinian point; in accordance with natural selection, hominin brain expansion clearly happened, but it happened opportunistically. Once a diet of high enough quality was consumed by enough people over a long enough period of time (probably at least tens of thousands of years), together with the genetic capacity of the primate brain to expand, actual expansion of the hominin brain could begin to occur. For a very long time, this imperceptible enlargement would have offered no added benefit or advantage for survival. Gradually, some individuals would benefit from a somewhat deeper insight into the benefits or risks of certain activities, or somewhat better hand–eye coordination, and so the ball started rolling slowly, almost imperceptibly, toward higher cognitive capacity, speech, language, and so on. In the meantime, the high quality diet had been in place long before the first flickering sparks of improved cognition.

NEUROCHEMICAL AND NUTRITIONAL EVIDENCE

In this book, we present evidence that increased availability of a relatively small cluster of *brain-selective nutrients* (most importantly docosahexaenoic acid [DHA], iodine, and

Human Brain Evolution, Edited by Stephen C. Cunnane and Kathlyn M. Stewart

iron) from freshwater and coastal food resources essentially controlled the cognitive destiny of hominins. Brain-selective nutrients are present in higher amounts in plants, invertebrates, fish, eggs, and amphibians found along freshwater and saltwater shores of lakes, riverbanks, marshes, estuaries, and intertidal zones than anywhere else. The fact that iodine had to be forcefully added (by government legislation) to table salt is a simple but graphic demonstration of the fact that when these nutrients, accompanied by others (see Chapter 3), are present in insufficient amounts, humans today simply cannot achieve their neurodevelopmental potential. If humans had continued to eat fish and seafood, this problem would never have arisen. Hence, using iodine as an example, inadequate intake of a single nutrient can mobilize governments worldwide to take action and prevent suboptimal brain development, hence avoiding the resulting unsupportable burden of hundreds of millions of unproductive adults. Inadequate dietary iodine availability then surely would have been an impediment to the cognitive evolution of our species much beyond that of Lucy some 4 million years ago. Fortuitous access to a lake or river margin diet changed all that.

What follows here is a brief encapsulation of the arguments made by the contributors to this book. The contributed papers provide compelling nutritional and neurochemical evidence for hominin diets being enriched in brain-selective nutrients; the consequences for inadequate supply of brain-selective nutrients; and the fossil, dietary, and paleoenvironmental evidence that documents how hominins exaptively (opportunistically) shifted to a diet containing freshwater and marine fish and invertebrates.

THE FOSSIL EVIDENCE

The papers in the first half of this volume make the case that hominin brain expansion occurred opportunistically, yet only under conditions in which the available supply of brain-selective nutrients could support it. The second half of the volume documents the archaeological and fossil evidence showing that meeting such nutritional requirements was possible on lake, river, and coastal margins in many areas of eastern and southern Africa. Indeed, for some hominins, the transition to wetlands vegetations and occasionally fish and invertebrates was highly plausible in the Plio-Pleistocene. Catfish remains are present even in marginal freshwater habitats and are prominent in many archaeological sites. They are good sources of brain-selective nutrients and easily caught. Invertebrates are similarly accessible and nutritious. What does confirmed consumption of these foods establish about the lifestyle or habitat of hominins 2 million years ago? First, it very definitely puts some of them on the margins of lakes and rivers during at least part of the year. Knowledge of fish behavior and ecology, and careful selection of fish in "good condition" (as some African fishers do today) requires a presence on the lake and river margins. Frequenting these ecotones for extended periods provides the opportunity to exploit other aquatic food resources besides fish, including shellfish, turtles, and birds.

Better than any other widely available food resources, fish and/or invertebrate consumption in *Homo habilis* and *Homo erectus* would have supported the metabolic and nutritional requirements of brain expansion (see Chapter 3).

In Chapter 8, Stewart argues that in pre-*Homo sapiens*, most fish and aquatic invertebrates came from freshwater sources, while *H. sapiens* and possibly late *Homo heidelbergensis* began to consistently exploit marine food resources. Consumption of fish and aquatic invertebrates at both freshwater and coastal marine sites in Africa in the Plio-Pleistocene is described in Chapters 7–10.

PLAUSIBILITY, PREDICTION, AND PARSIMONY

Constructing human brain evolution around lake and river margin ecotones and diets, and later coastal ecotones, is plausible from the fossil record, leads to certain testable predictions, and is parsimonious. It is *plausible* because the hominin and early human fossil record in several parts of Africa supports purposeful consumption of these foods, particularly shellfish and catfish, dating back almost 2 million years. The fossil record further supports consumption of marine invertebrate and later fish resources at least 165,000 years ago (Chapters 7–10). Furthermore, compared with all other diets, the lake/river margin and coastal marine food supply is richest in brain-selective nutrients (Chapter 3, Tables 3.4 and 3.5).

The *prediction* that follows is that a functional deficit in human brain development will occur if humans consume inadequate amounts of fish and/or mollusks. The World Health Organization's description of the widespread debilitating vulnerability to iodine and iron deficiency during childhood is compelling confirmation of this prediction. Low accumulation of fat on the fetus in humans, whether through intrauterine growth retardation or prematurity, also confers an increased risk of subnormal neurological development. The concept that humans are best adapted to freshwater and marine environments also leads to the prediction that adult humans would be vulnerable adverse effects of no longer occupying this niche. Elaine Morgan (1990) described these problems, particularly involving the joints and back, as *scars of evolution*. A great deal of epidemiological evidence as well as numerous clinical trials demonstrates that humans consuming a lake or river margin diet are less vulnerable to these scars (see Chapter 5). Many studies now indicate that Alzheimer's disease is more prevalent in populations consuming low amounts of fish and aquatic invertebrate nutrients. (Cunnane et al., 2009).

It is the ongoing developmental vulnerability of the infant brain that is the most damaging scar of all. Humans do not need to become divers or swimmers, nor do we need to become spear-wielding hunters to avoid these scars, but we do need moderately frequent access to a wetlands diet. The key trick to improving cognitive function during human evolution was to avoid exposing the brain's increasing developmental vulnerability as it expanded. This masking of developmental fragility was essential to refining brain function and was something that no other land-based species accomplished. Present-day vulnerability of the brain during infancy makes it abundantly clear that humans have not yet distanced themselves from nutrition based on freshwater and marine foods.

Other theories of human brain evolution do not have this predictive ability: There is no evidence that low intake of alternative sources of dietary energy, such as meat or nuts, is associated with impaired brain function during either early development or aging. Hence, diets that are not freshwater- or marine-based may have contained sufficient energy to meet the requirements of hominin brain expansion, but they had and still have two serious inadequacies for human brain development: (1) they are more likely to create nutrient deficiencies, particularly of DHA, iron, and iodine; and (2) plant-based diets contain *antinutrients* such as phytate and goiterogens that exacerbate deficiencies of nutrients such as zinc and iodine, respectively (see Chapter 3).

Parsimony is also important in evolutionary theory; if a transformation during evolution can plausibly be completed in one stage instead of two, then like Occam's Razor, the one-stage version should be favored unless there is overwhelming evidence for a more complex process. Adaptation to a wetlands environment is the simplest explanation for human brain evolution because all the evidence points to difficulties with brain development and function when humans adopt other habitats. Hominins needed these environments to support the process of evolving toward a human brain. As the failure of brain development shows today when the supply of brain-selective nutrients is inadequate, optimal human brain function is still absolutely dependent on freshwater food-derived nutrients. Hence, the use of wetlands environments and foods is not only compatible with the major events in prehuman and early human evolution; it also explains these events more parsimoniously than other theories.

SALIENT POINTS

We propose here a scenario of human brain evolution with several interlocking components:

1. *Exaptation:* Human brain evolution depended on a process of *exaptation*, that is, fortuitous discovery and exploitation of a cluster of food resources that, relative to other diets, was rich in brain-selective nutrients.
2. *Brain-selective nutrients:* Brain-selective nutrients include but are not limited to DHA, arachidonic acid (AA), iodine, iron, zinc, copper, and selenium. They are found in higher amounts in freshwater and coastal food resources than in any other habitat. Such food resources include a wide range of plants, amphibians, fish, mollusks, crustaceans, and eggs. Higher quality diets supplying more protein and energy but not more brain-selective nutrients would have been insufficient to support human brain evolution.
3. *Brain vulnerability:* Extensive evidence of potential developmental neurological deficits in humans today underlines the ongoing vulnerability of human brain development to inadequate intake of the same brain-selective nutrients as were required for its evolution. Most damaging among these nutrients is the inadequacy of iodine, a problem only corrected in the developed world because of government-legislated iodization of table salt. Of almost equal global significance are the roles of iron and DHA. In contrast to people living inland or in mountainous regions, those living in freshwater or marine habitats and consuming aquatic foods are largely protected against these nutrient deficits.
4. *Ketones and infant body fat:* Brain development requires ketone bodies for synthesis of structural lipids. Human brain development also appears to need ketones as an obligatory energy substrate complimenting glucose. Ketones are derived from fatty acids, principally those in fat stores. Therefore, among the primates, the evolution of subcutaneous fat depots only in human infants provided a unique means by which increased dietary energy could be stored and transferred toward the expanding brain.
5. *Thyroid hormone:* Wetlands environments also supplied some preformed thyroid hormone that would have facilitated the modification of gene expression and stimulatory effect of abundant brain-selective nutrients and energy supply toward evolutionary change in the development and function of the hominin brain.
6. *Habitat:* Such a wetlands environment not only provided abundant and accessible food resources rich in brain-selective nutrients but also may have facilitated evolution of neonatal body fat stores required to support the development of the expanding brain.
7. *Geographic and reproductive isolation:* By providing a degree of geographic isolation, wetlands habitats may also have provided sufficient reproductive isolation from other related primates to facilitate speciation toward *Homo*.

8. *Fossil evidence:* The hominin fossil record in eastern and southern Africa contains extensive evidence of exploitation of wetlands food resources, particularly catfish and shellfish, adding plausibility to the nutritional and metabolic reasons why increased availability of these foods was crucial for human brain evolution.

CONCLUSION

The events of significance in human evolution started with the evolution of fat babies and larger brains. Whether inland or on the seacoasts, these events were most plausibly a consequence of adaptation to and then exploitation of freshwater or marine resources. Most importantly, this scenario offers a clear rationale for the most important scar of human evolution – the ongoing developmental vulnerability that is peculiar to the human brain.

Wetlands environments provide not only the food, habitat, and lifestyle necessary for human brain evolution but also the geographic isolation that is generally more likely to foster speciation. Crucially, humans are *still in* this wetlands phase; we have explored other niches in the past and will continue to do so in the future, but we are still best adapted to, and dependent on, foods from a water/land interface.

REFERENCE

Morgan, E. 1990. *The Scars of Evolution.* Souvenir Press.
Morgan, E. 1997. *The Aquatic Ape Hypothesis.* London: Souvenir Press.

INDEX

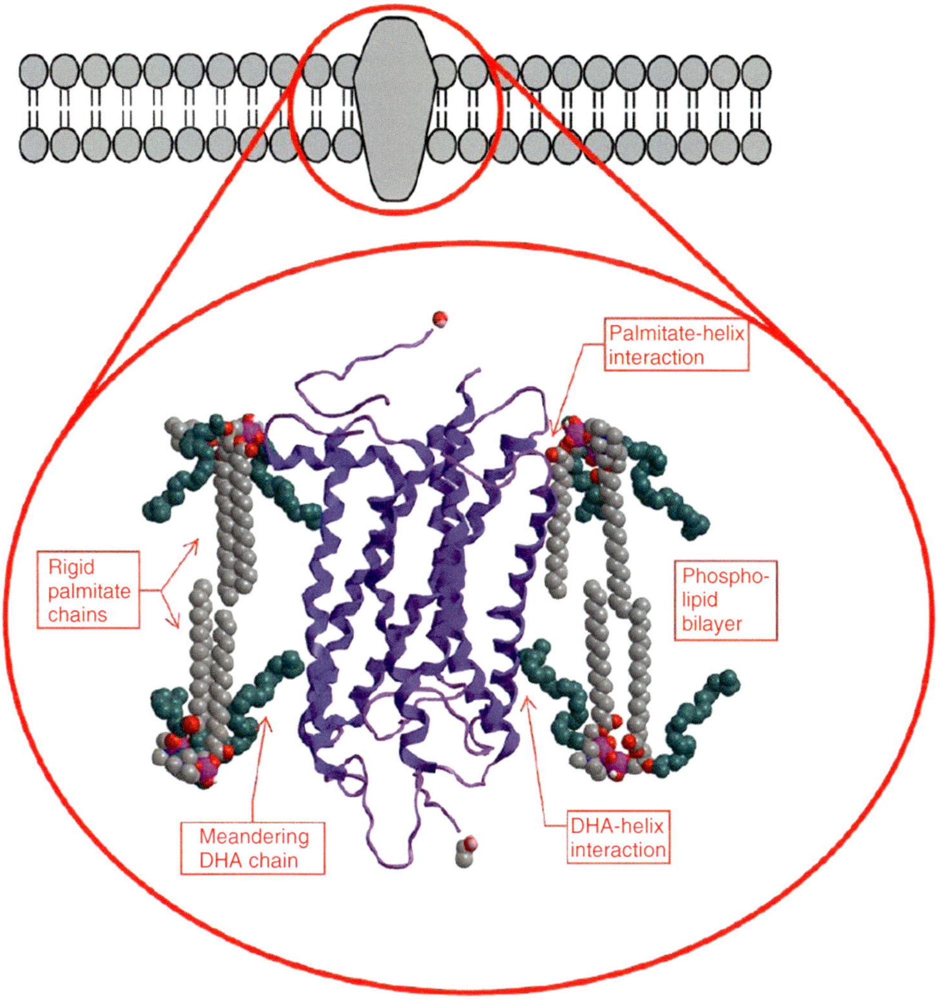

Figure 4.4 Top: Classical representation of a membrane bilayer embedded with a transmembrane protein. *Inset*: Schematic representation of rhodopsin (blue ribbon) embedded in a membrane. *Eight* identical phosphatidylcholine-docosahexaenoic acid (DHA)-palmitic acid molecules are arranged around the transmembrane rhodopsin protein. Palmitic acid (gray) forms straight chains that are rigid and interaction with one another between molecules. DHA (green) chains meander from the interior to the exterior surface of the membrane. Next to rhodopsin, the fatty chains interact very differently based on their shapes, with DHA thought to solubilize rhodopsin more strongly than palmitate or other saturates.

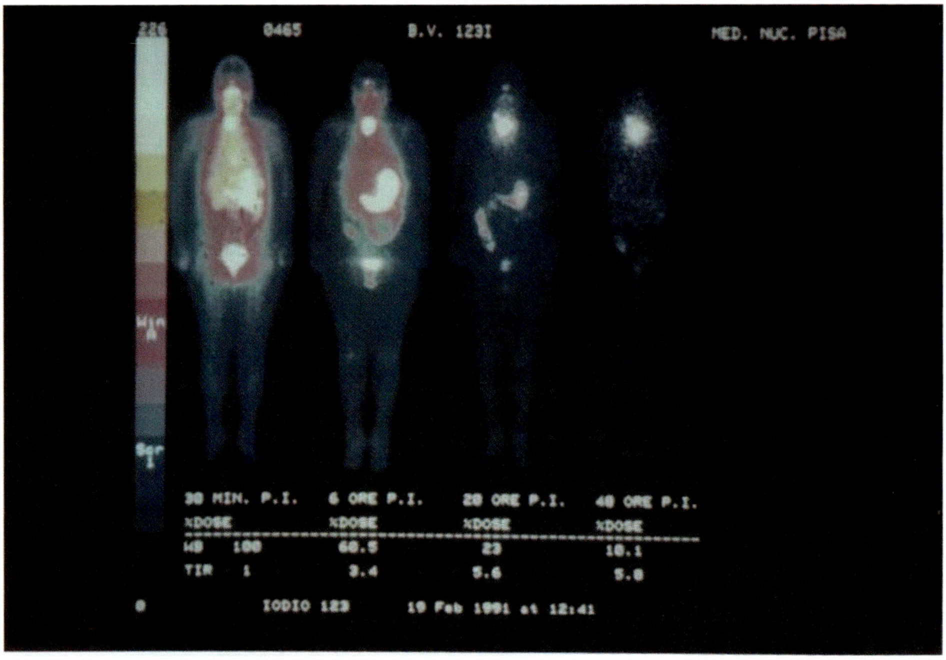

Figure 6.2 Sequence of 123-iodide total-body scintiscans of a woman after intravenous injection of 123-iodide (half-life: 13 h); (from left) respectively at 30 min, and at 6, 20, and 48 h. The highest and rapid concentration of radioiodide (in white) is evident in gastric mucosa of the stomach, salivary glands, and oral mucosa. In gastric mucosa of the stomach, 131-iodide (half-life: 8 days) persists in scintiscans for more than 72 h. In the thyroid, iodide concentration is more progressive, as in a reservoir (from 1% [after 30 min] to 5.8% [after 48 h] of the total injected dose). Here, iodide concentration by the mammary gland is not evident because this woman was not pregnant or lactating. A high excretion of radioiodide is observed in the urine.

Figure 8.1 Fisherman from the Republic of the Congo holds his catch of *Clarias*, a catfish. Photograph by Bruce G. Marcot.

Figure 9.2 *Aonyx capensis*, the African clawless otter, is a relatively large lutrine (12–19 kg) with a blunt head, powerful neck, and hypertrophied masticatory muscles. The digits of the hands are largely unwebbed, and those of the hindfeet are partly webbed. The distal phalangeal bones of *Aonyx* are reduced, and the keratinous claws are rudimentary or completely lost (*a-onyx* means "clawless"). *A. capensis* relies on its dexterous hands when foraging for crabs, frogs, and slow-moving fish in shallow water and semiterrestrial environments. Photo taken by Alan B. Shabel.

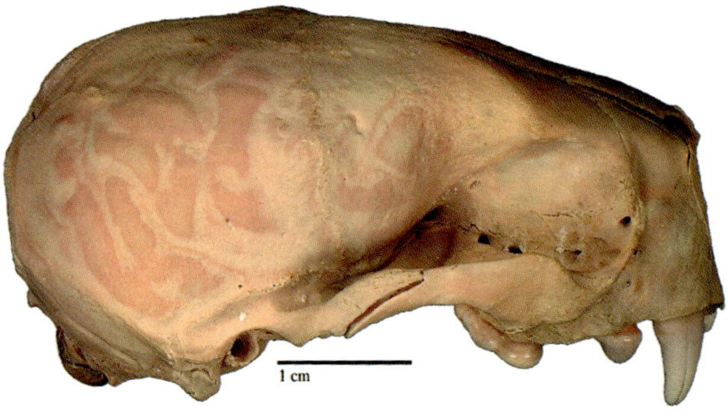

Figure 9.3 This specimen of *Enhydra lutris* (USNM 285430), a young sea otter from Amchitka Island, Alaska, illustrates the close connection between the brain and the cranial bones in a carnivoran. The purple-red stain on the braincase is probably related to the consumption of heavily pigmented sea urchins by these animals. Photograph by Alan B. Shabel.

Printed in the USA/Agawam, MA
September 16, 2026

910661.001